Space Shuttle Challenger

Ten Journeys into the Unknown

Space Shuttle Challenger

Ten Journeys into the Unknown

Ben Evans

Space Shuttle Challenger

Ten Journeys into the Unknown

Published in association with
Praxis Publishing
Chichester, UK

Ben Evans
Space Writer
Atherstone
Warwickshire
UK

SPRINGER–PRAXIS BOOKS IN SPACE EXPLORATION
SUBJECT *ADVISORY EDITOR*: John Mason, M.Sc., B.Sc., Ph.D.

ISBN 10: 0-387-46355-0 Springer Berlin Heidelberg New York
ISBN 13: 978-0-387-46355-1 Springer Berlin Heidelberg New York

Springer is part of Springer-Science + Business Media (springer.com)

Library of Congress Control Number: 2006935328

Printed in Germany

Cover design: Jim Wilkie
Project management: Originator Publishing Services, Gt Yarmouth, Norfolk, UK

Printed on acid-free paper

Contents

Preface . . . vii

Acknowledgements . . . ix

List of figures . . . xi

List of abbreviations and acronyms . . . xv

1 Flight of the Geritol Bunch . . . 1
Space cowboys . . . 1
Unexpected change . . . 2
Birth of Challenger . . . 5
Workclothes . . . 10
Time to operate . . . 13
The butterfly and the bullet . . . 17
Clean and spacious . . . 20
Hydrogen leaks . . . 23
Mental simulations . . . 26
Ride of a lifetime . . . 27
Switchboard in the sky . . . 30
"Extraordinarily exciting" . . . 36
Five-day high . . . 40

2 Ride, Sally Ride . . . 45
Astronauts wanted . . . 45
Four become five . . . 49
The new guys deliver . . . 56
Orbital ballet . . . 63
Scientific bounty . . . 66
A successful mission . . . 68

3 Weightlifters . . . 73
Funtime . . . 73
Off to work . . . 82
An eye on the future . . . 86
Night flying . . . 91

4 "At the peak of readiness" . . . 95
A touch of superstition . . . 95
Human satellites . . . 98
Lost, immovable and burst in space . . . 104
A bittersweet mission . . . 112
'Lucky' thirteen? . . . 114
Deployment of LDEF . . . 117
Repairing Solar Max . . . 123
Fire on the pad . . . 135
An experimental crew . . . 137
Earth watching . . . 139
Hazardous hydrazine . . . 151
"Five plus two equals seven" . . . 155

5 The Untouchables . . . 161
Unequal partnership . . . 161
'Routine' flights . . . 167
Seven men, two monkeys and two dozen rats . . . 170
"They left out the 'wow'!" . . . 175
Gravity gradient . . . 179
Going to Spain . . . 193
"We don't have time to talk about this" . . . 198
Solar observatory . . . 202
Particle 'sniffer' . . . 211
Multi-disciplinary mission . . . 216
Coming of age . . . 218
Suspended in a gondola . . . 223
Quiet time . . . 227

6 "Major malfunction" . . . 231
Ultimate field trip . . . 231
The golden age ends . . . 241
A preventable tragedy . . . 247
Missed warnings . . . 253
A full plate . . . 256
"The Sun kept on rising" . . . 266

7 Challenger missions 1983–1986 . . . 273

Bibliography . . . 283

Index . . . 287

Preface

Ironically, the loss of Challenger in January 1986 fired my interest in space exploration more than any other single event. I was nine years old. My parents were, at the time, midway through moving house and, luckily, the TV was one of the few domestic items still to be packed. I watched the entire horror unfold live on all of the network stations. Admittedly, my fascination with rockets and astronauts, stars and planets had begun several years earlier, but Challenger's destruction turned it from an occasional hobby to a fascination which has remained with me ever since. In September 1988, aged 11, I came home from school to watch STS-26 return the Shuttle fleet to orbital operations. Five years later, I gave a speech on the STS-51L disaster to my teacher as part of my GCSE English assessment. Another decade passed and, now a teacher myself, I returned to my school one cold Monday morning to explain to my pupils what had happened to Challenger's sister ship, Columbia, a few days earlier.

In some ways, the loss of Columbia affected me more deeply than Challenger. Aside from the fact that I was older, I had also corresponded with and interviewed many of the lost astronauts. Signed photographs of the STS-107 crew had adorned my bedroom walls. Personalised, hand-written notes and letters from my heroes had filled scrapbooks. I did not have the same personal link with Challenger's crew. However, in my mind, my passion for space exploration might not have endured were it not for STS-51L.

I find it depressing – even distressing – that, in the case of both Challenger and Columbia, most observers focus upon their final, fateful missions, avoiding the spectacular triumphs that both venerable orbiters achieved during their all-too-short lives. At the time of her loss, Challenger had flown more times than all three of her sister ships. She had rocketed into the heavens nine times successfully in less than three years, transporting nearly four dozen astronauts aloft – several of them on two occasions, one of them as many as three times – and had spent over two months in space. Twelve satellites had departed her payload bay, dozens of experiments ranging from studies of bees to sophisticated crystal growth facilities had flown aboard her, a snazzy jet-fed backpack had been tested and spacewalking astronauts had captured and deftly repaired NASA's crippled Solar Max observatory. In June 1983, the world got its first photographic glimpse of the 'whole' Shuttle in space, drifting serenely above a cloud-bedecked Earth: and the Shuttle in question was Challenger.

Of course, there were dismal times, including embarrassments and near disasters. Before she even undertook her maiden voyage, problems with her main engines enforced a delay of several months; then, when she finally achieved orbit, a booster malfunctioned and left a $100 million Tracking and Data Relay Satellite in a lower-than-planned orbit. On her fourth mission, two more satellites were lost and an experimental rendezvous balloon burst minutes after deployment. What should have been her seventh trip was cancelled six days before launch and her eighth mission experienced a harrowing on-the-pad main engine shutdown. When it finally set off, a main engine failed as she sped towards orbit at 15,000 km/h and a hairy abort landing in Spain was narrowly averted. On January 28th 1986, Challenger's luck finally ran out.

My intention in writing this book, as with its sister volume about Columbia, was to tell the story of Challenger from technical esoterica, press kits, reports, personal interviews, correspondences, newspaper and magazine articles from the time and original NASA sources. My goal was to present, in as much detail as I could possibly achieve, not a critique, but rather an appraisal of her many accomplishments. It will be up to the reader to gauge how successful I have been in this endeavour, but I hope that this book will prove interesting and informative when, a few years from now, the Shuttle fades into history, taking its place alongside Vostok, Voskhod, Mercury, Gemini and Apollo, and the next stage of space exploration begins.

Acknowledgements

I am indebted to a number of individuals for helping to make this book happen. Among them are several astronauts who actually flew Challenger – Norm Thagard, Vance Brand, Bruce McCandless, George 'Pinky' Nelson and Gordon Fullerton – who kindly took time to speak with me at length over the telephone or answer my many questions via email correspondence. Their insights into the similarities, differences and idiosyncracies of the orbiters, the development of the Manned Manoeuvring Unit, spacewalking procedures and early efforts to understand the causes of space sickness have proven invaluable. Thanks are also due to Roberta Ross and Beth Hagenauer of NASA and Linn LeBlanc of the Astronaut Scholarship Foundation for arranging interviews.

Once more, I am grateful to David Harland and Ed Hengeveld; to the former for reading the first draft of the manuscript, pointing out my errors and sharpening up the text, and to both for giving up their time to identify high-quality illustrations for this book. The enthusiasm of Clive Horwood of Praxis is acknowledged, as is the project's aptly-named copy editor and typesetter, Neil Shuttlewood.

Alongside the loss of Challenger, a unique group of friends from the Midlands Spaceflight Society helped foster my fascination with space exploration. In particular, Andy Salmon's infectious enthusiasm and encyclopaedic knowledge have been immensely helpful, as has the inexhaustible reservoir of astronaut and cosmonaut esoterica from Rob and Jill Wood. My family have constantly supported my interest and I must thank my fiancee, Michelle Chawner, for her endless love and encouragement. Additional grateful thanks go to my parents, Marilyn and Tim Evans, to Sandie Dearn, Ken and Alex Jackson, Malcolm Chawner and Helen Bradford and the ever-present distraction provided by a playful golden retriever named Rosie and a frog-catching kitten called Simba.

To Michelle – with love and thanks for your constant, unfailing support
In memory of Astronaut Chuck Brady (1951–2006)

Figures

Fully suited, Story Musgrave prepares for submersion in the WETF to practice his upcoming spacewalk 4
Challenger rolls from the Orbiter Processing Facility to the Vehicle Assembly Building for attachment to her External Tank and Solid Rocket Boosters 6
The airframe of STA-99 during construction at Rockwell's Palmdale plant in 1977 . 9
Story Musgrave translates hand over hand along the starboard sill of Challenger's payload bay during the first Shuttle spacewalk 15
Demonstrating the cramped nature of the Shuttle's flight deck, this image shows the four STS-6 astronauts during pre-mission training. 21
Challenger undergoes the first Flight Readiness Firing of her three main engines on December 18th 1982 24
Challenger thunders aloft on her maiden orbital voyage 28
With its communications gear folded up inside the large black solar panels, NASA's first Tracking and Data Relay Satellite is raised to its deployment position 33
Story Musgrave operates the Continuous Flow Electrophoresis System on Challenger's middeck. 37
Mounted atop a heavily modified Boeing 747 Shuttle Carrier Aircraft, Challenger flies over the Johnson Space Center in Houston, Texas, during her return journey from California to Florida in April 1983 42
Sally Ride in the Pilot's position on Challenger's forward flight deck 47
The STS-7 crew assembles in Challenger's forward flight deck for an in-flight photograph 50
Sally Ride glances through the overhead flight deck windows during simulator training on the Remote Manipulator System. 52
In addition to utilising his four crewmates for medical tests, Norm Thagard conducted a number of experiments on himself, including this series of observations of head and eye movements in microgravity 55
On June 13th 1983, with less than a week to go before the STS-7 launch and poised on Pad 39A, Challenger's payload bay is fully laden for her second mission. 58

Canada's Anik-C2 communications satellite, with a PAM-D booster attached to its base, departs Challenger's payload bay. 61
Deftly manipulated from within Challenger's flight deck, the RMS grapples SPAS-1 during the final phase of proximity tests. 64
Challenger touches down at Edwards Air Force Base on June 24th 1983. 70
Insat-1B is positioned inside its sunshield during STS-8 pre-flight processing 74
With the STS-8 stack looming behind them on Pad 39A, the five-man crew greets the media before their Terminal Countdown Demonstration Test. 77
A powerful electrical storm creates an eerie 'tapestry' of light at Pad 39A in the hours preceding the Shuttle's first nocturnal lift-off. 79
Turning night into day across a sleeping Florida, Challenger effortlessly carries out the Shuttle's first nocturnal lift-off. 81
Glinting in the sunlight, Insat-1B and its attached Payload Assist Module booster drift away into the inky blackness 84
The Payload Flight Test Article during 'dynamic' exercises involving Challenger's mechanical arm 87
Dale Gardner sleeps on Challenger's tiny middeck 90
Illuminated by powerful xenon floodlights, Challenger swoops into a darkened Edwards Air Force Base on September 5th 1983. 92
Spacesuited Bob Stewart and Bruce McCandless flank Ron McNair in the STS-41B crew's official portrait. 96
Stunning image of Bruce McCandless during his first flight with the Manned Manoeuvring Unit 99
A view of the 'rear' of the Manned Manoeuvring Unit during vacuum chamber tests at the Johnson Space Center in March 1981 101
A simulated Main Electronics Box for Solar Max is installed onto SPAS-1A by technicians in the weeks preceding STS-41B's launch. 105
Bruce McCandless, carrying a TPAD device similar to the one earmarked for the Solar Max repair, prepares to dock onto SPAS-1A 106
Bruce McCandless tests a mobile foot restraint, whilst attached to Challenger's Canadian-built mechanical arm 108
Thermal vacuum testing of the IRT at NASA's Johnson Space Center in December 1983 111
Surrounded by servicing vehicles, Challenger sits on the Shuttle Landing Facility, after becoming the first manned spacecraft to touch down back at her launch site 113
On March 29th 1984, after depositing the STS-41C stack on Pad 39A, the gigantic 'crawler' inches its way back to the Vehicle Assembly Building. 116
The Long Duration Exposure Facility in March 1984, shortly before installation aboard Challenger 118
Terry Hart – after recovering from his space sickness episode – prepares the IMAX camera to film aspects of the Solar Max repair effort. 122
George 'Pinky' Nelson, equipped with a Manned Manoeuvring Unit, makes his first unsuccessful attempt to dock onto the slowly spinning Solar Max 125
James 'Ox' van Hoften and George 'Pinky' Nelson pre-breathe pure oxygen using their launch and entry helmets before one of their two spacewalks 128
Crowded into Challenger's aft flight deck and framed by her overhead and payload bay-facing windows, the STS-41C crew celebrates their success. 130
Shot through the small circular window in the middeck side hatch, this view of the

Kennedy Space Center was taken by Kathy Sullivan moments before touchdown at the end of STS-41G . . . 132
The STS-41G crew at a pre-flight press conference . . . 136
Deployment of ERBS . . . 144
Challenger's robot arm is used to push one of the SIR-B antenna's leaves shut, prior to an orbit-lowering manoeuvre . . . 148
Hurricane Josephine as viewed from STS-41G . . . 149
Dave Leestma during his spacewalk . . . 154
Marc Garneau and Paul Scully-Power work with CANEX-1 experiments on Challenger's middeck . . . 156
Bob Crippen, photographed during Challenger's fiery re-entry on October 13th 1984 158
Challenger swoops like a bird of prey onto the KSC runway . . . 159
Vice-President George Bush talks with astronauts Owen Garriott and Ulf Merbold in the first Spacelab module, shortly after its dedication at the Kennedy Space Center's Operations and Checkout Building on February 5th 1982 . . . 164
When they finally sat down to eat breakfast on launch morning, April 12th 1985, Karol 'Bo' Bobko's crew would fly under a different mission number . . . 168
One of the two squirrel monkeys is prepared for STS-51B . . . 171
Perhaps trying to distract his attention from floating monkey faeces, Bob Overmyer undertakes Earth resources photography through Challenger's overhead flight deck windows, using a Linhof camera . . . 174
Silver team member Norm Thagard bails out of his sleeping bunk on Challenger's middeck, while gold team counterpart Don Lind works on the Autogenic Feedback Training experiment . . . 177
Diagram of the gravity gradient attitude adopted by Challenger during the bulk of STS-51B orbital operations . . . 180
Bill Thornton assists Taylor Wang, whose upper body appears to have been completely swallowed by the Drop Dynamics Module, during the repair of his experiment . . . 183
View of the Drop Dynamics Module during its operations later in the mission . . . 185
Don Lind observes the Vapour Crystal Growth System in the Spacelab-3 module . . 187
Granted considerably more 'space' than previous Challenger crews, the STS-51B astronauts stretch their legs in the Spacelab-3 module . . . 189
The abort mode switch on Challenger's instrument panel, ominously set at the Abort To Orbit option . . . 195
Challenger experiences a dramatic on-the-pad shutdown of her three main engines on July 12th 1985 . . . 197
Although both Coke and Pepsi dispensers were ultimately carried on STS-51F, following lengthy 'Cola Wars', the original 'first' sponsor, Coca-Cola, was actually the first to be sampled in space . . . 199
The Spacelab-2 hardware undergoes checkout in the Orbiter Processing Facility in March 1985 . . . 201
Spacelab-2's Instrument Pointing System, equipped with four powerful solar physics instruments, comes alive on STS-51F . . . 205
Blue team shift members Tony England and John-David Bartoe at work in Challenger's cramped aft flight deck area . . . 210
The drum-like Plasma Diagnostics Package in the grip of Challenger's mechanical arm during a series of experiments on August 1st 1985 . . . 212
The STS-51F crew poses for their official portrait in front of the Instrument Pointing System . . . 215

The record-sized crew of Challenger's last successful mission pose for an impromptu portrait in front of the Shuttle simulator . . . 218
The Spacelab-D1 payload is installed aboard Challenger in KSC's Operations and Checkout Building . . . 220
Working busily around their 'half' of the 24-hour clock, red team members Guy Bluford, Reinhard Furrer and Ernst Messerschmid tend to various research facilities in the Spacelab-D1 module . . . 221
Wubbo Ockels, whose shifts overlapped both the red and blue teams, climbs into his own sleeping bag in Challenger's airlock . . . 224
During their shift on the flight deck, Commander Hank Hartsfield and Pilot Steve Nagel keep watch on Challenger's systems . . . 226
Challenger returns from space for the ninth and final time . . . 229
Icicles on Pad 39B's launch tower on the morning of January 28th 1986 . . . 233
The STS-51L crew in the 'white room' on Pad 39B, during their Terminal Countdown Demonstration Test on January 8th 1986 . . . 234
The official mission emblem for the Spartan–Halley project, showing the small, boxy satellite and highlighting its quest to better understand the celestial wanderer . . . 235
First evidence of a puff of smoke from the right-hand Solid Rocket Booster, milliseconds after STS-51L's lift-off . . . 243
Almost 59 seconds into Challenger's ascent, an ominous flame from the breached right-hand Solid Rocket Booster begins to impinge on the External Tank . . . 245
Challenger's still-firing main engines, forward fuselage – containing the crew cabin – and residual propellants hurtle out of the fireball at T + 78 seconds . . . 246
Thirty per cent of Challenger's structure was recovered from her Atlantic grave . . . 248
Remains of O-ring seal 'tracks' and putty in recovered debris from STS-51L's right-hand Solid Rocket Booster . . . 251
Had STS-51L survived, these four men would have been Challenger's next crew . . . 257
Members of the Rogers Commission, including chairman William Rogers, arrive at the Kennedy Space Center on March 7th 1986, during the course of their inquiry into the Challenger disaster . . . 263

Abbreviations and acronyms

ACIP	Aerodynamic Coefficient Identification Package
ACOMEX	Advanced Composite Materials Experiment
AEM	Animal Enclosure Module
AFRSI	Advanced Flexible Reusable Surface Insulation
AFT	Autonomic Feedback Training
APE	Auroral Photography Experiment
APU	Auxiliary Power Unit
ASE	Airborne Support Equipment
ASEAN	Association of South-East Asian Nations
ASTP	Apollo–Soyuz Test Project
ATMOS	Atmospheric Trace Molecule Spectroscopy
ATO	Abort To Orbit
BMFT	German Federal Ministry of Research and Technology
BTS	Biotelemetry System
CFC	chloroflurocarbon
CFES	Continuous Flow Electrophoresis System
CHAMP	Comet Halley Active Monitoring Program
CHASE	Coronal Helium Abundance Spacelab Experiment
CRNE	Cosmic Ray Nuclei Experiment
CRRES	Cosmic Release and Radiation Effects Satellite
CRT	Cathode Ray Tube
CRUX	Cosmic Ray Upset Experiment
DDM	Drop Dynamics Module
DDU	Data Display Unit
DEMS	Dynamic Environment Measuring System
DFI	Development Flight Instrumentation
DFVLR	Federal German Aerospace Research Establishment
DPM	Drop Physics Module

DSO	Detailed Supplementary Objective
DSP	Defense Support Program
DTO	Detailed Test Objective
EIS	Experiment Initiation System
ELDO	European Launcher Development Organisation
EMU	Extravehicular Mobility Unit
EOIM	Evaluation of Oxygen Interaction with Materials
EOS	Electrophoresis Operations in Space
ERBE	Earth Radiation Budget Experiment
ERBS	Earth Radiation Budget Satellite
ESA	European Space Agency
ESRO	European Space Research Organisation
ET	External Tank
EUVE	Extreme Ultraviolet Explorer
EVA	Extravehicular Activity
FEE	French Echocardiograph Experiment
FES	Fluid Experiment System
FILE	Feature Identification and Location Experiment
FRF	Flight Readiness Firing
FSS	Flight Support Structure
GAS	Getaway Special
GFFC	Geophysical Fluid Flow Cell
GLOMR	Global Low Orbiting Message Relay
GPC	General Purpose Computer
GPWS	General Purpose Workstation
GRO	Gamma Ray Observatory
GSFC	Goddard Space Flight Center
GSOC	German Space Operations Centre
HiRAP	High-Resolution Accelerometer Package
HRTS	High-Resolution Telescope and Spectrograph
HUD	Heads-Up Display
HUT	Hopkins Ultraviolet Telescope
HXIS	Hard X-ray Imaging Spectrometer
IML	International Microgravity Laboratory
IPS	Instrument Pointing System
IRAS	Infrared Astronomy Satellite
IRT	Infrared Telescope
IRT	Integrated Rendezvous Target
IUS	Inertial Upper Stage
IWG	Investigators Working Group
JPL	Jet Propulsion Laboratory
JSC	Johnson Space Center
KSC	Kennedy Space Center
LATS	LDEF Assembly and Transportation System
LDEF	Long Duration Exposure Facility

LFC	Large Format Camera
LRSI	Low-Temperature Reusable Surface Insulation
MADS	Modular Auxiliary Data System
MAPS	Measurement of Air Pollution by Satellite
MAUS	Materialwissenschaftliche Autonome Experimente unter Schwerelosigkeit
McIDAS	Man–computer Interactive Data Access System
MDD	Mate–Demate Device
MEA	Materials Experiment Assembly
MEDEA	Materials Science Experiment Double Rack for Experiment Modules and Apparatus
MEM	Meteoroid and Exposure Module
MLP	Mobile Launch Platform
MLR	Monodisperse Latex Reactor
MMS	Multi-Mission Modular Spacecraft
MMU	Manned Manoeuvring Unit
MOMS	Modular Optoelectronic Multi-spectral Scanner
MPESS	Mission Peculiar Equipment Support Structure
MSE	Manned Spaceflight Engineer
MSFC	Marshall Space Flight Center
MSL	Microgravity Science Laboratory
NOAA	National Oceanic and Atmospheric Administration
NOSL	Day/Night Optical Survey of Lightning
NUSAT	North Utah Satellite
OMS	Orbital Manoeuvring System
OOE	"Out of the Ecliptic"
OPF	Orbiter Processing Facility
ORS	Orbital Refuelling System
OSP	Optical Sensor Package
OSTA	Office of Space and Terrestrial Applications
PAM	Payload Assist Module
PAPI	Precision Approach Path Indicators
PDP	Plasma Diagnostics Package
PEAP	Personal Egress Air Pack
PEC	Photoelectric Cell
PFTA	Payload Flight Test Article
PGU	Plant Growth Unit
POCC	Payload Operations Control Center
PSN	Pasifik Satelit Nusantara
RAHF	Research Animal Holding Facility
RCS	Reaction Control System
RME	Radiation Monitoring Experiment
RMS	Remote Manipulator System
RTG	Radioisotope Thermoelectric Generator
RTLS	Return to Launch Site

SAEF	Spacecraft Assembly and Encapsulation Facility
SAFER	Simplified Aid For Extravehicular Activity Rescue
SAGE	Stratospheric Aerosol Gas Experiment
SAL	Scientific Airlock
SFHE	Superfluid Helium Experiment
SIR	Shuttle Imaging Radar
SLF	Shuttle Landing Facility
SLS	Spacelab Life Sciences
SOS	Space Operations Simulator
SOUP	Solar Optical Universal Polarimeter
SPARTAN	Shuttle Pointed Autonomous Research Tool for Astronomy
SPAS	Shuttle Pallet Satellite
SRB	Solid Rocket Booster
SSBUV	Shuttle Solar Backscatter Ultraviolet
SSIP	Shuttle Student Involvement Program
STA	Shuttle Training Aircraft
SURF	Synchrotron Ultraviolet Radiation Facility
SUSIM	Solar Ultraviolet Spectral Irradiance Monitor
SWAA	Scientific Window Adaptor Assembly
TAL	Transoceanic Abort Landing
TCDT	Terminal Countdown Demonstration Test
TDRS	Tracking and Data Relay Satellite
TISP	Teacher In Space Project
TLD	Thermoluminescent Dosimeter
TPAD	Trunnion Pin Attachment Device
UARS	Upper Atmosphere Research Satellite
UIT	Ultraviolet Imaging Telescope
USML	United States Microgravity Laboratory
VAB	Vehicle Assembly Building
VCAP	Vehicle Charging and Potential
VCGS	Vapour Crystal Growth System
VHRR	Very High Resolution Radiometer
VPF	Vertical Processing Facility
WCDT	Wet Countdown Demonstration Test
WETF	Weightless Environment Training Facility
WUPPE	Wisconsin Ultraviolet Photopolarimeter Experiment

1

Flight of the Geritol Bunch

SPACE COWBOYS

Publicly, Paul Weitz' STS-6 crew was nicknamed 'The F-Troop'.

The nickname originated from a television series about an ageing cavalry unit and partly honoured their military backgrounds, as well as reflecting the fact that they were the sixth team of astronauts to fly the Space Shuttle. It was Weitz' idea and they even had 'official' F-Troop photographs and memorabilia produced.

"We had little T-shirts and pants," remembered Mission Specialist Don Peterson, "and bought cowboy hats. I had a sword that had once belonged to some lieutenant in Napoleon's army. We got a Winchester lever-action rifle, a bugle and a cavalry flag. Weitz was the Commander and sat there, very stern-looking, with the sword sticking in the floor. We had that picture made and were passing them out and NASA asked us not to do that. They thought it was not dignified, but I thought it was hilarious!"

Certainly, the aged cowboy image was apt, because when Challenger roared into clear Florida skies on April 4th 1983 to begin her first orbital voyage, Weitz, Peterson, Pilot Karol 'Bo' Bobko and Mission Specialist Story Musgrave may have inspired the movie 'Space Cowboys' as the oldest astronaut crew to date.

In fact, behind their backs and with tongues firmly embedded inside cheeks, fellow astronauts dubbed Weitz' team, somewhat less flatteringly, 'The Geritol Bunch'. Years later, Peterson would not recall that nickname with quite as much fondness. "Maybe that was something everybody said about us when we weren't around," he said, "but when we were in orbit, somebody was talking about 'how old you guys are'. We had a bunch of F-Troop pictures and I couldn't resist, so I said 'We're not going to show these to anybody under 35 when we come back, so you wises-asses won't see them!' "

It was true that the four men of STS-6, with a combined age of 191, were the oldest yet launched. Only Weitz had flown before – on a four-week mission to the Skylab space station in mid-1973 – and later assumed the mantle of deputy chief of NASA's astronaut corps. For his crewmates, it was their first flight, but all had vast expertise on the ground. Particularly notable was Musgrave, who had amassed extensive flying time as a US Marine Corps aviator and secured half a dozen degrees, yet waited an unenviable 16 years for his first trip into space.

Scientist, doctor, engineer, pilot, mechanic, poet and literary critic, Musgrave approached STS-6 with the characteristically philosophical outlook for which he was to become famous. "I got into this business to be on the intellectual and physical frontier," he said. "I wanted a transcendental experience – an existential reaction to the environment. I'm not talking about an illusion or seeing something that wasn't there, but a magical emotional reaction to the environment. That is what I've been after all my life: to experience and feel new sensations."

Many astronauts, without much hesitation, have labelled six-time spacefarer John Young as their hero, icon and one of the most outstanding pilots ever rocketed into the heavens; if that is true, then Musgrave is undoubtedly among the most spiritual. Indeed, Swiss astronaut Claude Nicollier – who flew with him on STS-61 in 1993 – likened Musgrave's intelligence to that of a super-developed alien. Others who worked with him over the years have paid tribute to his remarkable attention to detail and insistence on knowing every part of his mission, down to the tiniest aspect.

Musgrave has freely admitted that, even on his first flight, he exuded an aura of self-confidence "in myself and the mission. I knew what was going to happen – and it happened! I knew every valve, every switch and every number on this flight. It was sheer play for me to be able to so completely interact with my environment."

UNEXPECTED CHANGE

"Story was a fun guy to work with," remembered Peterson. "On the job, he was extremely dedicated and would do anything; he'd work 20 hours a day! He didn't argue about anything, but just did whatever needed to be done. It's really delightful to work with a guy like that." Musgrave would ultimately fly six times into orbit, but was already a rising star at NASA, having closely followed, virtually from conception, the development of the Shuttle's spacesuit. Entirely appropriately, on STS-6, he and Peterson became the first men to perform a 'real' spacewalk and show what it could do.

Their historic excursion came about by accident, rather than design.

A planned outing on STS-5 in November 1982, featuring astronauts Bill Lenoir and Joe Allen, was jinxed from the start when two members of Space Shuttle Columbia's crew suffered severe bouts of motion sickness. More trouble was afoot, however, when they finally donned their suits and ran through the laborious, pre-spacewalk checks. A problem was noted in Allen's ventilation fan; it sounded, said the crew, "like a motorboat". In effect, it was starting up, running unexpectedly slowly, surging, struggling and finally shutting itself down.

Nor was Allen's suit the only one causing headaches. The primary oxygen regulator in Lenoir's snow-white ensemble – which he would have used during a series of 'pre-breathing' exercises and throughout the spacewalk itself – failed to produce enough pressure; regulating to 3.8 psi instead of the required 4.3 psi. Several of the astronauts' helmet-mounted floodlights also failed to work properly. After fruitless attempts to troubleshoot the problems, the spacewalk was cancelled and deferred to Weitz' mission, scheduled for just ten weeks later.

"It didn't give us much time to train," recalled Peterson. "I didn't have much experience in the suit, but the advantage we had was that Story was the astronaut office's point of contact for the suit development, so he knew everything there was to know. He'd spent 400 hours in the water tank, so he didn't really have to be trained."

This water tank was known as the Weightless Environment Training Facility (WETF) in Building 29 of NASA's Johnson Space Center (JSC) in Houston, Texas. In anticipation of the immense spacewalking load required to build the International Space Station, in 1997 the WETF was superseded by the larger Neutral Buoyancy Laboratory. Since the mid-1960s, the use of 'neutral buoyancy' – placing astronauts, fully suited and laden with lead weights, underwater – has been recognised as an effective means of simulating the 'zero gravity' of low-Earth orbit.

Accompanied by scuba divers to ensure their safety, Musgrave and Peterson were thus able to rehearse both planned and contingency procedures for their spacewalk in the 7.6 m deep tank. Measuring 23.8 m long, 10.1 m wide and holding almost 1.9 million litres of water, together with a full-size model of the Shuttle's crew cabin, airlock and cavernous payload bay, the WETF's value was balanced by a number of lingering concerns.

"Its disadvantages," said astronaut Bruce McCandless, whose involvement in the suit's development closely rivalled Musgrave's own, "included the optical distortion caused by looking out through a concave helmet in the optically denser medium of water. That caused everything to look smaller, coupled with the viscous drag from the water and the risk of 'bends' if you stayed in too long and too deep."

Of course, during their training, neither Musgrave nor Peterson was truly 'weightless', and two key differences between operating in the tank and working in space were that they could still 'feel' the weight of their 125 kg suits and the effect of the water, which tended to make some tasks easier to perform on the ground. "The WETF lied to us," admitted astronaut Kathy Sullivan, explaining that pressure differences 'inside' and 'outside' the suits in orbit were greater than in the tank, meaning that fingers, arms and legs became harder to bend in space.

Nevertheless, even today, the concept remains the closest parallel to the real thing. During typical training runs, the scuba divers guided Musgrave and Peterson into the pool and fitted the weights onto their suits, to enable the pair to 'hang' in the water, neither rising nor sinking.

Inside the bulky ensembles, conditions became both uncomfortable and painful. The men's bodies were supported by the weight of their suits, which meant "the blood ran to your head when upside down," explained McCandless, "and your weight was supported on your collarbone!" This made a precise fit essential: both astronauts' heels had to be firmly planted against the backs of their boots, their shoulders against their harnesses and their heads touching the crowns of their helmets. Horror stories from earlier missions, in which astronauts lost untrimmed nails because of imperfectly fitting suits, also required gloves to be close to fingertips.

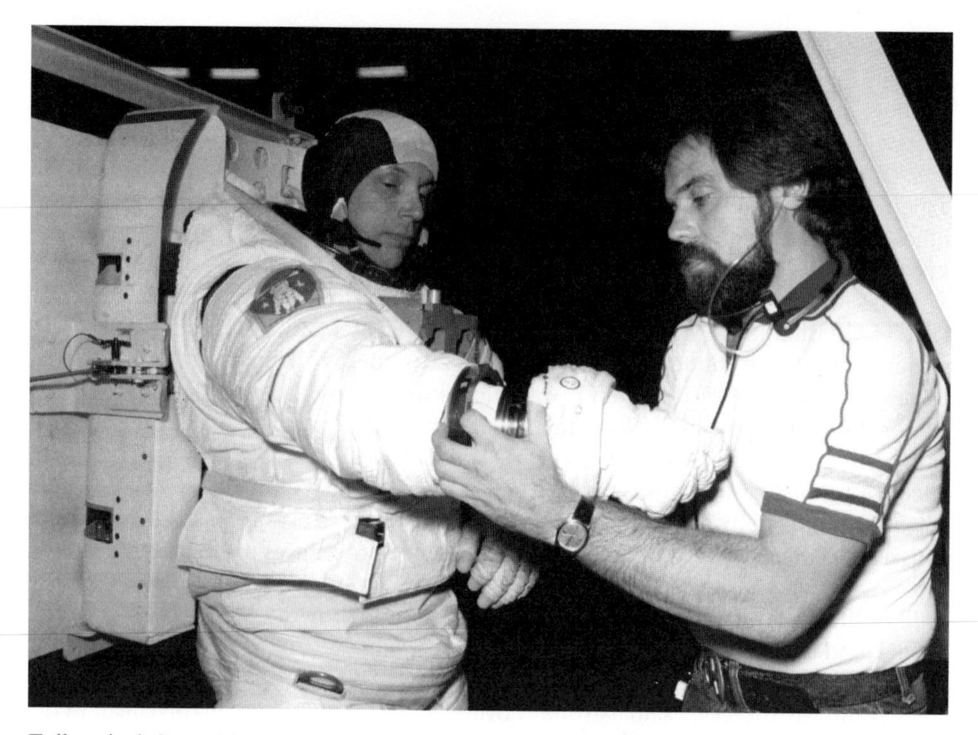

Fully suited, Story Musgrave prepares for submersion in the WETF to practice his upcoming spacewalk.

"It used to be a form of medieval torture to hurt people's fingernails," Peterson said, "but the gloves, if they're wrong, can be really bad." On the Apollo 15 expedition to the Moon in mid-1971, astronaut Dave Scott's gloves were so tight that he lost several fingernails and was forced to take aspirin to keep working. Too loose, on the other hand, and spacewalkers could lose the ability to feel and grip objects or make precise movements.

Astronaut George 'Pinky' Nelson, who worked on the development of the spacesuit before making two excursions of his own in April 1984, likened the effects of hand fatigue to "squeezing a balloon", adding that any irritation or pressure points could quickly lead to soreness in the fingers. Consequently, the gloves were manufactured in 15 sizes, permitting sufficient dexterity, according to Nelson, to pick up a coin the size of a penny, "given enough time!"

"My training was pretty rushed," recalled Don Peterson of his STS-6 preparations. "I was in the water 15 or 20 times, and that's not really enough to know everything you need to know. But all we were doing was testing the suit and the airlock, so we weren't doing anything critical to the survival of the vehicle. We were just testing equipment and the deal was that if something went wrong, we'd stop and come back inside. The fact that I wasn't highly skilled in the suit didn't matter that much."

Immediately after Columbia landed from STS-5, a task force was established, led by NASA's Richard Colonna, to investigate the spacesuit anomalies. The fault in Lenoir's suit was traced to two missing 'locking' devices – each the size of a grain of rice – in the primary oxygen regulator. The paperwork provided by its manufacturer, Hamilton Standard of Hartford, Connecticut, indicated that the devices had been fitted in August 1982, but actually they had not been fitted at all and inadequate checking failed to discover this. It was their absence that allowed the pressure in Lenoir's suit to drop back by half a pound.

The problem with Allen's suit was a faulty magnetic sensor in the fan electronics. Colonna's report, published in December 1982, pointed out that "even with no improvements, if the regulator were fabricated properly, the [backpack] would function properly". It also listed ways to test and inspect the regulators and motors, in addition to recommending checks inside the Shuttle's airlock on the day before launch. Additional plans were laid out to provide sensors with better moisture resistant coatings to future motors and to initiate new tests to highlight manufacturing defects, although these measures were not ready in time for STS-6.

BIRTH OF CHALLENGER

Musgrave and Peterson's moment of triumph would make them the first Americans to leave their spacecraft in orbit since February 1974 and give the world a glimpse of the new Shuttle suit in action. Today, it has become increasingly familiar as missions have routinely serviced the Hubble Space Telescope and begun constructing the International Space Station. Yet, originally, in the early developmental days of the Shuttle, an Extravehicular Activity (EVA) capability was considered unnecessary and was not provided.

"The NASA perspective of the Shuttle was an airliner," explained spacesuit engineer Jim McBarron, "and the people inside it wouldn't need suits. It was through prompting and questioning that Aaron Cohen, who was then the Shuttle's project manager, accepted a contingency capability for closing the payload bay doors – which was an issue they were faced with – to put in an EVA capability."

To understand the purpose of this payload bay, one must first comprehend the underlying reasons for the Shuttle's existence and the billions of dollars invested in what was undoubtedly the most advanced spacecraft yet to leave Earth. As well as being advanced, the four-strong fleet of orbiter vehicles were also to be reusable; capable of flying, supposedly, every fortnight to carry commercial satellites, scientific laboratories, space probes, astronomical instruments and – for the first time – ordinary civilians aloft. Plans were underway to send teachers, journalists and foreign nationals into space, with up to seven seats available on each mission.

The delta-winged Shuttle, which made its first orbital flight in April 1981 with a pioneering, two-day voyage by Columbia, was appropriately named: it would whisk people into space frequently, reliably, relatively cheaply and in conditions a world away from the cramped, one-use-only ballistic capsules of the 1960s. History has shown that only a few of these promises were fully realised.

Challenger rolls from the Orbiter Processing Facility to the Vehicle Assembly Building for attachment to her External Tank and Solid Rocket Boosters. Delta-winged and intended to be the spacegoing equivalent of a commercial airliner, the Shuttle never fully realised NASA's promise of 'routine' access to space.

Before such a complex machine could be declared to be operational, it had to be exhaustively tested. Much of this work took place during and after its construction and a series of high-altitude approach and landing runs were conducted in mid-1977 using an aerodynamic test vehicle called Enterprise, hauled aloft from Edwards Air Force Base in California by an adapted Boeing 747 airliner and released to glide back. Astronaut Fred Haise, one of her pilots, later called it "a magic carpet ride". Although she was never capable of flying into space and is now on display in the Smithsonian, Enterprise demonstrated the Shuttle's aerodynamic performance and ability to make precise landings on pre-determined runways.

Overall, Haise was happy with her flying characteristics. "It handled better, in a piloting sense, than we had seen in any simulation," he said later, "either our mission simulators or the Shuttle Training Aircraft. The term I use is that it was tighter. It was crisper in terms of control inputs and selecting a new attitude in any axis and being able to hold that attitude. It was just a better handling vehicle than we'd seen in the simulations."

Initially, it was hoped that, following her approach and landing tests, Enterprise would be extensively upgraded to make her capable of travelling into space. In March 1978, she moved to NASA's Marshall Space Flight Center (MSFC) in Huntsville, Alabama, and spent the following year undergoing further tests. The results prompted a number of design changes, after which she would be modified for her first orbital mission. However, as time has shown us, this never happened.

Lessons learned during her fabrication were subsequently incorporated into the design of Columbia and NASA realised in 1978 that Enterprise weighed too much to transport a full payload into orbit; she would need a new set of plans, quite different from those of her sister, to render her spaceworthy. Moreover, she contained no propulsion system, plumbing, fuel lines or tanks. Her three main engines were all dummies, her payload bay had no mounting hardware for cargo, its doors had no opening mechanisms or radiators and her thermal protection system was little more than black and white polyurethane and fibreglass.

In view of the fact that Enterprise would not be launched in the manner planned for her successors, nor fly in space, the instrument suite in her cabin was sparsely furnished with switches and dials, compared to later orbiters. She had no guidance equipment, such as star trackers or heads-up displays, and no indicators of the systems of an external fuel tank and twin solid-propellant rocket boosters that would support her violent climb into space. Elsewhere, she had no aft flight deck or overhead windows, no airlock, no middeck lockers, no galley and her fuel cells were high-pressure tanks, rather than cryogenic powerplants. Her landing gear was operated by explosive bolts, with no hydraulic mechanisms or manual backup systems.

She did, however, carry a pair of Lockheed-built ejection seats which would have fired her two pilots through a pair of aluminium panels in the ceiling in the event of an emergency. Modifying her for space missions, therefore, was envisaged to be a long, complex and costly process. Additionally, the new design specifications called for much stronger wings and mid fuselage than Enterprise possessed and several aluminium components would have been changed to titanium to save weight.

Transportation and modification funds to accomplish this were simply unavailable and, as 1978 wore on, NASA was already looking to a high-fidelity structural test article known as 'STA-99' as a cheaper option to upgrade for orbital service. Since the original Shuttle contracts were signed in July 1972, the reusable spacecraft's design had evolved under such weight-saving pressures that virtually all airframe components were required to handle significant structural stress. Furthermore, in view of the difficulties involved in accurately predicting mechanical and thermal loads on the vehicle using the limited 1970s-era computers, NASA opted to build STA-99 specifically as an engineering tool.

As a result, after its completion in February 1978, the structural test article underwent a year of intensive evaluation in a steel rig at Lockheed's Plant 42 facility in Palmdale, California. Originally built to test the Lockheed TriStar aircraft, the rig's 256 jacks subjected 836 load application points to pressures equivalent to those of launch, ascent, orbital flight, re-entry and landing. Even the tremendous jolt of main engine ignition was simulated by three hydraulic cylinders, each imparting a force of 450,000 kg. Additionally, cold nitrogen gas and thermal blankets were employed to recreate the frigid conditions of orbital flight and the intense heat of atmospheric re-entry. The decision to modify STA-99 as a 'true' orbiter came about because, unlike Enterprise, it was a bare, incomplete airframe and could be more economically upgraded.

Traditionally, manned spacecraft had been tested to 140 per cent of their design strength, but NASA engineers recognised that this might cause so much damage to STA-99 as to make it inadvisable to do so. Consequently, JSC's Thomas Moser and his team developed an analytical computer model to simulate over 3,000 measurement points on the airframe. Their results confirmed that it could easily withstand 140 per cent loads, with actual stress distributions in critical areas comparing favourably with pre-test model data.

On January 29th 1979, it was made official. Under a $1.9 billion contract between NASA and Rockwell International, STA-99 would follow Columbia into orbit as the second space-capable vehicle and two new orbiters would be built. On February 2nd the structural test article was renamed 'Challenger'.

Like Columbia (and, indeed, the subsequent vehicles), Challenger was named for a seafaring vessel that had made an outstanding contribution to exploration. The historical, nautical Challenger made a prolonged cruise from December 1872 until May 1876, gathering the equivalent of 50 volumes of information about the Atlantic and Pacific Oceans. Later, the name's proud heritage continued when the Apollo 17 crew chose it for their lunar module in December 1972.

However, ground evaluations, practice landings and structural test articles were no substitute for actually operating in space. Before she could be declared ready for flight, Challenger required substantial disassembly and rework – including the 'beefing-up' of her wings and installation of heads-up displays for her pilots – that got underway at Rockwell's Palmdale plant in November 1979. Her payload bay doors, aft body flap and elevons were removed and returned to their vendors for refurbishment, with her tailfin following in January 1980.

She had been built with a simulated crew cabin, which required the two halves of

The airframe of STA-99 during construction at Rockwell's Palmdale plant in 1977.

her forward fuselage to be 'cracked open' to remove it for modifications. In July 1981, after its own series of improvements, the aft fuselage returned to Palmdale. In physical appearance, the rebuilt Challenger looked similar to Columbia and Enterprise, at least at first cursory glance. External appearances, though, proved deceptive.

All three vehicles were similar in shape and approximate dimensions to the DC-9 airliner: roughly 36 m long with wings spanning 24 m from tip to tip. They comprised a two-tiered cockpit, cavernous, 18-m-long payload bay with clamshell doors and an aft compartment housing a cluster of three main engines, bulbous Orbital Manoeuvring System (OMS) pods and a vertical tailfin. Forty-four tiny Reaction Control System (RCS) thrusters in her nose and tail would provide additional manoeuvrability whilst in space.

Opening the graphite epoxy payload bay doors – the largest aerospace structures ever built, at that time, from composite material – was among the astronauts' first tasks in orbit, in order that radiators lining their interior faces could begin to shed excess heat from the electronic systems into space. The five-piece doors were hinged at either side of the mid fuselage, mechanically latched at the forward and aft bulkheads and thermally sealed at the centreline. Ordinarily, they were driven 'open' and 'closed' by electromechanical power drive units and gears, but if Weitz' crew had been unable to open the doors, flight rules dictated they return to Earth at the earliest opportunity. Conversely, if the doors would not close at mission's end, then Peterson and Musgrave were to go EVA and operate the mechanism manually.

WORKCLOTHES

As well as practicing how to manually winch the doors closed, plans for STS-6 called for Musgrave, designated 'EV1', with red stripes on the legs of his spacesuit for identification, and Peterson ('EV2') to rehearse procedures for the tricky recovery and repair of NASA's crippled Solar Max satellite, which was scheduled for the spring of 1984. Although the excursion was intended to last barely four hours, preparing for and conducting it consumed virtually the crew's entire work day on April 6th 1983.

Aided by spacewalk choreographer Bo Bobko, the two men rose early that morning to begin readying their equipment and Challenger's airlock, before spending three and a half hours 'pre-breathing' pure oxygen to wash nitrogen from their bloodstreams and avoid potentially lethal attacks of the 'bends'. Otherwise known as 'caisson disease', the bends are triggered by the formation and expansion of nitrogen gas bubbles in the blood when subjected to a rapid decrease in external pressure. The consequences can be dire: ranging from severe pain in the joints to paralysis and eventually death. Indeed, the name 'bends' comes from the fact that sufferers instinctively bend into a foetal position.

To sidestep this danger, in a procedure similar to that commonly followed by deep-sea divers, Musgrave and Peterson spent time in Challenger's 14.7-psi middeck, pre-breathing pure oxygen from facemasks to prepare themselves for operating inside the spacesuits at 4.3 psi pure oxygen.

Shortly before the onset of pre-breathing, the entire cabin had been reduced from its normal pressure to around 10.2 psi, while the percentage of oxygen in the atmosphere was slightly increased. As Musgrave and Peterson breathed through their masks, they were still able to attend to their other tasks on the middeck. At the end of pre-breathing, with sufficient dissolved nitrogen now cleared from their blood, they were at last ready to begin donning their spacesuits.

Most of this was done inside the airlock – a cylindrical structure about the size of a Volkswagen Beetle, situated at the rear of the middeck. Its inclusion within the cabin preserved the maximum amount of usable volume in the payload bay. It had two hatches: one for the astronauts to enter from the middeck and another through which they would venture into the payload bay. The interior was decidedly cramped and veteran spacewalker Michael 'Rich' Clifford lucidly described hanging in his bulky suit, barely able to even move his arms ...

Depressurisation was controllable either from the flight deck or within the airlock itself. Normally, two spacesuits were stored inside the chamber, although there was room for up to four, if needed. In fact, long after Challenger's tragic demise, many missions have involved four spacewalkers, working in two alternating pairs, and in May 1992 on STS-49 the airlock successfully demonstrated its ability to support three fully suited astronauts at the same time.

Since reaching orbit, Weitz, Bobko, Peterson and Musgrave had checked and rechecked their equipment for the long-awaited spacewalk: testing a third, 'spare' upper torso in accordance with flight rules, verifying that oxygen regulators and fans operated normally, inspecting for leaks and confirming that communications were satisfactory. In fact, the only problem raised was a need to replace some flat floodlight batteries. With everything in place, spacesuit donning began and, in true F-Troop fashion, it ran as crisply as a military campaign.

"You don't just put the suit on and open the hatch," Peterson explained. "You make sure everything's laid out properly and everything that you check is working properly. We were instrumented with little stick-on patches to measure our heart rates. Then we put on what looked like long underwear – a cooling garment – which had water tubes that ran through it and hooked through a connector to the suit."

This long underwear, officially known as the 'liquid cooling and ventilation garment', was a one-piece, zip-up suit, based on one developed for moonwalkers, composed of stretchable spandex fabric and laced with 91.5 m of plastic tubing. During the course of their spacewalk, cooling water would be pumped through this tubing to control Musgrave and Peterson's body heat, exhaled gases and perspiration. Next, anti-fog compound was rubbed onto the insides of their helmets. "I wore glasses," said Peterson, "and we rubbed this on the lenses so they wouldn't fog up, because I was inside a helmet and couldn't get my hands inside the suit."

To provide an additional measure of safety and prevent them from falling off, Peterson's glasses were tied to an elasticated strap around the back of his head.

Before actually clambering into the two-piece spacesuits, electrical harnesses were attached to their 'hard' upper torsos to provide biomedical and communications links through the backpack. A wrist mirror and spiral-bound, 27-page checklist were placed on each suit's left arm, followed by the insertion of a small fruit and nut food bar and water-filled drink bag. The next step was the connection of a black-and-white communications hat – famously nicknamed the 'Snoopy cap' since Apollo days – to the top of the torso.

Physically, the so-called Extravehicular Mobility Unit (EMU) was a $2.5 million miniature spacecraft in its own right, consisting of 'upper' (above-waist) and 'lower' (below-waist) segments, together with helmet, gloves and life-support backpack. Musgrave and Peterson firstly pulled themselves into the lower torso, which featured joints at its hips, knees and ankles and a metal body seal closure for connecting to a ring on the upper torso. It also included a large bearing at its waist, which offered greater mobility and allowed the astronauts to twist whilst their feet were held firmly in restraints.

After donning the trousers of the suit, their next step was to plug the airlock's service and cooling umbilical into a display and control panel on the front of the upper torso. This would provide cooling water, oxygen and electrical power from the Shuttle until shortly before they were scheduled to go outside, thereby conserving the limited consumables available in their backpacks. The two men finally entered the airlock itself, where the upper torsos 'hung' on opposing walls and, through a half-diving, half-squirming motion, manoeuvred themselves into the top halves of their suits.

With arms outstretched, and Bobko nearby to assist, they slipped themselves into the upper torsos and their waist rings were brought together, connecting the cooling water tubing and ventilation ducting of the long underwear and the biomedical sensors to their backpacks. Bobko then helped them to lock the body seal closure rings at their waists.

The hard upper torso was essentially a fibreglass shell under several fabric layers of a thermal and micrometeoroid garment. On its back, it held the life-support system and on its chest the display and control unit by which the spacewalker would manage his or her oxygen, coolant and other consumables; in fact, due to the difficulties in seeing 'down' to read labels on the unit, the mirrors on the suits' left wrists would help immeasurably. For additional ease, the labels were written backwards!

"The displays and controls in the suit are a challenge," said Pinky Nelson, who worked with Musgrave to develop them, "because you have to see them from inside the suit, looking down, so a lot of the old guys in the astronaut office couldn't read the displays as they were close to your face. We worked on [Fresnel] lenses and all kinds of ways to make the displays legible to people with old eyes!"

To aid his 49-year-old eyes, just before donning the helmet, Peterson tied his glasses securely onto his head and pulled on his Snoopy cap, equipped with microphone and headphones to provide two-way communications with his crewmates and Mission Control. Next came the gloves. Snapped into place on the wrist rings of the upper torso, these had silicone rubber fingertips to provide a measure of tactile sensitivity when handling tools in Challenger's payload bay.

Finally, the enormous polycarbonate bubble helmets were lifted over the astronauts' heads and clicked into place on the neck rings of their upper torsos. Over the top of each helmet was an assembly containing manually adjustable visors to shield their eyes from solar glare, together with two EVA lamps to illuminate work areas out of range of the Sun or the Shuttle's own payload bay floodlights. Mobility in the neck rings was unnecessary, because the helmets were easily big enough to allow the astronauts to move their heads around.

Unlike previous Apollo spacesuits, the modularised Shuttle ensemble, with its waist closure ring, eliminated the need for pressure-sealing zips and therefore had a much longer shelf life. Additionally, the use of newer, stronger and more durable fabrics enabled spacesuit engineers to design joints with better mobility, resulting in lower weight and a reduction in overall cost.

TIME TO OPERATE

Story Musgrave and Don Peterson, by now floating motionless in Challenger's tiny airlock, were, in effect, small spacecraft in their own right. However, they were not yet 'self-contained', as their oxygen, electricity and cooling water were still being provided by the Shuttle's systems; not until shortly before the two men ventured outside would they transfer to the life-support utilities of their suits' onboard consumables. Before they could do that, they had to lower the airlock's pressure to 4.3 psi in order to check their integrity, necessitating a final 45 minutes of pre-breathing.

At last, at 9:21 pm on April 6th 1983, Musgrave initiated the final depressurisation of the airlock and pushed open the outer hatch leading into the payload bay. The plan called for three hours of activities, but in order to accommodate delays he and Peterson had up to six hours' worth of consumables. Entering the overwhelmingly floodlit bay for the first time, one of his first comments, somewhat understatedly, was that "it's so bright out here!"

Their time outside was limited, but the experience would remain with both men for the rest of their lives. "You remember little things like sound," Musgrave told a post-flight press conference. "Even though there's a vacuum in space, if you tap your fingers together, you can hear that sound because you've set up a harmonic within the spacesuit and the sound reverberates within it. I can still 'hear' that sound today. But the main impression is visual: seeing the totality of humanity within a single orbit. It's a history lesson and a geography lesson – a sight like you've never seen."

Watching through the aft flight deck windows, Paul Weitz would later quip that "Story seemed like a butterfly coming out of a chrysalis – only he's not as pretty!" For Musgrave, who had spent years virtually designing the spacesuit that he now depended upon for his life, it was an intensely personal accomplishment. "This was to be only three hours of experience on top of 48 years," he said later, "but it's like a surgeon who's been training 16 years to operate. Sooner or later, a surgeon has to operate. Sooner or later, I knew I was meant to walk in space."

The poetic justice of being first to venture into space wearing the fruit of so many of his labours was clearly not lost on the intensely philosophical Musgrave.

Although somewhat different from the ensembles worn on previous Gemini and Apollo missions, they were designed with the same objective in mind: to leave the pressurised confinement of a spacecraft. However, the near-flawless procedures followed by Musgrave, Peterson and Bobko to don the suits masked a complex, tumultuous and near-tragic developmental history. "We had ten major replans," remembered Joe McCann, NASA's former EVA life-support systems manager, of its genesis. "The problems started pretty early with the upper torso and putting pivots in that, because Story couldn't get through it. He couldn't don the thing, because his elbows couldn't get close enough together, so we ended up putting gimbals and bellows on the suit to allow him to get in and be able to operate. That was a significant technical challenge to get that done, but it remained a key worry point because we had these pivots buried in the fibreglass and cases of them loosening up after a while. If they blew out, you'd blow the bellows and be dead! We got some early warnings of that happening in the WETF."

A multitude of other niggling glitches also characterised the spacesuit's early days: problems with CO_2-scrubbing lithium hydroxide cartridges, clogged sublimators, cycling water pressure regulators and battery failures. Pinky Nelson remembered it as a time of frustration and near-fatal consequences. "The suit actually blew up shortly before STS-1," he said of a harrowing episode on April 18th 1980. "I was home, working in my garden, and got a call that there had been an accident with the spacesuit. They were doing some tests in one of the vacuum chambers at JSC and going through the procedures of donning the suit, flipping all the switches in the right order and going through the checklist. There's a point at which you move a slider valve on the front and that pushes a lever inside a regulator and opens up a line that brings the high-pressure emergency oxygen tanks online. You do that just before you go outside. You don't need them when you're in the cabin, because you can always repressurise the airlock, but when you're going outside, you need these high-pressure tanks. It turned out that a technician threw that switch and the suit blew up! Not just pneumatically, but burst into flames, and he was severely burned. It was pure oxygen in there. The backpack is flammable in pure oxygen [and] it went up in smoke."

For Chester Vaughan, who was placed on NASA's investigation panel following the accident, there was a certain amount of debate over whether to publicly release the photographs of the badly charred spacesuit, "because most people might believe that we had someone inside". Fortunately, the technician did not actually need to be inside the ensemble on that occasion, but was required to reach over and activate switches on its chest panel. The resultant flash fire meant he spent that night in the Galveston Burns Center, downtown from JSC ...

Three years later, with these and other technical woes long since resolved, Musgrave and Peterson found that the suits functioned well as they read status reports to show all systems running near-perfectly. During their excursion, Bobko directed their every move from Challenger's aft flight deck, while Weitz photo-documented the historic event. The spacewalkers' first task was to tether themselves to slidewires running along the sills of the payload bay walls (one on either side, to prevent mutual interference) and move towards the aft bulkhead, in the process evaluating their ability to reach, pick up and handle tools.

The two slidewires, on the port and starboard sills, each ran for about 14 m to the aft bulkhead. During the spacewalk, they were used as part of a safety procedure to prevent Musgrave or Peterson from inadvertently floating away from the Shuttle. Meanwhile, the two men began conducting their first 'real' evaluations of the new

Story Musgrave translates hand over hand along the starboard sill of Challenger's payload bay during the first Shuttle spacewalk. Note the silvery lids of two Getaway Special canisters at bottom-centre and a tile-coated OMS pod at top-right.

suits: their comfort, dexterity, ease of movement and the performance of their communications and cooling systems and the payload bay floodlights.

One of very few concerns expressed by Peterson after the mission was that "the gloves are hard to work in – extremely stiff – and I had to get my hands strengthened with a little hand exerciser". Despite this, both astronauts reported that the suits' mobility enabled them to satisfactorily accomplish each of their tasks. Most of their work focused on identifying suitable locations from which future spacewalkers could best work on the malfunctioning Solar Max satellite, and on practicing some of the intended repair techniques.

This kind of work was essential, not only for the successful reactivation of the $240 million, Sun-watching observatory, but also for future servicing missions to the still-to-be-launched Hubble Space Telescope, at that time scheduled for the mid-1980s. Musgrave and Peterson finally evaluated the manual system for closing Challenger's payload bay doors in the event of a failure. This involved using a hand operated winch, attached to the forward bulkhead and was performed both with and without foot restraints.

It was during tests of the 7.3 m Kevlar winch line that they encountered difficulties. "Story got the rope hung over something," Peterson recalled, "and couldn't release the winch. It was under a lot of tension. There was some talk about how we could get this thing loose so we could get it restored. We couldn't just leave it where it was because it was on the rollers that were used to latch the doors down." After the crew's suggestion to cut the rope was rejected by Mission Control, Musgrave eventually pried it free with his gloved hands.

At one point in the spacewalk, during what NASA later labelled a "high metabolic period", Peterson received a 'high O_2 usage' warning on his chest display. Although the message cleared quickly and did not recur, it was attributed to flexure within the suit and his high work rate. "I was working with a ratchet wrench," he said. "We had launched a satellite out of a big collar mounted in the back of the orbiter and it was tilted 45 degrees. It had to be tilted back down before we could close the payload bay doors and come home. Instead of driving it with the electric motors, NASA wanted to see if we could crank it down with a wrench to simulate a failure. We had foot restraints, but it took so long to set them up and move them around that we didn't do that, so I just held on with one hand and cranked the wrench with my other. My legs floated out behind me. As I cranked, the suit's waist ring rotated back and forth, the seal popped out and the suit leaked badly enough to set off the alarms. I stopped and said 'I've got an alarm'. Story stopped what he was doing and came over. We were trying to check what was going on and the seal popped back in place and the leak stopped. Now, in those days, we didn't have constant contact with the ground. They weren't watching at the time that happened. By the time we dumped the data from the computers to the ground that showed the leak, we were back inside the orbiter."

The alarms, part of the suit's caution and warning system, had been another of Pinky Nelson's responsibilities. "It had a very crude computer," he said, "that monitored a number of different sensors and systems, so that you could tell how much air you had left in your tanks and how much life in your batteries and how much carbon dioxide was in the air. Then there was a caution and warning system attached to that, so if something went out of limits, it would ring a bell and scroll through a diagnostic program and offer advice on what to do."

It seemed that Peterson's alarm was caused by overworking and breathing excessively rapidly; this depleted his oxygen, forcing a higher feed level and triggering the warning. Biomedical data confirmed his heart rate was around 192 beats per minute whilst cranking the wrench, but Peterson doubted he could have worked so hard as to breathe enough oxygen to set off the alarm. Later, during a vacuum chamber test with another astronaut in 1985, a similar alarm sounded. Like Peterson's alarm, it was attributed to high oxygen usage, but turned out to be a leak caused by excessive flexure at the suit's waist.

In spite of the problems, both Musgrave and Peterson clearly savoured not only their first spaceflight experience, but also the opportunity to leave the vehicle in orbit. They were even able to look 'outwards' into the Universe. "The Shuttle flies with the payload bay towards Earth all the time," recalled Peterson, "but we thought it would be neat when we got on the dark side if we could look out at the night sky and see all the stars. We did better than that. When we were on the daylight side, we went into 'the

Ferris wheel mode'. Just like a Ferris wheel seat goes around and never changes attitude, we went around the world holding one attitude, so when we got on the dark side we faced exactly away. We got some beautiful pictures of Earth from different attitudes that we couldn't have done otherwise. We got on the dark side and Paul Weitz said 'Okay, guys, you asked for this. Now stop whatever you're doing and look'. We did – and, because there's a lot of light in the payload bay, we couldn't see anything! There was too much glare."

Musgrave asked that the floodlights be turned off, but Weitz would have none of it. He was concerned that, on Challenger's maiden mission, problems switching them back on again would have left the two spacewalkers struggling to find their way back to the airlock in pitch darkness. "We didn't get much of a view," continued Peterson, "but what we could see was pretty interesting."

"I guess I was hoping for a religious experience out there," Musgrave remembered of his first spacewalk. "I didn't expect anything particular. I was open-minded about it. Now, this is funny, but things went so smoothly that, in a sense, I was disappointed by what I felt! I never got that transcendental jolt. I never experienced a separation phenomenon. I had no sense of Earth being 'down'. In fact, I had no 'down' reference at all. My frame of reference was always the payload bay of the orbiter."

THE BUTTERFLY AND THE BULLET

After returning to the airlock, which was repressurised at 1:15 am on April 7th, the data on Peterson's alarm was pored over by flight controllers with dismay. "They were upset about it," he said later and, this being the suit's first outing in space, the astronauts would almost certainly have been directed back to the airlock had the problem occurred whilst in communication with Mission Control. At the time of STS-6, however, the Shuttle still relied heavily on a network of ground stations to relay its communications and data traffic during part – but, at just 20 per cent, by no means all – of each orbit.

That was set to change in time for Challenger's second mission in June 1983. It was ironic that on STS-6, which, in Peterson's mind, benefited from having gaps in communications with the ground, the first in a series of huge Tracking and Data Relay Satellites (TDRS) had been deployed to provide near-continuous communications with future Shuttle crews. It was optimistically hoped that, after the launch of a second, identical satellite on STS-8 later in 1983, it would be possible to talk to astronauts not only during the majority of their orbital time, but also throughout re-entry, eliminating the radio blackout of this phase of the mission.

Moreover, the existing network of 20-year-old ground stations – capable of supporting barely one or two low-Earth orbiting spacecraft at a time – could be effectively retired to save money. The TDRS system, on the other hand, could support the Shuttle and up to 26 other satellites simultaneously.

In the eyes of the world, Musgrave and Peterson's spacewalk was the defining moment of STS-6, but when Paul Weitz' crew was announced by NASA in March 1982 their key tasks were to evaluate Challenger's spaceworthiness and insert the first

TDRS into orbit. No spacewalk was planned. Originally, the mission was scheduled to fly on January 24th of the following year, but hydrogen leaks in the new orbiter's main engines led to two and a half months of delays.

After 13 years with NASA, a slightly longer wait proved of little consequence to Peterson. "It wasn't a huge big deal," he said. "I just figured, sooner or later, I'd get a chance to fly." His crewmates had waited an equally long time. Both Peterson and Bobko had been chosen in the mid-1960s by the US Air Force as research pilots to support its Manned Orbiting Laboratory. When that military space station, intended to conduct surveillance of the Soviet Union from polar orbit, was cancelled in June 1969, they were transferred to NASA as astronaut candidates.

"I had worked with Story before," Weitz said later, "because he was Joe Kerwin's backup for my Skylab mission and, of course, we would get together with the backup crew and compare notes, but I really wasn't that familiar with Bo or Don. Story had been an enlisted Marine, so he had been in the military and, of course, Bo and Don were Air Force pilots. That's one good thing about flying with military people: they understand chain of command. I'm not saying everything I said was God, because those guys would go off and we'd have crew discussions and I think we used a reasonable approach to accommodating different points of view on a certain aspect of getting ready to go fly." Unlike several previous astronauts, Weitz had no input in the selection of his crew; rather, he was simply told of their appointment by senior management. However, he added that he had full confidence in his colleagues' capabilities.

Musgrave, chosen as a scientist-astronaut in August 1967, stuck it out longer than Bobko and Peterson, and even Weitz waited almost a full decade between his first and second missions. By the time that the Shuttle undertook its maiden voyage in April 1981, it was three years behind schedule and what had been intended to be its highest profile flight – a delicate orbital ballet to re-boost Skylab and set it up for reoccupation – had been missed. Unexpectedly fierce solar activity in 1979 caused Earth's atmosphere to inflate, increasing air drag at orbital altitudes and the station disintegrated during re-entry in July of that same year.

The achievements brought by the Shuttle, including its widely publicised ability to make space travel 'routine', had not come about without problems. Since the original contracts to build the delta-winged spacecraft had been signed a decade earlier, its designers had faced setback after setback: frustrating problems with a patchwork of heat resistant tiles and blankets to shield it during its searing hypersonic descent and maddening explosions of its liquid-fuelled main engines. There was political fallout, too, with its powerful Congressional opponents questioning the need for a reusable manned spacecraft.

"Like bolting a butterfly onto a bullet" was how Musgrave described the unusual appearance of the Shuttle, its two behemoth Solid Rocket Boosters (SRBs) and giant External Tank (ET), the latter of which would feed Challenger's main engines with over 1.9 million litres of liquid propellants. It was an appropriate description. The 46.6-m-tall ET, reminiscent of an enormous aluminium zeppelin standing on end, was indeed bullet-like, but was actually far more than 'just' a container.

In fact, it comprised two tanks, one above the other. Separating the two was an 'inter-tank', which contained instrumentation and umbilical interfaces for the launch pad's purging and hazardous gas detection systems. Above the inter-tank, the liquid oxygen tank housed up to 542,640 litres of oxidiser and, below it, the liquid hydrogen tank held some 1.4 million litres of fuel. Both were then fed through two 43-cm-thick propellant lines into disconnect valves in the Shuttle's aft compartment and from thence into the main engines' combustion chambers.

"The main engine is very high performance," said Henry Pohl, NASA's former director of engineering and development, "with a very high chamber pressure for that day and time and very lightweight for the thrust that they were producing. I would say that we came out with that at the only time when it would have been successful. If we had waited another two years before starting development on the Shuttle, we probably would not have been able to do it, because the people that designed the main engine were the same that designed previous rocket engines. That group had designed and built seven different engines before they started the Shuttle development. A lot of them retired and so if we'd waited another two years, those people would all have been gone and we would have had to learn all over again on the engine development."

Built by Rocketdyne – formerly part of the Shuttle's prime contractor, Rockwell International, but now owned by Boeing – the engines burned for about eight minutes of Challenger's ascent and were shut down a few seconds before the ET was jettisoned, right on the edge of space. Each engine measured 4.2 m long, weighed 3,400 kg and was 'throttleable' at one per cent incremental steps from 65 per cent to 104 per cent rated thrust. This ability, which was controlled by the Shuttle's five onboard General Purpose Computers (GPCs), reduced stress on the vehicle during periods of maximum aerodynamic turbulence and also served to limit the g-loads in the final phase of ascent.

Despite the immense power generated by each engine and the colossal amount of propellant needed to run them for such a short length of time, they in fact provided Challenger with only 20 per cent of the muscle to reach space. The remainder came from the two 45.4-m-tall SRBs, which became the first solid-fuelled rockets ever used in conjunction with a manned spacecraft. Loaded with a powdery aluminium fuel and an oxidiser of ammonium perchlorate, the boosters, built by Morton Thiokol (now ATK Thiokol) of Utah, were mounted like a pair of Roman candles on either side of the ET.

Typically, during pre-flight preparations, the SRBs were paired in matching sets and loaded with propellant ingredients from identical 'batches' to minimise the risk of thrust imbalances during ascent.

This unusual combination, nicknamed 'the stack', was not wholly reusable and originated from financial and technical compromises back in the early 1970s. Each orbiter was designed to fly a hundred times before major refurbishment would become necessary, although none of the surviving vehicles will have achieved even half that number of flights by the time the fleet is retired in 2010. The SRBs were to be capable of flying 25 times apiece (although they would need to be stripped down and reassembled in between each mission), but the ET was discarded to burn up in the atmosphere. It would have proven more costly to recover and modify an ET than to simply build a new one for each mission.

Preparing for each Shuttle flight requires several years, but the actual bringing together of the components begins with setting up the boosters on a Mobile Launch Platform (MLP) in the gigantic Vehicle Assembly Building (VAB). This 160-m-tall structure – the world's largest engineering operations building, so vast that clouds once formed in its upper reaches before an air conditioning system was fitted – has dominated the marshy landscape of the Kennedy Space Center (KSC), just north of Cape Canaveral in Florida, for four decades. It was initially used to assemble the massive Saturn V rockets and, since 1980, the Shuttle.

Each booster comprises six blocks, called 'segments', each of which is positioned by overhead cranes with pinpoint grace, one atop the next and joined by a ring of bolts. To prevent a leakage of searing gases while operating, a series of rubberised O-rings seal the joints between the segments. After propelling the Shuttle and ET to an altitude of about 45.7 km, pyrotechnics separate the boosters from the supporting struts on the ET, explosive rockets at their nose and tail push them away and parachutes in the nose compartment are deployed to lower them to a gentle splash-down just off Cape Canaveral in the Atlantic Ocean. They are then recovered, stripped down, refurbished and reused.

When the assembly of the SRBs is complete, the ET is moved into position between them and connected by a series of spindly, but strong, attachment struts. Following checks of their mechanical and electrical compatibility, the Shuttle is moved from the nearby Orbiter Processing Facility (OPF), tilted by crane onto its tail and mated to the tank.

The transfer of the 1.8 million kg stack from the VAB to one of two pads at launch complex 39 – a distance of 5.6 km – takes six hours, with the aptly named 'crawler' inching the precious, $2.2 billion national asset along a track made from specially imported Mississippi river gravels. Once the stack is 'hard down' on the pad surface, further checks are conducted, payloads installed and the crew participates in a Terminal Countdown Demonstration Test (TCDT); essentially a full dress rehearsal of the last part of the countdown, followed by a simulated main engine failure and emergency evacuation exercise.

CLEAN AND SPACIOUS

By early December 1982, most of these preparations in support of STS-6 had already been completed; in fact, attached to her tank and boosters, and partially enshrouded in a gloomy midwinter fog, Challenger had crept to Pad 39A on November 30th. She had been delivered to Florida from California almost five months earlier and, despite her outward similarity to sister ship Columbia, appearances proved deceptive. Indeed, astronaut Gordon Fullerton, who flew both orbiters, described Challenger as much 'cleaner' and more spacious than her sibling.

One of the reasons for this spaciousness was the absence of two cumbersome ejection seats, which had been installed aboard Columbia in support of her first four test flights. After the completion of this orbital evaluation phase, the entire Shuttle fleet – not just Columbia – was declared 'operational' and many restrictions were relaxed; consequently, Challenger, like its new sisters in development, was designed without ejection seats. Instead, her astronauts were destined to journey into space in collapsible seats, clad in lightweight flight garments, rather than the bulky US Air Force pressure suits worn by Columbia's first four crews.

"From historical perspective," said Charlie Walker, a McDonnell Douglas engineer who flew three times on the Shuttle between August 1984 and December 1985, "that era was the equivalent of the 'white scarf' days of aviation. We were flying in blue coveralls: flight boots with steel heels and toes and a partial-pressure helmet. On the back of our seats was an oxygen-generating contraption that, if we could reach around and throw a switch, we'd have enough oxygen for four minutes to our helmets. We also wore fire retardant gloves, but no pressure suit. It was all intended for use with on-pad escape if there was an emergency before lift-off or on the ground after the vehicle had rolled to a stop."

Despite the ostensibly operational status conferred upon Challenger, even before her maiden flight, she was nevertheless fitted with a battery of test equipment to monitor her performance during launch, ascent, orbital operations, re-entry and landing. This included an Aerodynamic Coefficient Identification Package (ACIP),

Demonstrating the cramped nature of the Shuttle's flight deck, this image shows the four STS-6 astronauts during pre-mission training. At the front of the flight deck are Paul Weitz (left) and Karol 'Bo' Bobko. Story Musgrave, as 'flight engineer', is stationed behind and between the Commander and Pilot, with Don Peterson to his right.

which collected data on the various flight regimes – hypersonic, supersonic, transonic and subsonic – encountered during each mission, as part of NASA's ongoing plans to better define the spacecraft's capabilities. Elsewhere, mounted beneath the middeck and payload bay floors, the Modular Auxiliary Data System (MADS) took 246 pressure, temperature, strain and vibration measurements from different sections of Challenger's airframe.

Also aboard was a High-Resolution Accelerometer Package (HiRAP), which recorded aerodynamic motions along her principal axes during the early stages of re-entry into the atmosphere. In a similar vein to the research supported by ACIP, the goal was to determine Challenger's aerodynamic performance and handling qualities during some of the most dynamic – and least understood – phases of a mission. Such characteristics could not be reliably predicted on the ground through wind tunnel testing or computer modelling and needed 'real world' data to provide improved benchmarks.

Even loaded with this 'extra' equipment, the new orbiter weighed some 900 kg less than Columbia. This saving was achieved not only through the absence of ejection seats, but also through the replacement of several hundred thermal protection tiles with new insulating blankets, the removal of a number of tube supporting frames, the use of lightweight 'honeycomb' for her landing gear doors and tailfin and the incorporation of less weighty main engine heat shields.

It was not solely the operational nature of the Shuttle that precluded the use of ejection seats: on Columbia's fifth flight, her pilots were accompanied by two Mission Specialists, making history by rocketing as many as four astronauts into orbit on the same spacecraft. This posed a difficult and uncomfortable dilemma over how to eject from the vehicle in the event of a major malfunction. The problem was remarkably simple: only the pilots would have a chance of getting out alive.

There was only enough room at the front of the cramped flight deck to house two ejection seats and their rails for the pilots. Other crew members would have to make do with the collapsible seats at the rear of the flight deck or downstairs on the middeck. Installing four ejection seats with rails would be difficult and, as orbiter crews increased to a maximum size of seven, would become impossible.

"No-one wanted to fly with seven rockets in the cabin," former Shuttle manager Arnie Aldrich said, although others wanted to retain the seats. "That would have restricted the size of the crew," admitted former assistant Shuttle director Warren North, "because you couldn't put seven ejection seats in there, but we could leave the two pilot seats, then add four abreast that were of much lighter variety. The Yankee system, for instance, was a tractor rocket that pulled the pilot out in a prone position, where the seat pan collapsed and the pilot was pulled out head-first. That would have involved putting pyrotechnic escape hatches behind the flight crew and in the overhead payload deck, which would have involved redesigning the orbiter to some degree, including its wiring. It could have been done, but would have involved a time delay, been a little more expensive and added some weight. We made mistakes along the way. We've got a vehicle today that has a moderate escape capability, but not nearly what some of the crew would like."

No ejection seats would have made Peterson and Musgrave's chances of escape during an emergency extremely limited. Regardless, even if ejection seats had been practicable or available to Weitz' crew, they would not have offered greater advantages to the astronauts in terms of survivability. Then-chief astronaut John Young, who commanded the first Shuttle flight, joked darkly that, in an on-the-pad ejection, their parachutes would open "after we hit the ground!" With this in mind, perhaps, STS-5 skipper Vance Brand wanted nothing to do with them.

Consequently, Paul Weitz' launch on STS-6 would be very different from his Skylab mission ten years before, when he and his crewmates lifted off on a Saturn 1B booster. Weitz remembered, first-hand, the escape rockets atop the Apollo capsule, which were capable of whisking astronauts to safety in the event of a launch failure. The Shuttle's asymmetrical design made such systems impractical and rendered the whole vehicle potentially far more lethal than the Saturn had ever been.

HYDROGEN LEAKS

Following Challenger's rollout to the launch pad on the final day of November 1982, several milestones had still to be overcome before final preparations for STS-6 could commence. One of the most critical exercises was a Wet Countdown Demonstration Test (WCDT), which was scheduled to culminate on December 18th in a 20-second firing of her three main engines. This so-called Flight Readiness Firing (FRF) was necessary to demonstrate the engines' ability to throttle between 94 per cent and 100 per cent rated thrust and 'gimbal' under hydraulic command, just as they would be expected to do during the 'real' launch.

Similar 'wet' – or fully fuelled – tests had been performed before each Saturn V launch, although on those occasions the giant rocket's engines were not fired.

Preparations for the FRF proceeded in a manner not dissimilar to a real countdown: launch controllers started the clock by powering up the SRBs, ground support equipment and activating Challenger's flight systems. Four seconds before the simulated lift-off, at precisely 4:00 pm, the Shuttle's engines thundered to life at 120 millisecond intervals, reaching 90 per cent thrust within three seconds and hitting the 100 per cent mark precisely at T-zero. Three seconds later, controllers practiced retracting the ET's umbilical line and the boosters' hold-down posts; a further 12 seconds elapsed before shutdown commands were issued to all three engines.

As well as helping to validate Challenger's integrity under duress, the test also evaluated her ET, which was of a new, lighter design, weighing 4,500 kg less than earlier models. This had been accomplished by eliminating portions of longitudinal structural stiffeners – known as 'stringers' – and milling STS-6's tank with a thinner aluminium skin. Additional weight savers included replacing heavy SRB attachment points with lighter, yet stronger and cheaper, titanium alloy ones and removing an 'anti-geyser' line previously used to circulate liquid oxygen during the lengthy tank-filling process.

Moreover, the SRBs were also lighter, with walls 0.08 to 0.12 mm thinner than previous boosters were. This saved 1,800 kg in weight, although in the wake of STS-7 it was feared too much material had been removed and NASA reverted, for a time, to

Challenger undergoes the first Flight Readiness Firing of her three main engines on December 18th 1982. Note the unusual 'butterfly and bullet' combination of the Shuttle stack, consisting of the orbiter, External Tank and twin Solid Rocket Boosters.

the original thickness. Finally, Challenger's main engines were capable of achieving 104 per cent thrust – a four per cent increase over the capabilities of sister ship Columbia – which enabled her to transport larger and heavier payloads aloft. In fact, for each one per cent performance increase over Columbia's engines, the new Shuttle gained 450 kg of additional payload-to-orbit capability.

Challenger's higher thrust was accomplished by incorporating redesigned components into each engine. Such changes became necessary in anticipation of higher temperatures, pressures and pump speeds that they would encounter at greater thrust levels. An aggressive series of test firings, lasting over 62,000 seconds, were performed to validate the engines in readiness for their first orbital mission; additionally, the main injectors employed higher strength liquid oxygen posts and they sported modified fuel pre-burners to overcome turbine blade erosion and thicker tubes and coolant supply lines to handle higher aerodynamic loads at lift-off.

Their performance on the pad, however, was not fully successful.

As they blazed at 100 per cent power on December 18th, engineers detected levels of gaseous hydrogen in Challenger's aft compartment that significantly exceeded allowable limits. When it became impossible to pinpoint the cause or location of the leak, a decision was taken to perform a second FRF. New instrumentation was installed both inside and outside the aft compartment to determine if hydrogen was leaking from an internal or external source.

Suspicion focused initially on the latter possibility, because vibration and current had found their way into the aft compartment, behind the engines' heat shields. Extra sensors and a higher than ambient pressurisation level were duly installed to prohibit any penetration by 'external' hydrogen sources. However, the second test firing on January 25th 1983, during which the Shuttle's engines were run at 100 per cent for 23 seconds, again revealed the presence of leaking hydrogen gas.

Several more days of troubleshooting eventually identified a cracked weld in tubing leading to the uppermost (Number One) of the trio of engines, which was promptly removed on February 4th. A replacement arrived from the National Space Technology Laboratories at Bay St Louis in Mississippi, but initial inspections in the VAB uncovered a leak in an inlet line to its liquid oxygen heat exchanger. Before it could even be installed onto Challenger, the 'replacement' was itself replaced by a third engine. After more checks in Mississippi, including a 500-second test firing, it was despatched to Florida on March 3rd and fitted a week later.

Unfortunately, while this work was ongoing, painstaking efforts were underway to ensure that the other two original engines did not exhibit leaks – and the bad news seemed to be that they did! Towards the end of February, hairline cracks were found in one of the left-hand Number Two engine's fuel lines and borescope observations of the right-hand Number Three engine revealed a similar problem. Both were removed, returned to the VAB and repairs conducted. With the arrival of the replacement engine from Mississippi, all three were reinstalled by mid-March and verified as being ready for launch.

The leaks from the Number Two and Three engines were apparently caused by a generic 'seepage' in a 45-cm-long inconel-625 tube in their ignition systems. It apparently occurred underneath a protective sleeve brazed onto a small hydrogen line that sent fuel to the engine's augmented spark igniter. The sleeve had been added in Challenger's case to counter possible chafing. After practicing cutting off the sleeve on the Mississippi plant's test stand, Rocketdyne technicians proceeded to Florida and replaced it with a non-sleeved inconel-625 tube on each of Challenger's engines.

By the time their orbiter was finally declared 'flight ready', the four men of STS-6 had already performed their countdown demonstration test and, on February 5th, the TDRS-A satellite – attached to a US Air Force-funded Inertial Upper Stage (IUS) booster – was transferred to Pad 39A and inserted into the payload bay. This impressive combination, when accommodated in a doughnut-shaped 'tilt table', consumed three-quarters of the bay's 18 m length. After deployment, some ten hours into the mission, the two-stage, solid-fuelled IUS would boost TDRS-A into a geosynchronous transfer orbit, 35,600 km above Earth. When fully operational, the satellite would be numerically renamed as 'TDRS-1'.

In spite of the engine leaks, the payload had originally been transferred to the pad a couple of days after Christmas, but when it became increasingly clear that Challenger would not be flying in January 1983 and another FRF would be necessary, it was returned to KSC's Vertical Processing Facility (VPF) for temporary storage. By the end of the first week in February, however, it was back at Pad 39A and had been ensconced in the Payload Changeout Room of the Rotating Service Structure in readiness for installation aboard the Shuttle.

Then, on February 28th, strong winds whipped across the Merritt Island launch area and breached a weather seal between the changeout room and Challenger's payload bay, depositing a fine layer of particulate material on the $100 million satellite's solar array deployment springs. The result: an additional nine-day delay from March 26th until the beginning of April. After thorough inspections, TDRS-A was removed and carefully cleaned, before being replaced aboard the Shuttle on March 19th.

MENTAL SIMULATIONS

Now rescheduled for April 4th, the long delayed mission gave the astronauts and their control teams an opportunity to sharpen their skills. "This was about the most challenging job you could ask for," said STS-6 Flight Director Jay Greene. "The simulations were mental simulations that were as challenging as anything NASA has to offer. There were two things going on: one was the goal to train the crew to work with the control centre and, at the same time, train maybe a dozen different operators to the max extent possible. Instead of having one failure – which is about the most you'd expect during a launch – they'd try and give everybody something to play with and the flight director would have to co-ordinate everybody's problems and come out with a solution that got the crew safely to orbit or resulted in a successful abort and recovery. During the course of a day, we'd run maybe eight launch abort sims and every sim had maybe ten different faults that the [simulation supervisors] would put in. By the end of the day, you had somewhere between 80–100 problems that you dealt with."

Training, Don Peterson admitted, was tough, but he did not recollect a single cross word between himself and his crewmates during their 13 months together. "And that's unusual," he pointed out, "because the training's really intense and very demanding and you're working long hours and things go wrong and there are many delays." However, he remembered a characteristically unflappable Story Musgrave happily going off to fetch sandwiches for the entire crew after one gruelling session in the simulator.

"It changes your outlook a lot," said Bo Bobko of the training grind. "You've got to do it, so it just changes the way you look at things. It's not some thing out there in the mist; it's up close and personal. There's a lot to be learned for a spaceflight and it doesn't seem like there's ever enough time. You feel you want to learn it all and it gives you a lot of incentive to work hard and try to learn as much as you can and get things as squared away as you can."

Finally, on April 4th, after a decade and a half waiting for his first mission, which he jokingly said felt like being "a cosine wave in a sine-wave world", Bobko accompanied Weitz, Peterson and Musgrave out to Pad 39A for the real thing. Challenger's flight deck, despite being roomier than that of Columbia, was still cramped, with Weitz in the left seat and Bobko to his right. Sitting just between and behind them was Musgrave, to whom fell the job of primary flight engineer during the Shuttle's ascent and re-entry.

Shoulder to shoulder with Musgrave, and directly behind Bobko, was Peterson. After Challenger's hatch had been closed, after more than a year training together, after almost two decades apiece in the astronaut corps and even in light of Weitz' prior flight experience, all four men were rookies as far as flying the Shuttle was concerned. For launch, they sat on the ten-windowed flight deck – the main location for controlling the vehicle during ascent, re-entry and the bulk of orbital operations. Six windows wrapped, airliner-like, around the front, with two more in the aft roof and another pair looking into Challenger's payload bay.

Situated directly beneath the flight deck was the middeck, accessed in space by floating through a small, 66 × 71 cm opening; there were actually two openings, but normally only one was used. Essentially, the middeck provided a living area for the crew, including storage lockers for experiments and personal effects, sleep stations, a galley, toilet and the airlock providing access to Challenger's payload bay. Before launch and after landing, the astronauts entered and departed the Shuttle through a circular hatch in the middeck's port side wall.

RIDE OF A LIFETIME

Six seconds before 6:30 pm on April 4th, with a low-pitched rumble that soon grew into a thundering crescendo, the three main engines roared to life, causing the entire vehicle to rock perceptibly backwards and forwards. Then, as the countdown clock hit T-zero, came the ear splitting crackle of the SRBs and it was this punch-in-the-back ignition that really seized the astronauts' attention and convinced them that they were heading into space.

Challenger thunders aloft on her maiden orbital voyage.

For the first few seconds, as the stack cleared the tower and rose into the clear Florida sky atop two dazzling orange columns of flame from her boosters, the cockpit instruments were blurred, but readable. By the time Challenger rolled onto the correct flight azimuth for a 28.45 degree inclination orbit, then pitched onto her back under GPC control about ten seconds after lift-off, the vibrations had lessened to a point that allowed the astronauts to read their instruments without problems for the remainder of the ascent.

"As the main engines come up [to full throttle], you really feel the vibrations starting in the orbiter," said astronaut Jerry Ross, who flew the Shuttle a record tying seven times, "but when the boosters ignite, I describe it as somebody taking a baseball bat and swinging it pretty smartly and hitting the back of your seat. It's a real 'bam'. The vibration and noise is impressive. The acceleration level is not high at that point, but there is that tremendous jolt and you're off!"

"Thank goodness we'd got all this insulation in our helmets," added Charlie Walker, "because the acoustic level is [huge] in the crew compartment. We would readily be deafened if we didn't have the insulation of the helmets around our ears."

A minute into the flight, as Challenger approached an altitude of 15 km, she passed through a period of maximum aerodynamic turbulence, which required her GPCs to throttle the main engines back to just under two-thirds of their rated thrust. The passage through this phase was marked by an increase in the noise and vibration of the engines, although their performance remained within expected limits. The sound from the SRBs remained sporadic and decreased to virtually nothing as the time approached, two minutes and ten seconds into the climb, for their separation.

The booster separation was also accompanied by a harsh grating sound, which STS-1 veteran John Young likened to the noise made by the Saturn V rocket's final stage when it was firing. Both SRBs parachuted into the Atlantic Ocean, splashing down five minutes later, several hundred kilometres downrange of Cape Canaveral. With the severe vibrations caused by the cumbersome boosters now gone, Weitz and Bobko found it easier to flip switches as Challenger flew on for six more minutes, reaching Mach 19 – nearly 23,340 km/h – at which point her engines were throttled back to maintain a 3g environment in the cockpit and limit the loads on the ET's connecting struts.

"It feels like there's somebody heavy sitting on your chest and you're just waiting for this 3gs to go away," said Ross. "This is when the orbiter's main engines start reducing their power output so you don't exceed the structural limit of 3gs. You're getting lighter and lighter. Then, at the time that the computers sense the proper conditions, the engines go from around 70 per cent power on a 3g acceleration, then shut off and you're in zero-g. For me, I had the sensation of tumbling head over heels; a weird sensation."

For Bruce McCandless, who sat alone on the middeck during one of his launches, the auditory sensations and vibrations were his only cues to the titanic events going on outside. "I had not paid much attention to the lockers until we actually launched," he said of his somewhat dull view of a row of storage lockers in the middeck. "As the noise and vibration commenced, and the 'g-load' gradually built up, I developed a fervent hope that all of the quality control procedures associated with locker installation had indeed been followed to the letter! When the SRBs were jettisoned, suddenly everything became quiet and the ongoing sensation was that of a giant 'hand' pushing on your back – a relentless thrusting you onwards towards orbit. When the main engines shut down, I had the sensation that we had gone into reverse, for some reason. I pulled out a felt tip pen and released it in front of me. Instead of falling, it went nowhere, so I unstrapped and went to work."

At 6:38 pm, some eight minutes and 19 seconds after leaving Pad 39A, Challenger's main engines were finally shut down and several hundred kilograms of residual liquid oxygen was dumped into space through their nozzles. The last traces of unburnt hydrogen fuel, meanwhile, was expelled through a fill and drain valve on the port side of the aft compartment. This was then closed and, later in the mission, the crew 'vacuum inerted' the entire system by opening the oxygen (starboard side) and hydrogen (port side) fill and drain valves to vent remaining propellant into space.

Twenty seconds after the main engines went out, the ET was jettisoned to follow a ballistic, sub-orbital trajectory and burned up over a sparsely inhabited stretch of the Indian Ocean.

Weitz and Bobko pulsed the RCS thrusters in the Shuttle's nose and tail to push themselves away from the now-useless tank at about 1.2 m/sec; the separation, they reported, was noiseless and only indicated when the red main engine lights on the instrument panel winked out.

After firing the twin OMS engines on two occasions – one to achieve an elliptical orbit with a specific apogee, the second time, half a revolution later, to circularise it – and opening Challenger's payload bay doors one hour and 45 minutes into the mission, the first order of business was preparing TDRS-A for deployment. However, with three rookie spacefarers aboard, the experience of absorbing where they were proved overwhelming. "Every hour and a half," Musgrave said, "we made a complete orbit and it was like getting a crash course in world geography. Seeing entire continents with the naked eye is something special. We saw oil slicks off India, oil tankers in the Persian Gulf, the swirls in the Earth's crust where Iran, Pakistan and India collided millions of years ago and the mountains thrust upward by the force. We saw the White Nile and the Blue Nile converge in the Sudan, the dust storms in Mexico, thunderstorms over Africa and the tranquil beauty of the Bahamian islands."

SWITCHBOARD IN THE SKY

The astronauts' first fleeting glimpses of their home planet from orbit were, however, eclipsed by a hectic Day One schedule, which was already ticking down towards a first deployment opportunity for TDRS-A a little over ten hours into the STS-6 mission. Even two decades later, the network – which boasts six 'first generation' satellites and three updated 'second generation' ones – has proven instrumental in providing near-continuous voice and data contact between Mission Control and orbiting astronauts.

Unusually for a Shuttle mission at the time, no fewer than three Payload Operations Control Centers (POCCs) would follow TDRS-A during its manoeuvres to geosynchronous orbit. Throughout the pre-deployment checks, right up until the satellite drifted away from Challenger, Harold Draughton's team at JSC in Houston would assume responsibility. Command would then pass to Pete Frank at the US Air Force's Satellite Control Facility in Sunnyvale, California, until TDRS-A reached geosynchronous orbit. Finally, the Spacecom concern, based at White Sands in New Mexico, would take over the day-to-day running of the satellite on NASA's behalf.

Under an initial ten-year contract signed with NASA in December 1976, Spacecom had agreed to lease TDRS communications, tracking and data relay services to the space agency at a cost of some $250 million per annum. As a result, it represents the world's largest and most powerful privately owned tracking, communications and data relay system currently in orbit.

Housed inside Challenger's payload bay, TDRS-A was huge, even though still stowed in its 'launch' configuration with its communications payload hidden from view, its umbrella-like antennas closed and its two electricity generating solar arrays each folded into three parts. The satellite comprised three main segments: an equipment module, a communications payload and a battery of relay antennas. At this stage, however, it bore little resemblance to the enormous 'windmill' into which it would transform in geosynchronous orbit.

The TDRS concept was initiated in the early 1970s as a means of not only supporting Shuttle crews, but up to 25 other 'users', including a number of important scientific missions such as the Hubble Space Telescope and Gamma Ray Observatory. It was recognised by NASA that a system of relay platforms operated from a single ground terminal in New Mexico would provide more adequate and near-constant support than the worldwide network of tracking stations previously employed. In fact, as well as supporting low-orbiting missions, it could relay data from higher altitude satellites circling up to 5,000 km above Earth.

Since the dawn of human spaceflight, astronauts had been out of contact with Mission Control for up to 80 per cent of every orbit; furthermore, satellites had to tape record data and transmit it later, when they came within range of a tracking ship or ground station. As the Shuttle effort gained momentum in the mid-1970s, it was envisaged that two TDRS relays – one in geosynchronous orbit over the equator, just off the north-eastern coast of Brazil, a second over the central Pacific Ocean – would provide astronauts with space-to-ground voice and data links for between 85 per cent and 98 per cent of each orbit.

Yet, TDRS was no miracle worker. It was not capable of processing or adjusting communications traffic in either direction. Rather, it operated as a 'bent pipe' repeater, relaying signals and data between its Earth-circling users and the highly automated ground terminal. Signals processing, therefore, was done on the ground and the satellite's sophistication was devoted to its very high throughput. Located in the inhospitable New Mexico desert, White Sands provided a clear line of sight with both satellites and its limited amount of annual rainfall meant that weather conditions would not interfere with their Ku-band uplink or downlink channels.

Responsibility for deploying TDRS-A was shared by the entire crew, although the Mission Specialists led the effort, with Peterson stationed on the aft flight deck and Musgrave temporarily in the Commander's seat. There had already been some confusion in the weeks preceding lift-off. "We were in quarantine in the crew quarters at the Cape," remembered Peterson, "and a couple of nights before launch, two guys showed up from Boeing. It turned out that the software we'd trained on in the simulator was not exactly the same as the software that was flying and a lot of the codes were different. Story and I copied a bunch of stuff down with pen and ink and used that on orbit and that's really scary because we were taking these [Boeing] guys' words for it! We'd never seen some of this stuff in the simulator. Suppose what they told was not right and we messed up the payload? We'd never find those two guys again! They'd be gone and it'd be 'Why the hell didn't you guys do it the way you were trained to do it?' Story called somebody in Houston to confirm the codes, but it was pretty vague. That bothered us, because TDRS-A was extremely expensive and important to get working properly. When there were last-minute changes, we wondered if it had really been tested and thrashed out. The way we did commands was by dialling in a set of three numbers and then hitting a switch to execute them. The command was determined by what three numbers were set in there and they gave us numbers we'd never used before. We had no way of knowing whether they were right or not right."

Fortunately, the commands proved accurate, "so I guess the Boeing guys were right", said Peterson, and some eight and a half hours after launch the crew was in position to raise TDRS-A and its IUS booster to their pre-deployment angle of 29 degrees above the payload bay. This was followed by radio frequency checks and, finally, at 3:51 am on April 5th, a final "Go" was given for deployment.

Less than 20 minutes later, the IUS was switched onto internal power. Originally intended as a temporary substitute for a reusable 'space tug' when it was designed in the 1970s, the booster was dubbed the 'Interim' Upper Stage and later became 'Inertial' in recognition of its internal guidance system. Losing its 'interim' status also reflected a growing awareness, when the space tug was cancelled in late 1977, that the IUS' services would be needed throughout the 1980s. In fact, not until the early years of the present century did it fly for the last time as a 'standalone' booster.

Prime contractor for the IUS was Boeing, which began developing the two-stage vehicle in August 1976 and supported its first launch aboard a Titan 34D rocket six years later. Measuring five metres long and a little under three metres in diameter and weighing some 14,740 kg, the cylindrical booster – made from Kevlar-wound aluminium – was capable of hauling 2,270 kg payloads from low-Earth orbits to geosynchronous altitudes.

Its first stage carried 9,700 kg of solid propellant and a large motor, capable of firing for up to 145 seconds; this made it the longest burning solid-fuelled engine ever used in space applications. Meanwhile, the second stage carried 2,720 kg of propellant. Both the first and second stage nozzles, commanded by redundant electromechanical actuators, could steer the former by up to four degrees and the latter up to seven degrees. Although solid rockets were known to generate a harsh impulse, the separation mechanism between the first and second stages employed a low-shock ordnance device to avoid damaging TDRS-A.

Moreover, solid propellant was chosen over a liquid-fuelled booster because of its simplicity, safety, high reliability and low cost. Hydrazine-fed reaction control thrusters provided the IUS with additional stability during the 'coasting' phase between the first and second stage firings, as well as ensuring accurate roll control and assisting with the satellite's insertion into geosynchronous orbit.

Situated between the two stages was an equipment section loaded with avionics systems to provide guidance, navigation, control, telemetry, command and data management services to TDRS-A. Crucially, most critical components, except the bellows for the gimbal actuator, were fully redundant to assure reliability of more than 98 per cent. In the early days of the IUS' development, Boeing even proposed adding a smaller third stage to propel planetary missions out of Earth orbit, although the design would have been too large – with its payload attached – to fit comfortably aboard the Shuttle.

Mounted on the base of TDRS-A, the $50 million booster was held securely in Challenger's payload bay by the doughnut-shaped tilt table, alternatively known as the Airborne Support Equipment (ASE). As well as providing the crew with an ability to hoist the entire 14-m-long stack from a horizontal position to the deployment angle of 59 degrees, it incorporated electronics, batteries and cabling to enable Peterson and Musgrave to issue commands during the lengthy checkout of both the satellite and booster.

The ASE included a low-response spreader beam and torsion bar mechanism to reduce spacecraft dynamic loads to less than a third of what might be achieved without the system. It was secured into Challenger's payload bay by means of six standard, non-deployable attachment fittings, which mated to the ASE's forward and aft frames, and two payload-retention latch actuators.

Upon arrival in orbit, the Shuttle was oriented with her payload bay facing

With its communications gear folded up inside the large black solar panels, NASA's first Tracking and Data Relay Satellite is raised to its deployment position. Note the white Inertial Upper Stage mounted at its base and the 'tilt table' at the bottom of the frame.

Earthward, in order that the IUS and its precious satellite did not experience excessive thermal loads from the varying solar illumination during orbital flight. The entire payload, during this time, was supported by Challenger's onboard electricity generating fuel cells, although the ASE included its own batteries to take over in the event of a power interruption. After releasing the forward retention latch actuators and raising the stack to 29 degrees above the payload bay, telemetry checks were performed by the ground to ensure that both were ready for deployment and the IUS was transferred to its internal batteries.

By 4:18 am, payload-to-orbiter umbilical cables were released, TDRS-A and the IUS were positioned at their deployment angle of 59 degrees and, precisely on time, Musgrave flipped the switches to release the combination from the tilt table. At that point, the booster's ordnance separation device, fitted with compressed springs, physically ejected the payload from Challenger at a rate of just 12 cm/sec. Deployment occurred at 4:31:58 am, during the STS-6 crew's eighth orbit, and the stack swept silently into the inky blackness, directly over the flight deck, surprising Bobko as it did so.

"I learned how big it was when it came out," he said later. "I was up in the front, in the Pilot's seat, and [Musgrave and Peterson] both said something like 'Oh, my God!' when this big satellite came out over the cockpit."

The astronauts' surprise was short lived. Nineteen minutes after the combo left Challenger's vicinity, Weitz and Bobko fired both OMS engines to create a safe separation distance before the ignition of the IUS' first stage. When this two and a half minute firing got underway at 5:26 am, a complex ballet of celestial mechanics was set in motion to insert TDRS-A into geosynchronous transfer orbit. First stage separation was then timed to occur at 10:44 am and second stage ignition two minutes after that. By 11:01 am, it was expected that the satellite should have reached its final operational location.

Following the deployment of its solar arrays and communications payload, by 1:30 am on April 6th, TDRS-A should have been fully functional and ready to begin several weeks of testing. Unfortunately, problems with the booster set the satellite on a rocky path that would not see it rendered wholly operational until the winter of the following year.

During the IUS' development, obstacles had been encountered with its propulsion system, including burst cases, tacky liners, soft and cracked propellants and nozzle delaminations, together with problems with its onboard software and avionics. The first operational use of the booster was supposed to be on the Shuttle with TDRS-A, although delays and scheduling conflicts caused it to be leapfrogged by a pair of military communications satellites on an Air Force Titan 34D launch in October 1982. In spite of telemetry dropouts, that IUS mission proved successful. TDRS-A was not quite so lucky.

As the firing of the second stage was underway, a critical seal (a manifold in the baffle of the gimbal actuator) failed and the control system lost its ability to accurately point the motor nozzle. This 'canted' and caused both the IUS and attached satellite to tumble wildly through space; furthermore, unusually high levels of cosmic radiation disrupted the normal automatic sequencing activities. The result was that, instead of reaching a circular orbit, TDRS-A was left in an egg shaped orbit with a perigee of 22,000 km and an apogee of 34,000 km.

Worse, communications with the tumbling payload was also lost. Throughout the night, flight controllers tried to separate TDRS-A from the IUS.

"We started to transmit separation commands in the blind, not having any communications with the vehicle," said former Shuttle manager Glynn Lunney, "and we jacked up the power on the network to as high as we could get and we radiated commands to jettison the IUS. We never got any signal. The deployment happened at night and by the next morning, around breakfast time, we were getting ready to give up when the network picked up a signal from the satellite."

However, the ground team had first to regain control of TDRS-A, which was spinning at 30 revolutions per minute. Fortunately, it was stabilised. Then, starting on June 6th and lasting 58 days, in a procedure devised by NASA, White Sands terminal operator Contel and the satellite's manufacturer, TRW Defense and Space Systems Group of Redondo Beach, California, three dozen firings of its hydrazine-fuelled reaction control thrusters raised it to a circular orbit at the planned altitude, in the process consuming 400 kg of propellant (two-thirds of its total supply) that would otherwise have been used for station-keeping during TDRS-A's ten-year operational life span.

Overheating thrusters caused further headaches, but on July 6th 1983 the satellite was finally switched on to commence testing. After three months, one of its Ku-band single access diplexers failed; followed, later, by the loss of a Ku-band travelling wave tube amplifier. Despite these obstacles, which meant that TDRS-A could only provide one Ku-band single access forward link – thus postponing its ability to support the Shuttle's middeck text and graphics machine – the satellite was nonetheless declared operational for communications purposes in December 1984.

Already, by this time, it had supported data traffic on the first Spacelab mission, launched in November 1983. When fully deployed in orbit, the newly renumbered 'TDRS-1' resembled a colossal, 2,270 kg windmill, measuring 17.4 m across its fully unfurled solar panels, which extended from a hexagonal 'bus'. The dual panels generated 1,800 watts of electrical power, supplemented by onboard nickel–cadmium batteries when in Earth's shadow to support its decade-long life span. Inside the bus, the communications payload was capable of transmitting in a single second the entire contents of a 20-volume encyclopaedia.

Mounted atop the bus were the satellite's antennas, capable of receiving transmissions from White Sands, amplifying them and re-transmitting them to the 'user' spacecraft and vice versa. The main space-to-ground link was a circular, two-metre-diameter antenna, which operated across the Ku-band frequency, while data from other spacecraft was routed through one of two umbrella-shaped, 4.9 m diameter dual feed S-band/Ku-band single access parabolic dishes. Constructed from gold-clad molybdenum wire mesh, both had transmission rates in the order of 300 megabits per second, capable of handling heavy traffic from Hubble, the Shuttle and, specifically, the 1982-launched Landsat-4 Earth resources platform.

For multiple access service, an S-band 'phased' array of 30 helix antennas was mounted directly onto TDRS-1's body. This incorporated a forward link, which transmitted command data to the 'user' spacecraft, and a return link to relay signal outputs directly to White Sands. Upon receipt, the ground terminal 'de-multiplexed' the signal and distributed it to 20 sets of beam-forming equipment, which discriminated among the 30 signals to select the unique signatures of individual users.

However, the success story that TDRS had since become could not have been further from NASA's collective mind during those first few weeks of April 1983 and, before Challenger had even landed, an investigation board was put in place to establish the cause of the IUS failure. Co-chaired by Donald Henderson of the embarrassed US Air Force and Thomas Lee of NASA, the board's initial work obliged the agency to delete TDRS-B from the eighth Shuttle mission in August 1983. Not until the booster's performance had been satisfactorily demonstrated would a second such satellite ride the Shuttle into orbit.

A number of modifications were implemented, including improvements to prevent hot gases impinging on the oil-filled Techroll seal in the region at which the booster's nozzle was connected to its solid rocket motor. It was the failure of this seal, 83 seconds into a planned 107-second firing, that had prevented the insertion of TDRS-A into its appropriate orbit. Additionally, a new nose cap to protect the forward part of the seal was thickened in the IUS' second stage and higher density carbon was employed in a Grafoil seal between the Techroll and nozzles in both the first and second stages. Tests of the repaired booster were conducted from March 1984 onwards and, in January of the following year, it successfully launched a top-secret Department of Defense payload.

Ironically, and tragically, the next TDRS was part of the ill-fated cargo aboard Challenger's final mission early in 1986. In fact, not until the autumn of 1988 would another fully functional satellite be inserted into orbit to complement TDRS-1.

"EXTRAORDINARILY EXCITING"

Despite these problems, the STS-6 crew could have done nothing to prevent the fault in the IUS' second stage and, indeed, the remainder of their five-day mission proceeded extremely well. "Technically, the mission was extraordinarily exciting," Musgrave told the post-flight press conference, "because we accomplished everything we set out to do – launching TDRS, performing the EVA, conducting medical experiments and bringing the Shuttle home in great shape. It was also personally fulfilling, because I've been waiting for this for a long time. For some reason, I immediately oriented to weightlessness. I was totally at home in zero gravity and felt extraordinarily comfortable in a 'no down' environment. I trained myself not to expect to see a 'down'. I was prepared to tell myself that the floor of the spaceship was 'down' and to keep myself oriented that way, but I found that I didn't need a 'down'! To me, the Earth was neither 'down' nor 'up'. It was just ... there! Some people are different and get confused by all the sensory inputs telling them that 'down' should be there, but – wait a minute – it should also be there and the two don't match. I had done a lot of work on integrating the vertical and horizontal parts of the spaceship and I had no need to see or feel an 'up' or a 'down'. Since I had no fixed notion of 'down', it never bothered me to see things that should have been 'down'."

In fact, as Musgrave dozed off each night on Challenger's flight deck, alongside Paul Weitz, climbing into his sleeping bag was the only time when he actually missed the presence of full terrestrial gravity. "I tied it horizontally and slept horizontally, up near the Commander, just to keep him company, since Bo and Don were sleeping on the middeck. I'm a side sleeper and like to change to different positions throughout the night, but since there's no 'up' or 'down' in space, I really couldn't sleep on my side. No matter what position I tried to take, the zero gravity would keep me locked in a neutral position; neither 'up', 'down' or 'sideways'. I couldn't twist and turn or hold a new position. I was tempted to take a strap and lock my knees in a crouched position, just to get some variety, but I never did. It's amazing that man – a creature genetically coded to live in gravity – can survive in zero gravity. When the space programme began, there were people who said that man wouldn't be able to breathe or swallow in zero gravity!"

One of Musgrave's responsibilities during STS-6 was a machine known as the Continuous Flow Electrophoresis System (CFES), assigned to him by Weitz in light of his expertise as a medical doctor. The 'electrophoresis' technique worked by passing an electric field through a fluid as it moved from one end of a processing chamber to the other. Akin to a prism splitting white light into its constituent colours, the 1.8 m tall CFES device – situated on Challenger's middeck – had the ability to separate cells

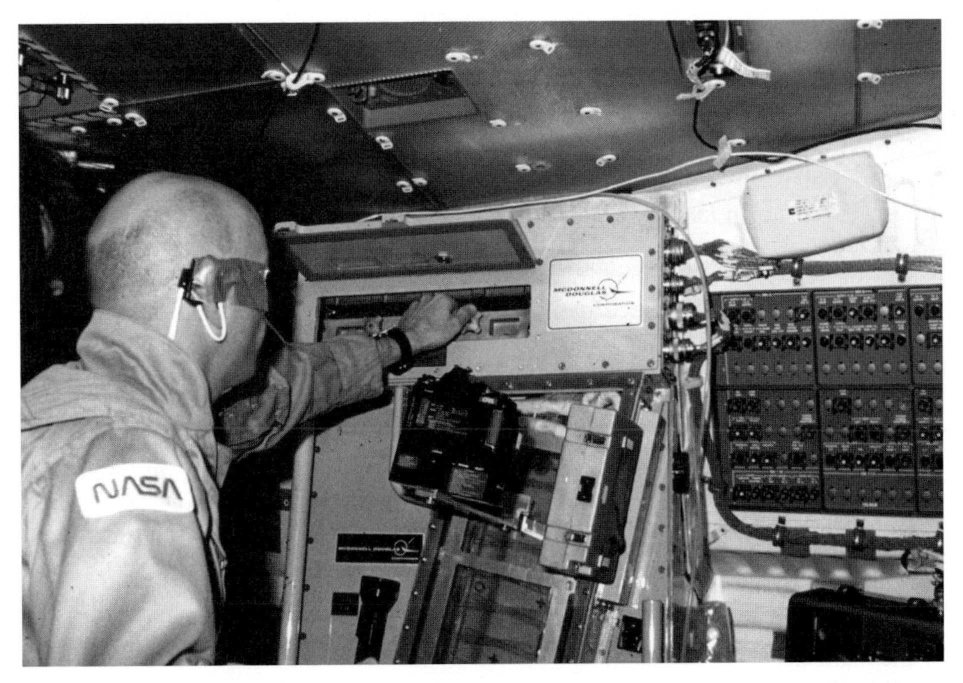

Story Musgrave operates the Continuous Flow Electrophoresis System on Challenger's middeck.

and proteins, but its effectiveness on Earth was limited by the gravity induced effects of convection and sedimentation.

In low orbit, where these influences were a million times lower, it was anticipated that such separation processes could achieve higher levels of perfection and purity. For example, components of some biological substances could become hotter or colder than on Earth, enabling greater control over the solidification process. Additionally, the lack of appreciable gravitational force meant particles could flow freely, avoiding the risk of contamination caused by contact with the container. On STS-6, during a pair of seven-hour-long experiment runs, CFES processed 700 times more biological material than was achievable in operations on Earth. Moreover, their purity was some four times higher.

"We saw the prospect of purifying medical materials like hormones and enzymes, which make up the basic components of treatments for a variety of diseases," explained Charlie Walker, formerly an engineer for CFES's sponsor, the McDonnell Douglas company. "In space, you can do that purification to a degree that's impossible here on Earth. We did the mathematics of the purification process using fluid dynamic equations and theorised [that] purities of four or five times [those that] could be done in the best processes on Earth could be achieved by taking this process into space."

First flown aboard Columbia on STS-4 in the summer of 1982, McDonnell Douglas and NASA intended to jointly carry the 250 kg electrophoresis machine six times, culminating in the development of a much larger processing facility for carriage in the Shuttle's payload bay. Unlike the middeck-borne CFES unit, which had only one processing chamber, the payload bay version – known as Electrophoresis Operations in Space (EOS) and weighing 2,270 kg – had two dozen. By the beginning of 1986, it was slated to fly for the first time aboard Challenger on the STS-61M mission in July of that year.

The promise of significant breakthroughs in pharmaceutical research seemed just around the corner in January 1986, when Challenger lifted off on what was expected to be a routine, six-day mission. Her destruction 73 seconds later, coupled with the deaths of her seven astronauts, put McDonnell Douglas' plans on indefinite hold. Subsequent plans implemented by NASA to limit the commercial utilisation of the Shuttle meant that neither the CFES, nor the payload bay mounted EOS facility, ever flew again. Moreover, the development of new gene-splicing techniques towards the end of the 1980s rendered electrophoresis effectively redundant.

Attitudes were different in 1983, with the promise of regular, fortnightly missions. Significant improvements, including software changes, better cooling and a greater separation capability, had been implemented during the interval between STS-4 and STS-6. "You don't fly a second time to do the same things you did the first time," said Charlie Walker. "You advance. You prove what you wanted to prove or find out you've got a problem and then, on the next flight, you make plans to go further in terms of the scientific or technical investigation. You 'stretch the envelope', as we say in the flight field."

Samples of rat and egg albumin and cell culture fluid had been successfully separated during STS-4 and, on Challenger's first mission, Musgrave tended high concentrations of haemoglobin to evaluate its flow profile in weightlessness. He also monitored a mixture of haemoglobin and polysaccharide to investigate the separation of different molecular configurations. Each sample was satisfactorily processed, although post-flight removal indicated that the refrigerator had been inadvertently turned off. The condition of the samples, nonetheless, was considered "acceptable".

"It showed great potential," explained scientist astronaut Don Lind, who worked closely with McDonnell Douglas and the experiment before flying aboard Challenger in the spring of 1985. "If we ever get to the point where we can have a guaranteed schedule – so companies can know if they send their samples up on Tuesday, they'll get them back three weeks later – there are a number of companies that have very productive manufacturing experiments in space. Electrophoresis just happened to be one of the first ones."

Of all the secondary experiments aboard STS-6, which Paul Weitz called "this other penny ante stuff to fill up the other four days", the CFES machine was by far the largest. Elsewhere, the astronauts conducted studies of lightning storms using the Day/Night Optical Survey of Lightning (NOSL), which comprised a 16 mm motion picture camera and cassette recorder. Typically, lightning discharges were detected by a photocell mounted on the camera, which generated electronic pulses and stored them on the cassette.

Since lightning events, often visible as single flashes, usually included many separate discharges, the photocell's ability to distinguish each one meant that individual strokes could be carefully scrutinised. Furthermore, during daytime activities, the crew used the camera to record cloud structures and the convective circulation of storm systems. Already, during observations on STS-4, lightning bolts forming a huge 'Y' shape and illuminating 400 km^2 were photographed. Images of South American thunderheads yielded recordings of powerful, 40-km-long lightning bolts. It was hoped that such data could lead to a clearer understanding of the evolution of storm systems.

Other 'candidate' storms were determined by scientists at the Marshall Space Flight Center in Huntsville, Alabama, using an advanced weather monitoring device known as the Man–computer Interactive Data Access System (McIDAS). This provided accurate co-ordinates for developing storm systems to Mission Control, which were then passed up to the orbiting Shuttle crew.

Elsewhere in the middeck was the Monodisperse Latex Reactor (MLR), intended to study the kinetics involved in the production of uniformly sized latex beads in the microgravity environment. The results of this experiment would subsequently find their way into world markets, with hundreds of adverts selling 'made in space' particulate spheres from the MLR. The tiny beads – invisible to the naked eye – were processed in four small reactors, each of which contained a chemical latex-forming recipe. In orbit, the experiment was heated to 70 degrees Celsius, which initiated the chemical reactions leading to the formation of larger beads.

Despite initial success in producing five-micron-sized beads on Columbia's STS-3 mission in March 1982, the reactor's next flight in June malfunctioned and processing was not completed. During STS-6, three reactors operated satisfactorily, with the final one not performing to completion; consequently, not all of its beads were produced.

However, particles in the ten-micron size range were achieved. It was hoped such 'monodispersed' beads could lead to medical and industrial benefits, for example by measuring the size of pores in the walls of the intestines for cancer research, assisting in glaucoma research and transporting drugs for the treatment of tumours.

Additional experiments included three Getaway Special (GAS) studies, which were part of a NASA drive, initiated in late 1976, to encourage universities, government agencies, foreign nationals and even private individuals to develop scientific investigations for carriage aboard the Shuttle. The three dustbin-sized GAS canisters aboard Challenger for STS-6, all attached to the starboard wall of the payload bay, included a Japanese study of producing artificial snow crystals, an experiment holding 11 kg of flowers, fruit and vegetable seeds to determine how best to package them for long duration missions and a series of US Air Force materials investigations.

The 40 varieties of fruit and vegetable seeds, ranging from potatoes to sweetcorn, were sealed in Dacron bags or airtight plastic pouches to be germinated upon their return to Earth. These were positioned around the edges of the GAS canister and at its centre, to demonstrate the effect of minimum and maximum cosmic radiation exposure on the seeds. Post-flight inspections would reveal no significant damage to any of the specimens – all were alive – and no reduced plant vigour or mutations were subsequently noted. Meanwhile, the military experiments investigated new techniques for purifying and electroplating metals and monitored the influence of weightlessness and space radiation on a number of micro-organisms.

FIVE-DAY HIGH

After what Musgrave called "a five-day high", with the investigation board just getting its teeth into resolving the IUS failure and TDRS-A shortly to begin limping to geosynchronous orbit, Challenger's crew prepared for their return to Earth on April 9th. Although this was Weitz' second re-entry, it was his first in the Shuttle. Years later, he paid particular tribute to the Shuttle Training Aircraft (STA) – a modified Grumman Gulfstream – which provided the most accurate analogue for Challenger's aerodynamic performance in the low atmosphere.

First delivered to NASA in mid-1976, the Gulfstream near-perfectly mimicked the orbiter's handling characteristics during subsonic flight, approach and landing. This was accomplished through the independent control of six degrees of freedom, achieved by the use of its flight surfaces and coupled with auxiliary direct lift, side force control surfaces and in-flight reverse thrust. The aircraft's computer could be programmed with the exact flying characteristics – even down to the weight and centre of gravity constraints – of a fully-laden orbiter to enable pilots to precisely simulate their return from space.

Typically, an instructor sat next to the astronaut 'pupil' and flew the Gulfstream to cruising altitude using the aircraft's standard controls. At this point, the astronaut took over with 'his' set of controls – which differed from those of the instructor in that they mirrored the displays in the Shuttle's cockpit – and guided the aircraft towards touchdown. Screens attached to the Gulfstream's windows provided a realistic field of view for the orbiter and, seconds before landing, the instructor again took control, returning to cruising altitude for another attempt.

"It's an excellent simulator," Weitz said later, "much better than the moving base simulators we have, because you're moving. You're not trying to 'fake yourself out' by emulating or simulating accelerations by just moving the fixed base up and down and around." He and Bobko also considered the inclusion of a new Heads-Up Display (HUD) in Challenger's cockpit as "a great aid, in my mind, of performing a landing".

Akin to the displays used in military and civilian aircraft at the time, the HUD projected instantaneous data on velocity, descent rate, altitude and other critical parameters onto a transparent viewing glass above the pilots' cockpit windows. This provided them with the ability to assimilate data from both the 'heads down' world of instrument flying and the 'heads up' domain of looking directly through the windows at the approaching runway.

Astronaut John Blaha worked on developing the HUD at manufacturer Kaiser Electronics of San Jose, California, and remembered it being cluttered with information and disliked by many pilots in the corps. "The biggest challenge," he said, "was the older, established, astronauts had not flown military aircraft with a HUD. The younger guys had all flown aircraft with a HUD, so there was some resistance." Ultimately, Blaha concluded, 'old heads' John Young and Dick Truly ended up liking the display enough for it to be declared fully operational, in a less cluttered form, on STS-8.

During Challenger's re-entry, quite contrary to standard operating procedures, Musgrave unstrapped and stood up inside the flight deck. It would lead to a reprimand from chief astronaut John Young after landing. Weitz admitted to Musgrave's indiscretion at the post-flight press conference. "Sure," he told journalists, "Story did it on the spur of the moment, but we all knew what he was doing and nobody quarrelled with him – at least until now." For his part, Musgrave would explain his reasoning as a desire to show that an astronaut could indeed stand during the transition from weightlessness to terrestrial gravity.

"I had my Hasselblad camera and was taking some photos," he said. "Also, I wanted to prove that I could do it. That's important if an astronaut ever has to leave the flight deck and go below to throw a switch or circuit breaker. I wanted to show that the cardiovascular system doesn't have any problem going back into gravity and you don't have to be strapped down. My standing was smooth and steady and it shows that the Shuttle is maturing. Standing up throughout re-entry, instead of being strapped down, was the perfect end to a perfect trip."

From the perspective of the four STS-6 astronauts, the second half of the hour-long hypersonic dive through the atmosphere was akin to hurtling through a blast furnace. "During the dark time of your approach," Bobko recalled years later, "the plasma sheath around the Shuttle recombines over the top and there's a big tongue of flame following you down."

Swooping into Edwards Air Force Base, deep in California's inhospitable Mojave Desert, Challenger alighted on concrete Runway 22 at 6:53:42 pm, a little more than five days after leaving Earth. "We landed on the solid surface, rather than the lakebed," Bobko recalled. "If the lakebed is dry, it gives you a little more latitude.

Edwards has large runways but, luckily, on this flight, nothing went wrong, so we didn't take that extra margin in any way. Landing on the concrete was just fine."

Since the beginning of the Shuttle era, it has ordinarily been the Pilot's job to both arm and deploy the landing gear, some 15 seconds before the Commander performs the actual touchdown. Like a conventional airliner, the 'business end' of the orbiter's landing hardware is arranged in a tricycle fashion, with two fixed main gears in the belly and a nose gear slightly aft of the nose cap. Normally, the gear is deployed at an altitude of around 75 m above the runway, whilst travelling at a ground speed no higher than 550 km/h.

Brake, axle and wheel damage suffered by Columbia at the end of STS-5 had already led to the incorporation of successful 'saddle' modifications. However, Challenger's landing was not as perfect as expected. During post-landing disassembly, six cracks were detected on three stators in her right-hand inboard brake. Subsequent investigation revealed an undersized machining template had caused expansion slots in the stator disks to be produced 'undersized'; it was possible, NASA's report said, that similar problems had arisen on STS-5, although on that mission the stators were so ruined that it was difficult to prove.

Inspections also highlighted damage to the Advanced Flexible Reusable Surface Insulation (AFRSI) thermal blanketing on Challenger's OMS pods. These consisted of silica tile material sandwiched between sewn composite quilted fabric which were much lighter than the Low-Temperature Reusable Surface Insulation (LRSI)

Mounted atop a heavily modified Boeing 747 Shuttle Carrier Aircraft, Challenger flies over the Johnson Space Center in Houston, Texas, during her return journey from California to Florida in April 1983.

tiles used on other sections of the orbiter's airframe. In total, the new blankets took the place of more than 600 LRSI tiles. Challenger's AFRSI damage ranged from missing outermost sheets and insulation to broken stitches and, in the severest cases, was even attributed to "some type of undetermined flow phenomena" during re-entry.

To further examine the problem, a set of four AFRSI 'test' blankets would be carried on the wings, upper surface and side fuselage on Challenger's next but one mission, STS-8, in the summer of 1983.

These blankets, capable of shielding sections of the airframe from temperatures of up to 650 degrees Celsius, were incorporated after the completion of Columbia as one of the features that helped shave weight from later orbiters, including Challenger herself. They were more durable, cheaper and faster to produce and fit than LRSI tiles and varied in thickness – depending upon their expected heating load – between 1.1 and 2.4 cm. Elsewhere, other thermal protection included white tiles, black tiles capable of withstanding up to 1,260 degrees Celsius and grey reinforced carbon–carbon panels for the nose cap and leading edges of Challenger's wings.

The orbiter's shielding experienced varying levels of degradation and discolouration, but, in general, NASA's second orbiter had returned in good condition from her maiden voyage.

As Challenger came down, efforts were underway to manoeuvre the TDRS satellite up to its correct orbital position, which ultimately allowed it to provide communications and data relay support for the first Spacelab mission. In view of the dramatic reduction of its station-keeping hydrazine supply by two-thirds and problems with its own Ku-band system, it is remarkable that it worked solo to provide near-continuous communications coverage of each Shuttle mission until the second TDRS reached orbit at the end of 1988.

Further underlining its importance, in early 1992, by which time the network had expanded to four satellites, TDRS-1 – then in a state of semi-retirement – was called upon at short notice by NASA to support its Gamma Ray Observatory, whose data recorders had failed. Engineers quickly assembled a ground station at Tidbinbilla, near Canberra in Australia, to minimise scientific loss and TDRS-1 was repositioned with line of sight of the new terminal. The result was that the observatory was granted a downlink capability over previously inaccessible portions of its orbit.

Other firsts achieved by a doddery, yet venerable, old satellite included the first live web cast from the North Pole and the first pole to pole phone call in April 1999. Due to its orbital inclination, TDRS-1 became the first satellite to 'see' both poles (though not at the same time) and, in co-operation with the National Science Foundation, it was commissioned to support ongoing research in Antarctica. In 1998, NASA abandoned plans to retire it and instead allowed scientists at the Amundsen–Scott base to employ it as a relay for transmitting research data to the continental United States.

Additionally, in 1998, it supported a medical emergency at McMurdo station, allowing scientists to conduct a telemedicine conference with doctors in the United States; this enabled a welder to be guided through a real operation on a woman diagnosed with breast cancer. Two years later, it aided an extended scientific expedition, jointly funded by the National Science Foundation and the US Coast Guard, to the Gakal Ridge, just below the North Pole.

Although the arrival of TDRS-1 undoubtedly made the world a more interconnected place, it was to Story Musgrave's intense regret that his fellow humans seemed far more interested in building barriers than bridges. "I'm an optimist," he said quietly at the post-flight press conference. "I like to think positive, but man is not a social animal. One of my biggest disappointments is the absolute failure of the human being as a social animal. You get back here on Earth and open the newspapers and, every week, there are ten or 12 new wars breaking out all over the world. When I was in space, I never thought about war. I never had one negative thought. It was an incredibly positive experience – there was no time or inclination to think of war or problems, disease or death. I had absolute confidence that this mission would go as smoothly as it did," he continued. "This is my career and though I'm not scheduled for another flight as yet, I hope I don't have to wait another 16 years."

He wouldn't.

2

Ride, Sally Ride

ASTRONAUTS WANTED

A strange paradox occurred in the summer of 1976. For nearly a year, no Americans had ventured into orbit; nor would they do so for at least another four years. The space ambitions of the United States were by no means directionless, but its 30-strong astronaut corps faced a crisis: no missions were available in the foreseeable future, yet more astronauts were urgently required. By the time the Shuttle entered operational service sometime in 1980, NASA optimistically hoped that missions would be launching as often as once every fortnight.

In other words, more crews would rocket into the heavens during its first couple of years than had previously ridden every American spacecraft since May 1961. A corps of less than three dozen could not support such an ambitious flight rate, obliging NASA, in a July 8th 1976 press release, to announce plans to hire "at least 15 Pilot candidates and 15 Mission Specialist candidates" for the Shuttle effort. Crucially, and totally at odds with previous astronaut candidates, this group would specifically include both ethnic minorities and women.

Pilots, declared the agency, were required to possess a bachelor's degree in engineering, biological or physical science or mathematics, with advanced qualifications desirable. Moreover, they needed to have accrued at least a thousand hours of pilot-in-command time in high-performance aircraft, with flight test experience preferable. For the Mission Specialists, similar academic credentials – plus three years of related professional expertise or advanced degrees – were demanded, although flight experience was not mandatory. These requirements remained in force for NASA's most recent astronaut selection in May 2004.

More than 8,000 applications were made by the closing date of June 30th 1977. Later that year, in groups of less than 20, the most promising candidates were summoned to the Johnson Space Center (JSC) in Houston, Texas, for screening. One of the first to arrive was a 36-year-old naval officer named Rick Hauck. "I was a project test pilot for the Navy's carrier acceptance trials on the F-14 [Tomcat fighter]," he recalled, "and at the end of that tour of duty, I went to Air Wing 14 on the USS Enterprise, which was the first time the F-14 was deployed overseas. It was a concentration of some of the up and coming pilots in the Navy. During my second cruise on Enterprise in 1977, there was a [leaflet] from NASA, saying they were looking for applicants for the astronaut programme to fly the Shuttle and, in fact, four of us on Enterprise wound in my astronaut class: myself, Robert 'Hoot' Gibson, Dale Gardner and John Creighton. Three of the 15 Pilots were from that air wing! Dale was a Mission Specialist, which is interesting. Twenty per cent of the Pilots came from that ship."

Two weeks after Hauck's screening, a 26-year-old PhD physics student from Stanford University, named Sally Ride, arrived as part of another group of candidates. "It was a group I'd never met before," she said, "and I didn't meet any of the other 180 who were interviewed. The only ones I met were the ones in my little group of 20. We spent a week going from briefing to briefing, from dinner to medical evaluations, psychological exams and individual interviews with the astronaut selection committee."

A month and a half later, two others – a 38-year-old US Air Force test engineer named John Fabian and a 34-year-old physician and former enlisted US Marine Norm Thagard – came down to Houston for screening. Little did these four candidates know at the time that, not only would they be chosen by NASA on January 16th 1978 as part of the agency's eighth group of astronauts, but that they would fly together aboard Space Shuttle Challenger a little more than five years later. For Ride, though, the media attention at becoming one of six female candidates was especially intense.

"The impact started before I left for Houston," she remembered years later. "There was a lot of attention surrounding the announcement, because not only was it the first astronaut selection in nearly ten years, it was the first time that women were part of a class. There was a lot of press attention surrounding all six of us. Stanford arranged a press conference for me on the day of the announcement! I was a PhD physics student. Press conferences were not a normal part of my day! A lot of newspaper and magazine articles were written, primarily about the women in the group, even before we arrived. The media attention settled down quite a bit once we got to Houston. There were still the occasional stories and we definitely found ourselves being sent on plenty of public appearances."

By the middle of July 1978, the 35 candidates had effectively more than doubled NASA's existing astronaut corps. However, unlike previous selections, the new arrivals were positively welcomed by the 'old heads' from the Gemini and Apollo era. "They seemed to accept us pretty well," said Ride. "We had them outnumbered, so I'm not sure they had a choice! It was clearly very different for them. They were used to a particular environment and culture. There were a few scientists among them, but most were test pilots. Of course, the entire astronaut corps had been male, so they were not used to working with women. There had been no additions to the astronaut corps in nearly ten years, so even having a large infusion of new blood changed their working environment. However, they knew this was coming and they'd known it

Sally Ride in the Pilot's position on Challenger's forward flight deck.

for a couple of years. By the time we actually arrived, they'd adapted to the idea. We really didn't have any issues with them at all. It was easy to tell, though, that the males in our group were really pretty comfortable with us, while the astronauts who'd been around for a while were not all as comfortable and didn't quite know how to react. But they were just fine and didn't give us a hard time at all."

The selection committee, co-chaired by chief astronaut John Young, was looking specifically not only for academic and technical talent, but also for the ability of men and women to work effectively together. "And they succeeded," added Ride. "It was a congenial class and we really didn't have any issues at all within our group. They were very respectful and incorporated us as part of the group from the beginning. We all walked in as rookies; as neophytes in the astronaut corps. None of us knew anything about what was going to happen to us and so, as you can imagine, we were a pretty close-knit group. None of the astronauts who applied did it for publicity. Everybody applied because this is what they wanted to do, so the males in the group didn't really want to be spending their time with reporters – they wanted to be spending their time training and learning things. Frankly, the women would have preferred less attention."

The new astronauts were desperately needed, said Hauck, due to natural attrition from the corps since the end of the Apollo missions. Over a period of just a year, by mid-1977, four veteran pilots – Stu Roosa, Gene Cernan, Ron Evans and Gerry Carr – had retired from NASA to pursue other interests. The remainder, Hauck explained, "wanted to get us as smart about the Shuttle's systems as soon as they could. We got a year of training that involved, for virtually eight hours a day, lectures about the systems or observations from space or visiting one of the NASA centres. Then, eventually, we got 'in', so we were assigned on-the-job training, assigned specifically to one of the old guys," Hauck continued, "and I was assigned to Dick Truly. Everyone was very hospitable to us, bending over backwards to make us comfortable and telling us how much they needed us."

Training co-ordinators from those heady days would recall that the classes for the new astronauts were more like briefings on ascent and re-entry aerodynamics, space physics, tracking techniques and physiology, followed by practical experience in various simulators. Kathy Sullivan, one of the six women in the group, admitted that there were few formal tests, but each astronaut was keenly aware that they would someday need everything they had learned, potentially, to keep them alive.

"The first few months were spent in more of an 'observer' mode," said veteran Skylab astronaut Ed Gibson, who co-ordinated the Mission Specialists' training schedules. "After that, they'd be assuming responsibility the same as anybody else in the office." Some candidates, such as Fred Gregory, were detailed to work on enhancing the Shuttle's cockpit instrument suite, while others, including George 'Pinky' Nelson, Anna Fisher and Jim Buchli, worked on procedures for donning and doffing spacesuits.

Elsewhere, fellow newcomer Dan Brandenstein described the training as an incredibly intense learning experience. "A common joke was that training as an astronaut was like drinking water out of a fire hose," he said, "because it just kept coming and coming and coming! Probably the good point was you weren't given written tests, so they could heap as much on you as possible and you captured what you could."

By the time they completed initial training in the summer of 1979, they began working hand in glove with the 'Devil's Advocates' – the team of instructors who dropped fault after fault and failure upon failure into each mission simulation, testing their knowledge to the limit and proving themselves as the astronauts' worst enemies and closest friends, all rolled into one. Years later, many would look back warmly on the Devil's Advocates as having prepared them to be able to respond to almost any emergency during a 'real' mission.

"My first big project for Dick Truly," recalled Hauck, "was to develop the emergency procedures for flying the Shuttle. I was supposed to be a co-ordinator for the flight crew in how [the procedures] would be formatted, how they would read and what kind of book they would be in. This project was to put in one document all the procedures that would have to be acted on quickly, either during launch or re-entry. Many of them would be on cue cards Velcroed to the panels around the cockpit, but there wasn't enough space to Velcro all the procedures of the Shuttle. It was much more complicated in terms of crew interaction than any of the previous vehicles, so I looked at the existing T-38 jet trainer emergency checklist and proposed a certain format and certain flip pages, how they'd be tabbed, how it would be organised and what it would look like. Dick and I tried several versions and that's what became the emergency procedures checklist. Then my job was to work with the flight controllers, who would develop the specific reactions to emergencies, put them in words, try to format them in a way that could be used – in not too much detail – by the flight crew and that was a massive effort. Eventually, we got to the point where we had an ascent and re-entry pocket checklist because, depending on the environment, you had different reactions to the same problem."

Despite the hard work, the newcomers bonded exceptionally well; so well, in fact, that two astronaut marriages resulted from Group Eight. One was between Ride and Steve Hawley, another between Hoot Gibson and Rhea Seddon. Years later, Gibson, who flew Challenger in February 1984, remembered the group was so large that it had to be split into two halves, both of which frequently entered into friendly competition through 'red' and 'blue' football matches. They organised happy hours on Friday nights, Christmas parties and New Year celebrations; turning, said Gibson, into an extended family as much as a spacefaring flight squadron.

To highlight the distinction between themselves and the Grizzled Veteran Astronauts already in Houston since the 1960s, they gave themselves the nickname 'Thirty Five New Guys', designing TFNG patches and T-shirts to foster closer camaraderie. Mike Mullane, another of the 1978 arrivals, has remarked that military pilots also knew of an obscene double entendre with the same acronym – 'The F***ing New Guys' – but that, as far as the outside world was concerned, TFNG reflected solely the number of candidates in the class ...

Judy Resnik, who died aboard Challenger in January 1986, came up with the design for the group's T-shirt: a forward-facing view of the orbiter, literally overflowing with 35 astronauts crammed, sardine-like, into every available orifice, and proudly displaying the Shuttle's 'We Deliver' motto that would later become world famous.

FOUR BECOME FIVE

In spite of the teamwork, however, each member of the TFNGs knew that one day their performance in the simulators and through their technical assignments would drive the decision as to which of them would fly first. Then, in April 1982, a few weeks after Space Shuttle Columbia landed from her third test flight, Sally Ride was called into George Abbey's office at JSC. As the agency's director of flight operations, Abbey had chaired the selection committee and it was he who gave final approval on the choice of astronaut crews.

His power in determining the fate of many a spacefarer has become the stuff of legend and subject of considerable praise and criticism over the years, but Abbey's influence in crew selections is indisputable. For Ride, being summoned to his office, alone, that spring day in 1982, was unusual. "The Commander is the first to know about a flight assignment," she remembered. "Bob Crippen, who would be the

The STS-7 crew assembles in Challenger's forward flight deck for an in-flight photograph. From left to right are Sally Ride, Bob Crippen, Rick Hauck, Norm Thagard and John Fabian.

Commander of my crew, had already been told, but then usually the rest of the crew is told together; at least, that was the way it was done then. In this case, Mr Abbey told me first, before he called over the other members of the crew. He took me up to JSC Director Chris Kraft's office, who talked about the implications of being the first American woman astronaut. He reminded me that I would get a lot of press attention and asked if I was ready for that. His message was 'Let us know if you need help. We're here to help you in any way and can offer whatever help you need'. It was a very reassuring message, coming from the head of the space centre."

Ride's colleagues on the STS-7 mission would be Crippen, a veteran of the first Shuttle flight, joined by Rick Hauck in the Pilot's seat and fellow Mission Specialist John Fabian. They were destined to train for a year, with a tentatively scheduled launch date sometime in April 1983 aboard Challenger to deploy two commercial communications satellites and release and subsequently retrieve a free-flying platform using the Shuttle's Canadian-built robot arm. Little did they know at the time that their crew would ultimately expand to five members with the inclusion of a third Mission Specialist, Norm Thagard.

The $100 million robot arm, officially known as the Remote Manipulator System (RMS), was Canada's contribution to the Shuttle effort – a contribution that dated back to 1974, when Spar Space Robotics Corporation was contracted by the country's National Research Council to build a mechanical device for deploying and retrieving satellites from orbit and, ultimately, assembling the components of a space station.

The challenges involved in building an arm of such complexity and dexterity were enormous: it needed to operate autonomously and under manual control and meet strict weight and safety requirements. Moreover, nothing quite like it had ever been built or used in space before, which made Spar's task yet more difficult. Although a functional floor rig was built to test its joints, the first real demonstration did not come until the RMS was actually uncradled in orbit on Columbia's STS-2 mission in November 1981.

Measuring 15.2 m long, which thus enabled it to reach the far end of the payload bay, it consisted – just like a human arm – of shoulder, elbow and wrist joints, linked by two graphite epoxy booms. Other components were constructed from titanium and stainless steel. To protect it from thermal extremes in space, the arm was covered in white insulation and fitted with heaters to maintain its temperature within required limits. Without a payload attached, it could move at up to 60 cm/min, but this was reduced to a tenth of that speed when fully loaded.

Ingeniously, the means by which the arm could 'pick up' and 'put down' objects was achieved by the so-called 'end effector' – essentially a hand that employed a kind of three-tie wire snare to capture a prong-like grapple fixture attached to deployable or retrievable payloads. Already, the Hubble Space Telescope, at that time scheduled for launch in the mid-1980s, had an in-built grapple fixture that would enable it to not only be deployed, but also retrieved and serviced, by future Shuttle crews. So too did the Solar Max satellite and, indeed, Landsat-4, which were specificially designed to be serviced by astronauts.

During operational missions, like STS-7, astronauts would use two television cameras on the arm's wrist and elbow to guide the end effector over a target's grapple fixture, before commanding the three metal ties of the snare to close around it at precisely the right instant. When this was done, it would impart a force of 500 kg onto the grapple fixture, thus enabling the RMS to move the target. Although the arm was controlled by the Shuttle's General Purpose Computers (GPCs), its movements were directed by an astronaut using a joystick on the aft flight deck.

As the astronauts issued each instruction, the GPCs examined them and determined which joints needed moving, their direction and their speed and angle. Meanwhile, the computers also looked at each joint at 80-millisecond intervals and, in the event of a failure, automatically applied a series of brakes and notified the crew.

"One of my first assignments was on the RMS," Ride said. "I was one of a couple of astronauts that became heavily involved in the work to verify that the simulators accurately modelled the arm: to develop procedures for using the arm in orbit, to develop the malfunction procedures, so astronauts would know what to do if something went wrong. There weren't any checklists when we started; we developed them all! We also helped with the testing of the hardware itself at the contractor's facility in Canada. Until you actually start using something, it's very difficult to make predictions on how well it's going to work, what it's used for and how to accomplish the tasks that it's designed to accomplish. Many of the recommendations came in the form of the procedures that we developed. We did a lot of development of the visual cues. The astronaut controlling the arm looks at it out the window and also monitors its motion using several cameras. Often, critical parts of the view are blocked or the arm is a long way from the window or the work is delicate. In those cases, the astronaut needs reference points to help guide the direction he or she moves the arm. How do you know, exactly, that you're lifting a satellite cleanly out of the payload bay

Sally Ride glances through the overhead flight deck windows during simulator training on the Remote Manipulator System.

and not bumping it into the structure? What limits should be put on the use of the arm to make sure that it's kept within its design constraints? We did a lot of work on that. It was rewarding work, because it was at a time when the system was just being developed and nobody had paid attention to those things yet.''

John Fabian, too, had amassed considerable expertise in developing the techniques for operating the RMS. It was entirely appropriate, therefore, that on STS-7 – which involved the first ever deployment and recovery of a free-flying platform by the Canadian arm – its movements would be under the watchful gaze of both astronauts. Designated the Shuttle Pallet Satellite (SPAS), the $13 million platform was laden with ten scientific and technical experiments, funded by the then-Federal Republic of West Germany, the European Space Agency (ESA) and NASA.

Famously, it would also take the first photograph of the entire Shuttle in orbit.

Before STS-7 even left Earth, however, the most famous aspect of Challenger's second mission was Sally Ride herself. In some of the more cynical areas of the media, journalists speculated that she had been added to the crew as a public relations ploy, in response to the Soviet Union's launching in August 1982 of its second female cosmonaut, Svetlana Savitskaya. Bob Crippen vehemently disagreed. "She is flying with us because she is the very best person for the job," he told the press. "There is no man I would rather have in her place."

Years later, Rick Hauck agreed, adding that, despite a few awkward occasions, training and execution of the first American mixed-sex spaceflight went without many problems. "There were situations," he acquiesced, "where, maybe in the potty training, I'd never been involved in professional discussions with women about those! It was uncomfortable in a few situations, but the discomfort disappeared easily. Sally was great and Crip set the right tone in terms of what his expectations were of the crew. We just did it."

Awkwardness was also a problem faced by NASA's male-dominated engineering community, who decided the female astronauts were bound to require a makeup kit! "So they came to me," laughed Ride, "figuring that I could give them advice. It was about the last thing in the world that I wanted to be spending my time training on, so I didn't spend much time on it at all. There were a couple of other female astronauts who were given the job of determining what should go in the makeup kit and how many tampons should fly as part of a flight kit. I remember the engineers trying to decide how many tampons should fly on a one-week flight and there were probably other issues, just because they had never thought about what kind of personal equipment a female astronaut would take. They knew that a man might want a shaving kit, but they didn't know what a woman would carry."

Four people confined in an area the size of a camper van for six days made for cramped accommodation. Then, eight months into their training, the STS-7 quartet became a quintet. When Vance Brand's STS-5 crew rocketed into orbit on November 1982, one of their objectives had been to perform the first-ever Shuttle spacewalk. That was cancelled due to unrelated equipment failures in the two spacesuits. However, the day before these problems materialised, another area of concern – space sickness – reared its ugly head.

Alternatively known as 'space adaptation syndrome', or 'stomach awareness', the problem, according to a NASA news release of June 1982, was akin to a nauseous motion sickness and could not be resolved entirely through ground-based medical research. As a result, a series of Detailed Supplementary Objectives (DSOs) were timetabled into several early Shuttle missions, comparing in-flight observations and crew-completed questionnaires with medical data acquired before launch and after touchdown. It was hoped, eventually, that this might enable doctors to identify unique parameters and predict which individuals would be especially susceptible to the condition.

During their training, these early Shuttle crews did everything from filling in the questionnaires to having motion sickness artificially induced in the unforgiving rotating chair of NASA's neurophysiology laboratory. This allowed doctors to provide each astronaut with a 'data point' against which their predicted in-flight susceptibility could be compared. Ultimately, medication was provided in the form of Dexedrine and Scopolamine tablets, taken minutes after arrival in orbit, and bags of salted water were drunk shortly before re-entry to lessen the punishing effects of the onset of terrestrial gravity.

"The symptoms were not only occasional nausea, but also what the docs called 'episodic vomiting'," explained sufferer Charlie Walker. "The only other symptoms were a malaise and slightly sweaty palms, like symptoms that others have with a cold or the flu. You feel low in energy, a little stuffy in the sinuses and I felt, with weightlessness, that the blood was rushing to my head, which is exactly the case! I still had these symptoms for about 72 hours and then they went away, just like that. Three days into the flight and I felt fine."

NASA considered the impact of space sickness to be minimal, highlighting the fact that only four previous crews – those of Apollo 9 in March 1969, Skylab 3 in the summer of 1973, STS-3 in March 1982 and Vance Brand's mission – had been directly affected by the condition. On Apollo 9, a spacewalk was postponed by a day to allow astronaut Rusty Schweickart to recover, the Skylab crew's workload had to be reduced for their first 36 hours and, on STS-3, Jack Lousma and Gordon Fullerton had experienced symptoms shortly after reaching orbit.

None of these instances of space sickness proved detrimental to the satisfactory conduct of mission objectives – the STS-5 excursion having been cancelled due to equipment malfunctions – but they did reflect a problem that had affected ten per cent of all American manned flights to date. Another key issue raised by the studies was that, although 'cross-trained' Shuttle astronauts could accommodate a sickness-stricken colleague for a limited time, serious obstacles could arise if the syndrome affected the entire crew.

Consequently, following the STS-5 incident, NASA decided that, in addition to its ongoing efforts to define the physiological and behavioural mechanisms responsible for space sickness, it would add a pair of medical doctors to STS-7 and STS-8. Norm Thagard, the physician joining Crippen's crew, was already well known to Rick Hauck. "He and I had first met when we were both on the USS Lake Champlain, learning to land airplanes on aircraft carriers," recalled Hauck. "In order to try to learn more about space sickness, NASA generated a bunch of tests and I was one of the guinea pigs! As soon as we got on orbit, Norm had these visual, spinning things that I had to watch and, boy, I felt miserable. They sure accomplished the purpose! At one point, I said 'Hey guys, I've had it. I'm going to go into the airlock', which was a nice place to hide. I said 'I'm going to close my eyes and please don't bother me until I come out'. I didn't know whether I was going to throw up. It was after about four hours that I started to come out of it and that resolved itself."

"The principal experiments," Thagard explained, referring to the tests that proved to be Hauck's nemesis, "involved eye-tracking changes. Since space adaptation syndrome is thought to be caused by visual–vestibular conflict, looking at changes in eye motion during tracking studies was thought to be important. Also, the auditory portion of the eighth – vestibular – nerve was studied using audio-evoked potentials." However, he added, "no reliable predictors of susceptibility were uncovered by either [STS-8 physician] Bill Thornton's or my work."

At the time of Thagard's assignment to the crew – just four days before Christmas 1982 – the STS-7 launch was still scheduled for April of the following year, which also provided NASA with invaluable data for how long astronauts needed to fully prepare themselves for missions. Eventually, due to hydrogen leaks that pushed Challenger's

In addition to utilising his four crewmates for medical tests, Norm Thagard conducted a number of experiments on himself, including this series of observations of head and eye movements in microgravity.

maiden voyage from late January until early April, Bob Crippen's team found themselves rescheduled for mid-June. Despite his late addition, Sally Ride recalled that – aside from the TFNG family ties they shared – Thagard blended in exceptionally well. "We didn't spend every waking hour together," she said, "but we did spend almost all our time together, either as an entire crew or in groups of two or three. I was spending almost all my time with Crip and Rick in launch and re-entry simulations or with John and Norm in orbit or RMS training. Also, because we had things that required the whole crew, we did a lot of training together. We got to know each other very well. We never had any issues at all and got to be very good friends through the training."

Thagard, too, felt comfortable joining STS-7. "I was already assigned to support the crew," he told me in a March 2006 email correspondence, "[and] I had been working with them for months before being added to the crew, so I was familiar with the mission before my assignment. I performed a lot of the photography, was EV1 for contingency spacewalks and operated the RMS to capture the SPAS. Except to add my space sickness activities to the mission, there was little change to the pre-existing flight data file. As I was the physician most familiar with STS-7 and my previous technical assignments included rendezvous and proximity operations similar to those involved in releasing and recapturing the SPAS satellite, as well as operations of the Canadian-built robot arm, I was the obvious choice to fly."

THE NEW GUYS DELIVER

By the time Paul Weitz' crew brought Challenger swooping into Edwards Air Force Base on April 9th 1983, the STS-7 launch date had slipped to "no earlier than" June 18th. Even as the new orbiter slowed to a halt, two-thirds of the components for her next mission were already in place on the other side of the United States. In High Bay Three of the Vehicle Assembly Building (VAB), the fully stacked Solid Rocket Boosters (SRBs) had been attached to their External Tank (ET) on March 2nd.

When Challenger finally returned to Florida on April 17th, atop the heavily modified Boeing 747 airliner, it was a race against the clock to ready her for a mid-June lift-off. In a quite remarkable turnaround that was trumpeted by NASA in its STS-7 press kit, Challenger was overhauled, the payloads from her last mission removed and support hardware for her next set of equipment installed and she was rolled into the VAB on May 21st. At just 34 days, turnaround was accomplished a week faster than the previous record holder, STS-4 in June 1982.

The speedy processing flow was, however, a worrying harbinger of future problems and would be one of several issues highlighted during the inquiry by the Rogers Commission into flawed decision making that contributed to Challenger's loss in January 1986.

Many of the time savings were achieved by deleting the need to repeat tests of systems that had operated perfectly throughout STS-6. Other important tasks included repairing damaged areas of the two Orbital Manoeuvring System (OMS) pods with around 170 white tiles; similarly, sections of Challenger's elevons – the flap-like assemblies at the rear of her wings – required replacement with new thermal protection material. Elsewhere, an additional seat for Thagard was installed on the middeck and the RMS (not carried on STS-6) was fitted alongside the port side payload bay sill.

Following attachment to her boosters and tank on May 24th and rollout to Pad 39A two days later, preparations to insert Challenger's cargo began in earnest. Although SPAS, with its rendezvous commitment, was among the most visible of the STS-7 payloads, the commercial focus was a pair of drum-shaped communications satellites: one belonging to Canada, the other to Indonesia. Both arrived by aircraft at Cape Canaveral Air Force Station on November 30th 1982 and were transferred to the Vertical Processing Facility (VPF) for checkout.

Less than six months later, on May 23rd 1983, the satellites and their attached solid rockets – each encased in a lightweight sunshade to protect it from temperature extremes in low-Earth orbit – were moved to the pad and loaded aboard Challenger. Looking 'down' on her payload bay, it seemed that two oversized versions of Pacman were sitting there, for when the protective sunshades opened to expose the satellites shortly prior to deployment, they looked just like a pair of jaws from the children's game. Fortunately, unlike the real Pacman, these jaws were designed to release something, rather than gobble it up.

Each cradle was composed of a series of machined aluminium frames and chrome-plated steel longeron and keel trunnion fittings, covered with Mylar insulation and measuring 2.4 m long and 4.6 m wide. At the base of the cradle was a turntable that used two electric motors to impart the required spin rate, which varied between 45 and 100 revolutions per minute, depending on the stability needs of the payload, together with a spring ejection system to release the satellite and its booster. During ascent, two restraint arms held the precious satellites steady inside their sunshades and, shortly after reaching space, the Pacman jaws were closed to protect them from the thermal extremes of low-Earth orbit. At operational geosynchronous altitudes, on the other hand, they would rotate to even out thermal stresses.

The 620 kg Anik-C2 satellite was built by the Hughes Aircraft Company at its El Segundo plant in California, but owned by the Ottawa-based Telesat Canada concern, between which fabrication contracts had been signed in April 1978. It offered, for the first time, rooftop-to-rooftop voice, data and video business communications, together with Canadian pay television and other broadcasting services. With this in mind, it seems fitting that the word 'anik' translates to 'little brother' in Inuit.

After their construction, Anik-C1 and C2 were placed into storage until suitable dates could be established to launch them both. By coincidence, Anik-C3's completion occurred at the same time as Telesat's first contracted flight opportunity on the Shuttle, so it was decided to take it straight from the factory to the launch pad. Anik-C3 thus rode aboard Columbia on STS-5 in November 1982 – the first 'commercial' Shuttle mission – followed by Anik-C2 on Crippen's flight and Anik-C1 aboard the orbiter Discovery in the spring of 1985. Telesat reportedly paid NASA somewhere between nine and ten million dollars to launch Anik-C2 alone.

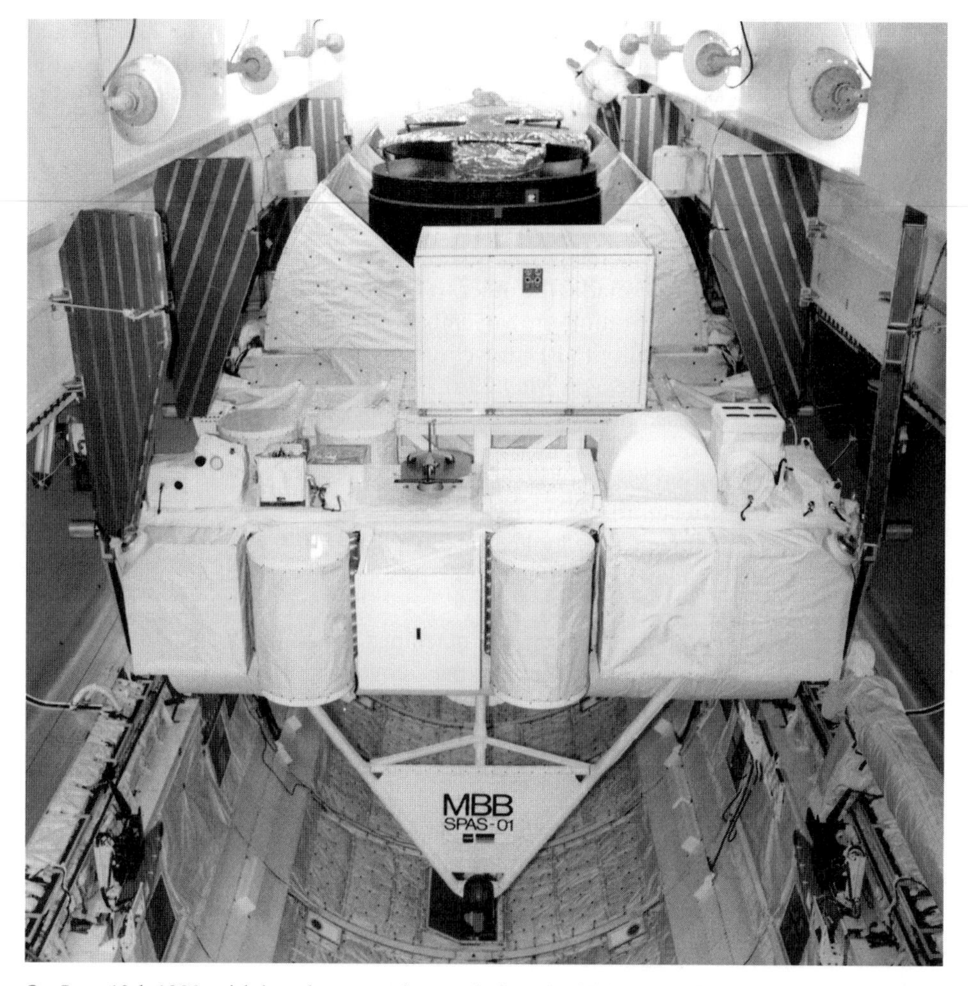

On June 13th 1983, with less than a week to go before the STS-7 launch and poised on Pad 39A, Challenger's payload bay is fully laden for her second mission. Clearly visible are SPAS-1 and the open sunshields of Palapa-B1 and Anik-C2.

Both it and the Indonesian satellite, Palapa-B1, which sat behind it in Challenger's payload bay, close to the aft bulkhead, were of the Hughes 'HS-376' bus type. These cylindrical, spin-stabilised drums measured 2.8 m tall and 2.1 m wide when stowed, but increased to more than twice that height in their final operating configurations. Both carried two concentric, telescoping solar panels – comprising 14,000 solar cells in total – which generated 1,100 watts of DC power to support ten-year life spans and carried their own 100 kg supplies of hydrazine fuel. Each also had an onboard power system, including rechargeable nickel–cadmium batteries, to run their communications payloads.

Attached to the 'top' of each satellite was a 1.7 m diameter shared aperture grid antenna with two reflecting surfaces to provide 'transmission' and 'reflection' beams. Together with its sister satellite, already in orbit, the new Anik-C2 would focus four transmit beams to cover virtually all of Canada, including its remote northern regions. Moreover, when joined by Anik-C1 in April 1985, the trio operated exclusively at 12 and 14 GHz in the microwave Ku-band, with 16 transponders apiece, each capable of carrying two colour television channels and their associated audio and control circuits.

Its combination of high-power and microwave band usage meant that much smaller antennas, just 1.2 m across, could now be routinely situated on rooftops or office blocks. This marked a significant reduction in size from the 3.6 m C-band reception dishes used previously, which had been viable only for hotels and major office buildings. Eventually, however, when all three satellites were in orbit, anticipated communications traffic did not grow as much as expected. Moreover, Anik-C1 – whose main role had been to back up the others in case of failure – became surplus to requirements, was placed into 'orbital storage' and eventually sold.

Physically identical, yet serving a very different, scattered nation on the opposite side of the planet, Palapa-B1 was first in a second-generation series of satellites to offer regional telecommunications across Indonesia's 6,000 inhabited islands, 150 million inhabitants and 250 languages. Unlike the earlier Palapa-A series, launched by NASA in the 1970s to provide telephone, television and fax services, the newer version was four times more powerful and extended communications to the Philippines, Malaysia, Thailand, Singapore and Papua New Guinea. Appropriately, the word 'palapa' translates to 'fruits of labour' and, in Indonesian political ideology, has symbolised harmony and unity for centuries.

Interestingly, the oath 'amuktl palapa', in ancient Javanese, literally means 'relaxation after exertion'.

Displaying a similar theme of unity as a crew, and heralded by signs that screamed 'Ride, Sally Ride', Crippen led his team out of the Operations and Checkout Building into the glare of flashbulbs in the early hours of June 18th 1983. Their countdown and lift-off at precisely 11:33 am was one of the smoothest ever conducted. Challenger's three main engines shut down on time, eight minutes and 20 seconds into the mission, and by 12:19 pm, Crippen and Hauck had completed the second OMS burn needed to circularise their 28.45 degree inclination orbit at an altitude of 260 km.

The picture perfect ascent demonstrated NASA's seemingly effortless ability to fly on time and within very short 'launch windows'. Only five minutes were available to the STS-7 crew for their first opportunity on June 18th and only two minutes for a second shot at 12:24 pm. The shorter than normal window was dictated by three considerations: Earth horizon sensor constraints on Anik-C2 for a deployment during Challenger's eighth orbit and on Palapa-B1 some 11 circuits of the globe later, together with a requirement for adequate lighting conditions at Edwards Air Force Base in California, should an emergency landing become necessary.

For the four rookies on the crew, their years of training had paid off.

"Physically, the simulator does a pretty good job," Sally Ride said of its closeness to the real thing. "It shakes about right and the sound level is about right and the sensation of being on your back is right. It can't simulate the g-forces that you feel, but that's not too dramatic on a Shuttle launch. The physical sensations are pretty close and, of course, the details of what you see in the cockpit are very realistic. The simulator is the same as the Shuttle cockpit and what you see on the computer screens is what you'd see in flight."

There, however, the similarities ended.

"The actual experience of a launch is not even close to the simulators," Ride exclaimed. "The simulators just don't capture the psychological and emotional feelings that come along with the actual launch. Those are fuelled by the realisation that you're not in a simulator – you're sitting on top of tons of rocket fuel and it's basically exploding underneath you! It's an emotionally and psychologically overwhelming experience; very exhilarating and terrifying, all at the same time."

During ascent and re-entry, Ride served as the flight engineer, seated behind and between Crippen and Hauck and helped them keep track of Challenger's systems. "My job was primarily to keep track of where we were in the checklists and be prepared with the malfunction checklists should anything go wrong," she remembered. "I was the one that was expected to be first to find and turn to the procedures should anything go wrong. I was also monitoring systems and status on the computer screens. My main job, though, assuming nothing went wrong, was to read the checklist and tick off the milestones. One of the first things that I was supposed to do – seven seconds after booster ignition – was, once the Shuttle started to roll, to say 'Roll program'. I'll guarantee that those were the hardest words I ever had to get out of my mouth. It's not easy to speak seven seconds after launch!"

Meanwhile, in the right-hand Pilot's seat, Rick Hauck recalled seeing the sky outside his cockpit window change colour as Challenger climbed higher. "Seeing the sky turn from blue to black in a fraction of a second was amazing," he said later, "because as you leave the atmosphere, the Sun's rays are no longer being scattered by the air molecules. I remember as I was glancing out the window, startled, Crip said 'Eyes on the cockpit!' Back to work. Watch all the gauges. I guess that's one thing that stands out in my memory. Everything about it was thrilling."

After unstrapping, all five astronauts had little time to contemplate their new surroundings. The main objective of their first day in space was the deployment of Anik-C2, performed under the auspices of Fabian and Ride. Three hours into the mission, updated computations of Challenger's orbital path – including her altitude, velocity and inclination – were radioed to the two Mission Specialists. Then, about 40 minutes before deployment, Crippen and Hauck manoeuvred the Shuttle into the correct attitude with its long axis 'horizontal', one wing down, and the open payload bay doors facing into the direction of travel.

The restraint arms pulled away from the $160 million satellite and the astronauts flipped a switch on Challenger's aft flight deck to open the Pacman jaws and impart a spin rate of 50 revolutions per minute on the payload. This steady rotation would help to stabilise Anik-C2 during its deployment. Next, at 9:01:42 pm, nine and a half hours after leaving Florida and flying high over the Pacific, Fabian and Ride fired and released a Marman clamp that held the satellite and its booster in place. Almost in slow motion, the payload left the bay at just 90 cm/sec.

Canada's Anik-C2 communications satellite, with a PAM-D booster attached to its base, departs Challenger's payload bay. Note the open 'jaws' of the insulation-enshrouded sunshield. In the foreground is SPAS-1, which would be deployed by Sally Ride and John Fabian later in the mission.

Fifteen minutes later, Crippen and Hauck backed the Shuttle away to a distance of around 40 km, aiming their spacecraft's belly towards the satellite to protect their delicate topside from the exhaust of the Payload Assist Module (PAM)-D booster. At 9:46 pm, as the combination hurtled over Africa, an onboard timer automatically fired the motor for approximately 100 seconds to push Anik-C2 into a highly elliptical geosynchronous transfer orbit.

The PAM-D – formerly known as the 'spinning solid upper stage' – was part of a family of three small boosters destined for use by the Shuttle. It was essentially a portable launch platform, developed in the mid-1970s by McDonnell Douglas on NASA's behalf, to enable the reusable spacecraft to transport satellites weighing up to 1,250 kg aloft, mostly bound for 35,600 km geosynchronous transfer orbits. Two others – a PAM-D2, capable of hauling 1,880 kg, and a more muscular PAM-A to support 2,000 kg payloads – were also in the pipeline for Shuttle missions at the time of the Challenger disaster. McDonnell Douglas' decision in the 1970s to develop this new booster came about following apparent dissatisfaction from the Shuttle's largest customer base – the commercial communications satellite industry – with Boeing's Inertial Upper Stage (IUS), which had been used on Challenger's maiden mission to deploy the first segment of a NASA tracking and data relay network. Most commercial communications satellites, including Anik-C2 and Palapa-B1, were 'spin stabilised' and McDonnell Douglas opted to utilise a turntable capable of spinning and ejecting payloads from the Shuttle, after which a much smaller, lighter and more compact booster than the IUS could propel them on to geosynchronous transfer orbits.

The performance of both the PAM-D and its titanium-skinned, Thiokol-built Star-48 solid-fuelled rocket motor were described as "satisfactory" on STS-7, with the only minor concern being a slight hesitation of Anik-C2's Pacman sunshield during closure. Post-flight inspections revealed that a small Teflon rub strip, laced into one of its insulating panels, had inadvertently pulled itself loose.

After insertion into geosynchronous transfer orbit, Anik-C2 raised its omni-directional antenna and, over the following three days, employed its own hydrazine-fed apogee motor to position itself into the required near-geosynchronous slot at 112.5 degrees West longitude, south of central Alberta. Initially used by an American provider, the GTE Satellite Corporation of Stamford, Connecticut, in support of one of the world's first direct-to-home pay television services until December 1984, it supported Anik-C3 and ultimately handled educational broadcasts and the Trans-Canada Telephone System. American involvement stemmed from an agreement with Telesat to purchase temporarily surplus satellite capacity for US companies.

Launching of the Indonesian Palapa-B1 followed much the same routine. Once more under the watchful eyes of Fabian and Ride, it was sent spinning out of the payload bay at 1:36 pm on June 19th. Forty-five minutes later, its own PAM-D ignited to insert it perfectly into an accurate transfer orbit. Commanded from an Indonesian ground station at Cibinong, near Jakarta, Palapa-B1 was then manoeuvred, in a similar manner to Anik-C2, into its operational slot at 108 degrees East longitude. In doing so, when it became operational on July 30th 1983, it eventually replaced the earlier Palapa-A1 satellite.

Operating on C-band (4/6 GHz) frequencies with 24 transponders, Palapa-B1 increased telecommuncations coverage of small rural satellite terminals in remote locations over its eight-year operational life span. Its advertised capabilities included carrying a thousand two-way voice circuits or a colour television channel on each of its transponders. Initially used by the Indonesian government's Perumtel organisation, it was replaced in April 1990 by the newly re-launched Palapa-B2R satellite and later sold to PT Pasifik Satelit Nusantara (PSN) for its 'inclined satellite' business.

ORBITAL BALLET

"I didn't really know what to expect, because there isn't a way to train for being weightless," said Sally Ride of her first experience of life off the planet. "It's so far removed from a person's everyday experience that even hearing other astronauts describe it didn't give me a clue how to prepare for it. What I discovered was that, although it took an hour or so to get used to moving around, I adapted to it pretty quickly. I loved it! I really enjoyed being weightless."

It was a pity that physician Norm Thagard, with his battery of space sickness tests to operate, could not have applied some of his expertise to the third deployable payload aboard Challenger. For, had the Shuttle Pallet Satellite been a 'human' crew member, its manoeuvres in space during the second half of the STS-7 mission would undoubtedly have rendered it somewhat queasy. The aim of flying the research platform was to demonstrate the Shuttle's ability to conduct close range 'proximity' operations, including rendezvous, station-keeping and retrieval.

Such operations on STS-7 would provide critical, real world data in support of one of Challenger's most important assignments planned for the spring of 1984: the recovery and repair of NASA's crippled Solar Max satellite. To further underline the link between these two flights, and highlighting his knowledge of the orbiter's computers, rendezvous and navigation hardware, Bob Crippen would command both missions.

Indeed, the SPAS operations were, admitted Rick Hauck, one of the most challenging aspects of STS-7. "It was going to be the first time that the Shuttle had flown in close proximity to another object," he explained. "We knew that the Shuttle had a lot of capabilities that had been designed into it and one of our major objectives was to flight test the ability to do the last stages of rendezvous and fly very close to another object when you're both going at [28,000 km/h]. The objective was, using the RMS, Sally was to lift it out of the bay and release it. Crip would fly the Shuttle with it just sitting there, because we could always drift relative to each other. We needed to make sure we could fly close to it comfortably, then back away, fire the jets to go back to it, eventually up to [300 m], fly around it and see if we could fly it without having the reaction jets upset the satellite."

Designed and built by the West German aerospace firm Messerschmitt–Bolkow–Blohm (MBB), under a June 1981 agreement with NASA, it was designed to accommodate scientific and technical experiments provided by fee-paying customers. Roughly triangular in shape, it measured 4.2 m across, 0.7 m wide and 1.5 m high

Deftly manipulated from within Challenger's flight deck, the RMS grapples SPAS-1 during the final phase of proximity tests.

and weighed 1,500 kg when fully laden. During missions, it could operate in the payload bay – secured by one keel and two longeron trunnions – or be deployed for up to 40 hours in autonomous free flight.

On STS-7, it was equipped with six experiments provided by the West German Federal Ministry of Research and Technology, two from ESA and three from NASA. In fact, the US space agency's decision to carry it was part of a deal with MBB to help with ongoing tests of Challenger's robot arm. In return, the West German concern received significant discounts for the satellite's maiden flight. Although crammed with experiments – ranging from studies of metallic alloys to a state-of-the-art remote sensing scanner – it became most famous for its NASA-provided cameras, which yielded the first pictures of the full Shuttle in space.

Beginning on June 20th, the first of two phases of SPAS activities got underway with initial testing in the payload bay. During this time, seven of its ten experiments were switched on and allowed to run continuously for 24 hours. Then, next day, with the satellite held securely by the RMS, Crippen and Hauck pulsed Challenger's Reaction Control System (RCS) jets to evaluate movements within the arm. Again, Sally Ride found her months of practice on the ground had prepared her amply for operating the real thing in orbit. "The simulators did a good job," she said later. "It was a little easier to use the arm in space than it was in the simulators, because I could look out the window and see a real arm! Although the visuals in the simulator were very good, there's nothing quite like being able to look out of the window and see the real thing. It felt very comfortable and familiar. The simulators had prepared me very well."

Early on June 22nd, the second phase – releasing SPAS into space – got underway. Shortly before 9:00 am, under John Fabian's control, it was released from the arm. The crew reported that the satellite's handling characteristics were exactly as expected and the RMS imparted no appreciable motion. For the next nine and a half hours, the astronauts tested the arm, fired off RCS plumes to deliberately disturb the satellite and practiced the rendezvous and proximity operations needed during the Solar Max repair.

During deployed proximity operations, Challenger flew 'down' and 'forward' of SPAS to a distance of 300 m, during which time the crew used a newly fitted Ku-band antenna as a 'rendezvous radar' to track the satellite. Crippen then approached SPAS and Fabian retrieved it, before releasing it again and recapturing it as it rotated. An hour later, it was deployed yet again, this time under Ride's control, and Challenger flew forward, 'up' and 'down' to a distance of 60 m.

Later, at closer gaps of ten and 30 metres, Crippen and Hauck fired the RCS jets at nine different locations to evaluate plume effects on the satellite. Subsequent tests included releasing the satellite with the RMS in automatic mode, before finally reberthing it in the payload bay and deactivating it.

As this celestial ballet was ongoing, the remaining experiments aboard SPAS, costing around ten million dollars, were activated. One of the most important was the West German Modular Optoelectronic Multi-spectral Scanner (MOMS), which acquired high-resolution imagery and conducted thematic mapping of ground-based targets, including arid regions, coastal areas, islands and mountains. This proved extraordinarily successful for geological mapping, mineral exploration, hydrology and monitoring of renewable resources for agriculture, forestry and urban or regional planning. Twenty-six minutes of high-resolution imagery validated the concept and cleared the way for MOMS' inclusion on a West German-dedicated Spacelab mission in the autumn of 1985.

Additionally, the University of Bonn provided a double focusing magnetic mass spectrometer to measure the intensity and composition of gaseous contaminants in Challenger's payload bay. An experimental heat pipe, a yaw sensor package and a variety of developmental solar cells were also affixed to the satellite, together with a number of investigations associated with STS-7's fourth major cargo: the Office of Space and Terrestrial Applications (OSTA)-2 payload.

Proximity operations with SPAS were aided immeasurably by the maiden flight of the Shuttle's steerable Ku-band communications antenna, which provided a rendezvous radar capable of 'skin tracking' satellites, passively or actively detecting a target's relative position in space. Since the antenna could only be effectively supported by having an active Tracking and Data Relay Satellite in geosynchronous orbit, STS-7 marked the first time the high-data-rate device had flown aboard the Shuttle; on STS-6 in April 1983, the slower S-band communications link was used instead.

The 91 cm Ku-band dish, mounted on the starboard wall of the forward payload bay, enabled high-data-rate communications to be transmitted to Mission Control. Although the S-band link could operate through TDRS, it could only do so at a lower data rate, since it did not have a high enough signal gain to support high-rate traffic. A drawback of the Ku-band system, however, was its narrow 'pencil' beam, which rendered it more difficult for TDRS antennas to lock onto its signal.

Consequently, the larger bandwidth S-band was used to 'locate' TDRS-1 and lock the Ku-band dish into position. When this had been achieved, the latter's signal was switched on and the device conducted a three-minute-long spiral conical scan to detect the satellite. During proximity operations, the dish provided Crippen and Hauck with target-angle and range data to update Challenger's navigation software. The only problems were occasional communications 'dropouts' from the payload interrogator, which provided a telemetry link between the Shuttle and SPAS.

Meanwhile, the crew described the retrieval – both in stable and slowly rotating attitudes – as easy to perform, "but the act of going up and capturing it was a little scary," admitted Ride. "What if we couldn't capture this satellite? It was easy in the simulators, but was it going to be easy in orbit? The experience was different because it was real! In the simulator, it wasn't that important and if you missed, it was just a virtual arm going through a virtual payload. In orbit, it really mattered that I captured the satellite." Fortunately, the retrieval went perfectly.

SCIENTIFIC BOUNTY

Although the most visible elements of STS-7 were the carriage of three satellites and the presence of Sally Ride, a vast amount of valuable research was being monitored and conducted autonomously by NASA's Office of Space and Terrestrial Applications (OSTA)-2 payload. Although this was the second time the office had flown a set of experiments aboard the Shuttle – the first was aboard Columbia on STS-2 in November 1981 – the 1,448 kg OSTA-2 marked the first use of the Mission Peculiar Equipment Support Structure (MPESS) in the payload bay.

Developed jointly by NASA's Marshall Space Flight Center (MSFC) of Huntsville, Alabama, and West Germany's Aerospace Research Establishment, it included the Materials Experiment Assembly (MEA), which included three studies of advanced semiconductor crystal growth, metallurgy and containerless glass technology. Elsewhere was the Materialwissenschaftliche Autonome Experimente unter Schwerelosigkeit (MAUS), consisting of three cylindrical Getaway Special (GAS) canisters laden with investigations into the melting and solidification of metals, alloys and industrial glasses.

It was already known that, on Earth, gravitational effects influenced the formation of materials in ways that yielded undesirable effects such as 'sedimentation' – the 'settling' of melts of composite materials whose constituents had different densities. Other results of terrestrial gravity included hydrostatic pressure and convection currents, both of which were known to cause 'stirring' in fluids. Even the walls of the containers in which such materials were solidified could cause stresses and imperfections. As a result, samples processed on Earth were often flawed in structure and composition and much less suitable for advanced technologies than more 'homogeneous' materials would be.

Until the flight of OSTA-2, it had proven extremely difficult to observe some aspects of the theoretical properties of specific materials because near-freedom from Earth's gravitational constraints could only be achieved for a few seconds at a time in ground-based laboratories. Skylab experiments in the early 1970s had already hinted at the effectiveness of the microgravity environment for advanced materials processing. During typical OSTA-2 experiment runs, Challenger was placed into a 'gravity gradient' attitude – a stable orientation with her tail pointing towards Earth – which achieved the minimum quantity of vehicle-induced g-forces and restricted the required number of thruster firings.

The development of MEA began in 1977, when NASA issued an announcement for proposals of materials investigations for carriage aboard the Shuttle. It was anticipated that the reusability of the system – which flew again aboard Challenger as part of the West German-sponsored Spacelab-D1 payload in late 1985 – would provide a cost-effective means of getting experiments into space for longer periods of time than had been possible aboard sub-orbital rockets.

Activated by Fabian and Ride from instrument panels on the aft flight deck, the rectangular, box-like MEA began operations on June 19th, barely 24 hours into the STS-7 mission. During the course of the next five days, it processed samples of germanium selenide – which, it was hoped, could ultimately lead to benefits in the semiconductor industry – as well as mixing liquid metals and exploring the viability of producing high-temperature, containerless glass-forming substances.

Meanwhile, one of the MAUS canisters operated for almost its full programmed duration of 80 hours, while the second prematurely shut down at the end of its first processing run. Two of the MAUS experiments also had components installed aboard SPAS and were conducted during its free flight alongside Challenger to minimise the impact of Shuttle-induced mass and stabilisation movements.

Sponsored by the West German Federal Ministry for Research and Technology, the SPAS-mounted MAUS experiments explored the processing of a new permanent magnetic alloy using the properties of two metals – bismuth and manganese – which have proven difficult to mix uniformly on Earth. A second investigation measured oscillatory convection in fusion processes, while a third determined the effects of gravity on ground-based pneumatic conveyor systems.

Seven additional GAS canisters, mounted along Challenger's payload bay walls, conducted a variety of academic and government-funded experiments. Six of these were attached to the port and one to the starboard sill. One West German study focused on conducting crystal growth in liquid solutions, manufacturing metallic catalysts and exposing plant seeds and eggs to cosmic radiation. Elsewhere, investigators from Purdue University explored seed germination, fluid dynamics and traced the movement of high-energy particles.

The strange effect of microgravity on a colony of carpenter ants was the subject of an experiment provided by Camden High School in New Jersey. Housed in a special 'farm', along with television and movie cameras inside the GAS canister, the aim was to assess the impact on the ants' social structure. Other investigations supplied by the Naval Research Laboratory, which utilised a motorised 'door' atop the canister for the first time, measured ultraviolet emissions and the effects of the Shuttle's payload bay environment on ultraviolet-sensitive films.

Meanwhile, tended in Challenger's middeck were the Monodisperse Latex Reactor (MLR) and Continuous Flow Electrophoresis System (CFES), both of which had been aboard STS-6. The former, designed to produce large quantities of ten-micron-sized latex beads as part of ongoing studies, operated extremely well, using all four MLR reactor chambers. Already, the National Bureau of Standards had indicated its interest in using such mass-produced beads for calibration standards in ground-based medical and scientific equipment.

Meanwhile, the CFES machine had, on STS-6, successfully separated one sample containing haemoglobin and a second containing a mixture of haemoglobin and a complex sugar known as polysaccharide. During the STS-7 flight, polystyrene latex particles were carried to further investigate the concentration limitations of continuous flow electrophoresis in space and better calibrate the machine. Its success on both missions had already guaranteed manufacturer McDonnell Douglas a seat for one of its employees – engineer Charlie Walker – on a Shuttle flight, specifically to operate it on a full time basis, in the spring of 1984.

"As I remember it," said Walker, who was assigned to STS-12 in June 1983, "the initial agreement was for six flights of the proof-of-concept CFES. It was very limited in terms of the number of flights available. I think there was wording in the contract of optional additional flights, to be negotiated later, if the concept proved to be of merit to both the industry and NASA and there were future needs to move into." Little did he know that those "optional additional" flights would not only be added, but would lead to no fewer than three missions for himself.

A SUCCESSFUL MISSION

In general, only minor problems marred Challenger's second mission. The new text and graphics system, akin to an onboard fax machine, failed after printing a single page and the urine flow system proved erratic on the toilet. On the evening of June 20th, a Cathode Ray Tube (CRT) display on the flight deck went blank and refused to respond to tests, although, fortunately, it was not needed during re-entry. Post-flight analysis confirmed a power supply failure had occurred and corrective measures were put in place prior to STS-8.

One of the more worrying problems was a four millimetre pit in one of the Shuttle's six forward flight deck windows; caused, it turned out, by the impact of a piece of 'space debris'. First reported by the astronauts on June 20th, it became the subject of detailed energy discursive X-ray analysis after landing. Titanium oxide and small quantities of aluminium, carbon and potassium were found in addition to pit glass. The morphology of the impact was suggestive of an impacting particle – most likely a tiny fleck of paint – just 0.2 mm in diameter, travelling at six kilometres per second!

The six windows wrapped around the orbiter's cockpit represented the thickest ever manufactured as optical quality view ports, each consisting of no fewer than three individual layers. The innermost pane, measuring 15.8 mm thick and made from tempered alumino–silicate glass, helped to maintain the cabin's pressure. Next came a 3.3 cm thick sheet of low-expansion, fused-silica glass to provide high optical quality and excellent thermal shock resistance. Lastly, came the 15.8 mm thick outermost pane, also of fused silica, but containing a high-efficiency, anti-reflection coating and capable of withstanding temperature extremes up to 420 degrees Celsius.

It was fortunate that Challenger's windows were thus equipped with these three layers, for the outermost pane – primarily employed to provide thermal protection during the later stages of atmospheric re-entry – was the only one affected by the pit. However, post-flight inspections noted that significant structural weaknesses caused to the outermost panes by such minutely sized debris particles could lead to further problems during a particularly harsh re-entry.

Additionally, NASA's mission report added that more debris damage was experienced generally by Challenger's thermal protection system on STS-7 than any previous Shuttle flight. The damage was close to the left 'chine' – between the leading edge of the left wing and the main fuselage – and was caused, apparently, during ascent on June 18th, as breakaway foam and ice tumbled from the External Tank. More discolouration of her insulating blanketing was evident, compared with STS-6, and several tiles were lost, including a fragment from one belonging to the left main landing gear door.

Originally, STS-7 was scheduled to perform the first Shuttle landing at the Kennedy Space Center (KSC), a fact highlighted in the mission's press kit, which would have helped to reduce turnaround times significantly. "We were looking forward to that," remembered Sally Ride. "They had a red carpet ready to roll out for us and our families were all waiting for us in Florida." Unfortunately, the touchdown on June 24th, due to occur on Challenger's 96th orbit, was postponed by two further revolutions in the hope that conditions would improve or facilitate a landing attempt in California.

It was expected that bringing each Shuttle mission back home to the East Coast launch site would save around one million dollars and five days' worth of processing for the next flight. Moreover, KSC landings would remove the necessity to expose the two-billion-dollar orbiter to the uncertainties and potential dangers of a cross-country ferry flight from Edwards Air Force Base in California atop NASA's heavily modified Boeing 747. However, as Crippen's crew discovered that June day in 1982, the West Coast landing site exhibited far more stable weather conditions than Florida.

The KSC runway, known as the Shuttle Landing Facility (SLF), opened in 1976 and is located a few kilometres north-west of the VAB. Measuring 4.6 km long and 91 m wide, with 300 m overruns at each end, it is all concrete and slopes slightly from the centreline to facilitate drainage. In contrast, Edwards has a multitude of runways: several dry lakebeds and one concrete strip, the largest of which was over 12 km long. Much of the increased size of these runways became necessary not in view of the orbiter's size, but in view of its touchdown speed – roughly 350 km/h – and the consequent need for additional margins of safety.

Challenger touches down at Edwards Air Force Base on June 24th 1983.

Two options were available to Shuttle crews returning to KSC: they could either approach from the south-eastern 'end' of the landing strip, designated 'Runway 33', or the north-western 'end', known as 'Runway 15'. The decision over which runway to use was largely dependent upon wind speed and direction, but in STS-7's pre-flight press kit, Crippen was aiming for Runway 15. Sadly, not until February 1984 would a Shuttle crew make landfall in Florida, although Crippen, Hauck, Ride and Thagard would all make landings there later in their careers.

The resultant three-hour delay to STS-7's homecoming, therefore, gave the crew some much deserved free time and, said Rick Hauck, provided them with an opportunity to hold a makeshift, Earth-circling Olympics. "Someone said 'Okay, we'll time this'. Each person, in turn, had their hands coming up from middeck to flight deck through that opening on the port side, hands curled over the floor of the flight deck. On the count of three," Hauck explained, snapping his fingers, "we went as fast as we could up into the flight deck, down through the starboard entryway, down through the middeck and back up. We gave out five awards. Sally won the fastest woman. John Fabian won the competitor that caused the most injuries; no-one got hurt, but I think his leg hit Crip coming around at one point. I think Norm Thagard was the fastest man. Crip was the most injured!"

Eventually, after the hopes of an East Coast touchdown came to nothing, Crippen and Hauck duly fired Challenger's OMS engines at 12:56 pm to begin the hour-long glide to Earth. Sally Ride was pleased. "I remember being disappointed that we weren't going to land in Florida," she said later, "but I grew up in California and we'd spent a lot of time at Edwards Air Force Base. The pilots had done a lot of approach and landing practice at Edwards, so it almost felt like a second home. But there weren't many people there waiting for us!"

It was true. All of the astronauts' families were gathered at the viewing site at KSC. Nonetheless, Challenger's second touchdown in just over two months, occurring at 1:56:59 pm, was near perfect. Her systems had performed satisfactorily throughout re-entry and landing, but during towing operations a chattering noise was heard from one of the wheels on her right-hand main gear. The Shuttle had to be jacked up, the wheel removed, its brake assembly disassembled and the wheel remounted before towing could resume.

Detailed inspections revealed that the right-hand inboard brake had actually suffered major structural damage to two of its rotors, including the beryllium heat sink and carbon lining segments. Additionally, the right-hand outboard brake had two loose carbon pads with retainer washers missing. Cracked retaining washers were found in all brake assemblies and it was discovered that a similar situation might have occurred on previous Shuttle missions with no adverse effects. None, however, had been positively identified before STS-7.

It became clear that the washers had probably cracked during their manufacture or pre-flight assembly, with structural and thermal analyses confirming that neither the flight nor landing could have caused the damage. One of the main 'to-do' tasks on the list for Challenger's processing team at KSC before her next mission, STS-8 in August 1983, would be the replacement of all cracked or suspect brake washers.

The NASA convoy responsible for recovering STS-7 after touchdown was somewhat smaller than intended, due to the diverted landing site: instead of the 24 vehicles and 110 personnel normally in attendance, only six trucks and 24 people were at Edwards on June 24th. As with previous flights, they determined that residual hazardous vapours were below significant levels and began attaching purging and coolant equipment to Challenger's aft fuselage. These measures enabled them to remove re-entry heat from the Shuttle and better protect its electronic hardware.

Half an hour after landing, Crippen, Hauck, Fabian, Ride and Thagard departed Challenger, using an airliner-like mobile stairway. Ground personnel then replaced the astronauts on the flight deck to complete safing activities and prepared the vehicle for transfer to the enormous Mate–Demate Device (MDD) hoisting crane, which would later install her atop the Boeing 747 and attach a tail cone to protect her main engines and OMS pods for the return flight to Florida.

When the astronauts returned to Houston, the media frenzy was more intense than previous missions, although their opportunities to relax were limited. Almost immediately, Crippen was immersed in training to lead the high-profile Solar Max repair – a mission to which he was assigned before STS-7 lifted off – while Thagard had joined the Spacelab-3 flight. Before 1983 was out, Fabian was attached to a new crew and Hauck received his first command. Ride, too, would fly again, teamed with Crippen once more. In fact, by the end of 1984, 'Crip' would acquire the new and perhaps more fitting nickname of 'Mr Shuttle'.

3

Weightlifters

FUNTIME

Guy Bluford, the first black American spacefarer, laughed with excitement all the way into orbit on STS-8.

It was around midnight, local time, at the Kennedy Space Center (KSC) in Florida, on the rainy evening of August 30th 1983 when he and his four crewmates – Commander Dick Truly, Pilot Dan Brandenstein and fellow Mission Specialists Dale Gardner and physician Bill Thornton – left the Operations and Checkout Building, bound for Pad 39A. Sitting out at the launch complex, resplendent in the dazzling glare of powerful xenon floodlights, Space Shuttle Challenger was ready for her third orbital voyage in less than five months.

Admittedly, the reusable spacecraft was far from achieving NASA's vision of a flight every fortnight – a rate which presumed a six- or seven-strong fleet of orbiters, rather than the four ultimately built – but it was certainly beginning to prove its commercial worth. Tucked into the Shuttle's payload bay for the planned five-day flight was an Indian communications satellite called Insat-1B, which had netted the agency four million dollars in fees and which Gardner and Bluford would deploy a few hours after lift-off. Unfortunately, another major cargo element – the second Tracking and Data Relay Satellite, known as 'TDRS-B' – had already been deleted from STS-8's roster following the embarrassing Inertial Upper Stage (IUS) booster failure in April 1983.

Had TDRS-B remained aboard Challenger, alongside Insat-1B, for this mission, it would, said Bluford, have been the heaviest cargo complement yet ferried into orbit at over 29,000 kg. "There was very little weight-growth margin," he said later. "During the training, Dale and I made several trips to Boeing Aircraft Corporation in Seattle, Washington, to learn about the IUS. We were becoming well versed in the operation of the IUS when it malfunctioned on STS-6 and, because of that, NASA decided not to fly the TDRS on our flight until after the mishap was investigated."

Insat-1B is positioned inside its sunshield during STS-8 pre-flight processing.

The presence of two of these communications and data relay platforms in geosynchronous orbit – one at 171 degrees West longitude, above the central Pacific Ocean to the south of Hawaii, and another just off the Atlantic coast of Brazil, at 41 degrees West – was highly desirable to support the first Spacelab research flight in late 1983. A third orbital 'spare' would then be launched on Commander Hank Hartsfield's STS-12 mission and placed over the equator at 79 degrees West.

However, by May 27th, as investigators got to grips with finding out why an IUS booster had failed to inject TDRS-A into its 35,600 km orbit, NASA opted not to risk launching another one until the problems were resolved. Efforts were already underway to raise TDRS-A, which Paul Weitz' crew launched on Challenger's maiden mission, into its correct 'slot', but did so at the expense of using two-thirds of its valuable hydrazine station-keeping fuel. In place of TDRS-B on STS-8 would ride an unusual contraption that NASA originally wanted to fly in early 1984: the Payload Flight Test Article (PFTA).

Measuring 4.6 m long by 4.9 m high and weighing 3,900 kg, it was, in effect, a giant dumb bell to evaluate the performance and handling characteristics of the Shuttle's Canadian-built Remote Manipulator System (RMS) mechanical arm. The PFTA was constructed from aluminium and stainless steel and equipped with four grapple fixtures; two of which would be used on STS-8. Similar to the Shuttle Pallet Satellite (SPAS) tests undertaken on the last mission, it sought to acquire 'real world' data and develop crew expertise on elbow, wrist and shoulder joint reactions before the RMS was committed to the Solar Max repair.

The experience gained on STS-8 would thus help to prepare the Solar Max crew not only for the repair procedure, but also to deploy their own payload: a 9,750 kg monster of a satellite called the Long Duration Exposure Facility (LDEF). As a result, Dale Gardner's performance as lead RMS operator on STS-8 was being carefully scrutinised by NASA and Bob Crippen's next crew to ensure that the mechanical arm could indeed handle and manoeuvre large payloads with dexterity.

Yet it was the deployment and tracking requirements of their other payload – the Indian National Satellite, known as 'Insat-1B' – that brought about one of the most historic features of the mission: the first Shuttle launch in darkness. After returning from California to Florida at the end of STS-7, Challenger spent a little under a month in the Orbiter Processing Facility (OPF) and the PFTA was installed into her payload bay on July 21st. Following rollout to Pad 39A less than a fortnight later, Insat-1B was also loaded aboard.

In doing so, preparation for STS-8 snared a new record for the fastest processing time between missions so far – a mere 62 days – which was attained primarily by Challenger's personnel working around-the-clock to get her flight-ready. Seventy-six thermal protection tiles were replaced, as were the damaged brakes in her landing gear, the pitted flight deck windowpane and a failed Auxiliary Power Unit (APU). Other experiments, including a record dozen Getaway Special (GAS) canisters, were also affixed to her port and starboard payload bay walls.

When Dick Truly's crew arrived in Florida in their T-38 jet trainers on August 27th, they included among their number a trio of the Thirty Five New Guys (TFNGs) from NASA's 1978 astronaut class. Although they were assigned at the same time as Bob Crippen's STS-7 team, they would actually become the second subset of TFNGs to fly. Years later, Dan Brandenstein recalled the excitement of the call to George Abbey's office and reception of the sacred news.

"By April 1982, the first six Shuttle flights had been assigned and they were all experienced people that had been around a long time. Nobody from our class had flown, but it was hoping and guessing and rumblings, starting with STS-7, that they'd be picking up some of the new class. I got called over one day and they said that I was going to fly STS-8. One of the neat things about it was that it was going to be a night launch and a night landing. What drove that was we were launching Insat and, to get it in the proper place, we worked the problem backwards. They wanted the satellite 'here', so then we had to go back down our orbital mechanics and it meant we had to launch at night. The fact we launched at night meant that we would end up landing at night. Dick and I had both done night carrier landings and, judging from the way the Shuttle flies and doing that at night, we both looked at each other and said 'Oooh. This is going to be interesting!' We got very much involved in developing a lighting system to enable us to safely land at night. We didn't have enough time to focus 'just' on that, although we got involved because we were the ones doing it first. Astronauts Bo Bobko, Loren Shriver and Mike Smith were all involved in developing the night landing system, so we went through a rather long evolution of floodlights and flares, trying to develop some way to give us the visual cues we needed to make a successful night landing."

This nocturnal launch commitment was also simulated, to an extent, on the ground. "We concentrated on flying night launches and night landings in a darkened simulator," Bluford recalled. "We learned to set our light levels low enough in the cockpit that we could maintain our night vision and I had a special lamp mounted on the back of my seat so that I could read the checklist in the dark. The only thing that wasn't simulated was the lighting associated with the Solid Rocket Booster ignition and the firing of the pyros for SRB and External Tank separation."

STS-8's boosters had themselves changed from the set flown aboard Challenger's previous mission, since they contained new, high-performance motors, which expanded the initial thrust by four per cent. This improvement was achieved by lengthening the exit cones of their nozzles by 25.4 cm and decreasing the diameter of the nozzles' throats by 10.1 cm; the result was an increase in the velocity of solid fuel gases as they departed the booster. Moreover, some of the propellant inhibitor used in previous SRBs was removed, allowing the fuel to burn more rapidly.

By the late summer of 1983, the five astronauts had become a close-knit miniature family in their own right. Like Bob Crippen's crew, they started with just four members, picking up physician Bill Thornton in December 1982 as part of NASA's ongoing investigation into possible countermeasures for space sickness. Thornton, an astronaut since August 1967, had actually designed much of the experimental hardware used by Norm Thagard on STS-7 and would himself be accompanying it into orbit on Dick Truly's mission.

"You spend so much time working together," said Brandenstein, "and that's part of the process of crew selections. You don't put oil and water together. When I ran the astronaut office [from 1987–1992], I was responsible for the crew assignments and you specifically look for people that are compatible. I can't speak for assignments that were made on me before I was doing them, but it was obvious by even looking at it that NASA looked for a good mix. They looked for people with specialities that mesh with the mission requirements. STS-8 was a good crew. Dick Truly had been around a long time and was a good commander; he taught us a lot. Everybody had their strengths and their area of expertise and they focused on those and shared their experience and wisdom with the other folks. We got the job done."

Thornton's assignment had actually led to the creation of an extra, unofficial crew patch. Historically, astronauts avoided doctors like the plague, remarking that there were only two ways a pilot could emerge from their surgeries: either 'fine' or 'grounded'. None of the STS-8 astronauts was at risk of being grounded by Thornton, of course, but his experiments – which included a series of blood tests – resulted in a patch featuring his bespectacled eyes peering at a cluster of four pairs of frightened eyes in Challenger's flight deck.

With the STS-8 stack looming behind them on Pad 39A, the five-man crew greets the media before their Terminal Countdown Demonstration Test. From left to right are Dale Gardner, Guy Bluford, Bill Thornton, Dan Brandenstein and Dick Truly.

Behind the humour, however, there were serious concerns among NASA's senior management that space sickness could detrimentally affect future missions if crew members reacted severely to it. During a lecture in October 1991, Thornton admitted that it remained difficult to predict which individuals were susceptible, although he pointed out that Dale Gardner experienced the nauseous ailment, yet was still able to complete all of his assigned tasks, including the hours-long Insat-1B deployment. "You can't redesign the human body," Thornton said, "but human beings have learned and will continue to learn to adapt and work in zero gravity."

During STS-8, his investigations encompassed seven medical disciplines: testing aural sensitivity thresholds ('audiometry'), tracking his crewmates' general health ('biomonitoring'), recording electrical signals generated by their eye movements ('electro-oculography'), studying the effect of repeating physical movements ('kinesymmetry'), examining changes in their limb-volume circumference ('plethysmography'), measuring external tissue pressures ('tonometry') and photographing changes in leg volume throughout the mission.

Thornton's main conclusions were twofold: that none of the astronauts were directly 'motor control affected' by the condition and that symptoms had more or less disappeared within 72 hours of launch. Since the earliest reported instance by Soviet cosmonaut Gherman Titov in August 1961, around 40 per cent of space travellers have experienced the problem, although detailed investigations during the Spacelab-1 mission in late 1983 identified the practice of rolling or pitching the head as a helpful countermeasure. "Ambiguous visual cues", on the other hand, such as viewing a crewmate from an unusual orientation, generally exacerbated the sensation of malaise and sickness.

However, the near-impossibility of determining which astronauts were most likely to fall prey to space sickness came as a surprise to Dan Brandenstein. "I'd never been seasick, airsick or anything in my life," he said. "I don't understand half of those medical experiments, but during training they put us in a spinning chair and put a blindfold on each of us. They spun the chair and then they had us move our heads down, up, right, left, down, up, right, left. I was convinced I could never get motion sickness but, man, in about 30 seconds, I was a sick puppy!"

Only Dick Truly had flown before, as Pilot on STS-2, and had never experienced a night launch, so it was with an air of excitement and trepidation that Challenger's crew headed into a bewildering glare of flashbulbs in the opening minutes of August 30th 1983. Their lift-off, at 6:32 am (2:32 am local time), came 17 minutes into a half-hour-long window, due to thunderstorms in the area which lit up the sky of a slumbering Florida. Bluford, whose historic journey made him the first black American in space, vividly recalled being strapped into his seat directly behind and between Truly and Brandenstein.

For Brandenstein – one of 15 Pilot candidates chosen by NASA in January 1978 – it was his first opportunity to put more than five years of training in a variety of simulators into practice. "We got a full set of briefs on each system, so we knew how the electrical system worked and how the hydraulic system worked and the computers," he said. Part of his requirement for being on 'active' flight status was also having the ability to maintain proficiency in NASA's fleet of T-38 jet trainers.

These legendary – some observers have called them 'antique' – aircraft continue to be used by today's Shuttle astronauts for flight training and, literally, as personal taxis to reach appointments across the United States. Unfortunately, the ability of this sleek, supersonic dart to precisely mirror the handling characteristics of the stubby, delta-winged orbiter has long proven problematic: the lift-to-drag ratios of the two vehicles are quite dissimilar. In order to best simulate the steep-angled Shuttle approach to the runway, astronauts typically opened the T-38's speed brakes as wide as possible and deployed its landing gear at the very start of their descent.

A powerful electrical storm creates an eerie 'tapestry' of light at Pad 39A in the hours preceding the Shuttle's first nocturnal lift-off.

"There was an area, just outside Houston, over the Gulf of Mexico, where we could go out and do what we called 'turn and burn'," Brandenstein explained of his T-38 escapades, "which is do aerobatics and loops and rolls and chase around clouds and stuff like that. All the time, that's a way of maintaining your piloting skills. Obviously, it's a kick for people that had flown thousands of hours, but for somebody who had never flown before or had very little experience, it was a 'real' kick, because you could go supersonic, pulling 7 gs. All the Pilots had been test pilots before, so we'd go out and run the Mission Specialists over the wringer, showing them the various things you'd do if you're testing a new airplane. We'd do simulated combat runs and show them what it was like to have a dogfight and all those sorts of things." As mission-specific training got underway, Brandenstein and Truly found themselves practicing Shuttle landing approaches, at least once or twice per month, not only in Houston, but also at KSC, Edwards Air Force Base in California and White Sands in New Mexico.

Since the crew would be launching at night, it became necessary in the final week before the flight for them to enter quarantine and shift their sleep patterns into the daytime hours. "It took us about a week to get comfortable with that," recalled Bluford, who ended up 'sleep shifting' in readiness for three of his four missions. "Some of us slept at home, while others slept in the crew quarters at the Johnson Space Center (JSC) in Houston. We ate food prepared at the center and practiced in the simulators at night. About three to four days before launch, we flew to the Cape for the final launch countdown. On August 29th, we were awakened at 10:00 pm local time. We had breakfast and suited up for the mission, then headed downstairs for the van ride to the launch pad. I noticed it was raining. There was lightning in the area and there was some concern expressed by the launch control center about our safety as we proceeded out to the pad. Finally, they left it up to Dick to decide if it was safe for the crew to go to the pad. He made the decision for us to proceed and we went out to Challenger. As we climbed into the vehicle and completed our pre-flight checks with the launch control center, the rain began to subside and the clouds began to clear away. The ride into orbit was really exciting! We had darkened the cockpit to prepare for lift-off; however, when the SRBs ignited, they turned night into day inside! Whatever night vision we hoped to maintain, we lost right away at lift-off. The ride on the SRBs was noisy and bumpy as Challenger rotated to align us to a 28.45 degree inclination. The orbiter pitched down as we headed downrange, upside down. Approximately two minutes into the mission, we jettisoned the boosters. There was a large, momentary flash of light in the windows when the SRB pyros fired. We continued to ride on the three main engines for the next six and a half minutes and then jettisoned the External Tank at eight minutes and 45 seconds into the flight."

During ascent, it was Bluford's job to act as the flight engineer, checking off each stage of the Shuttle's violent climb to orbit and reading out procedures to support the pilots in the event of problems. Next to him on the flight, directly behind Brandenstein, was Gardner, while Bill Thornton sat alone in the darkened, locker-studded middeck. From his vantage point, the 54-year-old physician had little to see: the only window was a small circular one in the side hatch, although, craning his neck, he could see 'upwards' into the flight deck and through the overhead windows.

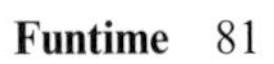

Turning night into day across a sleeping Florida, Challenger effortlessly carries out the Shuttle's first nocturnal lift-off.

At the instant of ignition, he recalled years later, the sensation was similar to "taking a fast ride on the London Underground". From his perspective, Thornton added, all was dark during the first two minutes of ascent, but as soon as Challenger shed her twin SRBs, the entire cockpit was eerily lit up. Upstairs, Dale Gardner's main view was through the overhead windows – and what he saw worried him sufficiently to call Brandenstein, one of whose tasks was to monitor the performance of the main engines.

"Obviously, Dick Truly and I were upfront, watching the instruments," recalled Brandenstein, "and Dale was looking back over his head out the window and back at the ground. At night, he could see how it lit everything up. During the first stage, it was really bright, because we had the boosters going. In fact, from the front cockpit, looking out, it was like we were inside a fire, because we didn't really see the flame, but we did see the reflection and the light. We weren't very far into the launch and Dale said 'Dan, how do the engines look?' I said 'Yes, look fine'. Thirty seconds later, he said again 'Dan, how do the engines look?' 'Fine'. I don't know how many times this happened going uphill. We didn't have a lot of time to chat about it, so finally we got all settled down on orbit [and] I said 'What was going on?' He said 'I was looking out the window', and when you watch a Shuttle launch the flame from the engine is solid. It comes out of the nozzle and just 'sits' there. During all those engine tests before STS-1, you'd have an engine running on the test stand and the flame would be solid and then, all of a sudden, the flame would 'flutter' and the engine would blow up! As you get higher in altitude, and from the perspective Dale had, the flames from the engines seemed to be fluttering, so his connection was that when the flames flutter, the engine blows up. You just have a different perspective as you get higher. The air pressure goes way down and you get into a vacuum, so basically what holds your flame real tight is the atmospheric pressure factors in that. When you get outside atmospheric pressure, they expand and flutter a little bit more."

After the mission, as the five astronauts listened back on their cockpit intercom tapes from Challenger's ascent, they were puzzled to hear someone chuckling all the way into orbit. It was Bluford. Years later, in an April 2003 interview, he remembered being so excited by the whole event that his only feeling at the time was not fear, but sheer elation.

OFF TO WORK

Bluford's journey into space had taken much longer than eight and a half minutes and represented an enormous leap for African Americans, who came to regard him as a new role model for their own aspirations and dreams. However, the astronaut himself has since remarked that it was never his intention to become the first black American in space and, fortunately, the media circus surrounding the achievement of STS-7's Sally Ride had kept him largely ignored.

"I recognised the importance of it," Bluford admitted, "but I didn't want it to be a distraction for my crew. We were all contributing to history and to our continued exploration of space." Nonetheless, he felt the Tuskegee Airmen – the United States' first all-black flight squadron in the Second World War – helped pave the way for his own achievement. When he was chosen as one of the Thirty Five New Guys in January 1978, Bluford was joined by two other black astronauts who would later ride Challenger into orbit: a Pilot named Fred Gregory and another Mission Specialist, physicist Ron McNair.

Seeing his ancestral homeland also proved profound for Bluford. "I still remember seeing the African coast and the Sahara Desert coming up over the horizon," he said later. "It was beautiful. Once we completed our Orbital Manoeuvring System (OMS) burns, I unstrapped from my seat and started floating at the 'top' of the cockpit. Like all the other astronauts before me, I fumbled around in zero-g for quite a while before I got my 'space legs'. However, it was a great feeling and I knew right away that I was going to enjoy this experience."

Upon reaching orbit, Dan Brandenstein discovered that, despite having been sick during the ground tests, he adapted to microgravity exceptionally well. "I'm one of the lucky ones in that I did a back flip out of my seat and never looked back," he said, "and never had a hiccup in any of my missions. It certainly makes your mission more enjoyable if you don't have to deal with that, but NASA was trying to decide what made people sick and how to prevent it and it turned out, after a while, they quit trying and there was no correlation. Some guys could ride the spinning chair until the motor burned up and didn't get sick and then got into orbit and, within ten minutes, they were as sick as could be. Ultimately, they found phenegren worked on almost everybody. Doctors use it on people that have had chemotherapy. So as soon as somebody would start getting a symptom of space sickness, you'd give them a shot and, in about 15 minutes, they'd be as good as new for the rest of the flight."

Bluford, too, did not recall any problems. "We had little sandwiches tied to our seats," he said later, "and when we got on orbit, a couple of crew members weren't feeling well as they adapted to space, so they 'passed' on lunch. I felt fine. I not only ate my lunch, but part of theirs, too!"

Despite concerns about space sickness and the fact that Dale Gardner, as lead crew member for both the Insat-1B deployment and RMS operations, suffered from the ailment, all five astronauts were able to conduct their prescribed tasks without problems. Releasing the $50 million Indian communications satellite and its attached Payload Assist Module (PAM)-D booster followed a similar protocol to that of the Anik-C2 and Palapa-B1 deployments on STS-7.

As its name implies, Insat-1B was the second in a series of multi-purpose geosynchronous platforms to provide telecommunications, television broadcasting, meteorology and search and rescue services to most of the Indian subcontinent and Indian Ocean. Its predecessor, Insat-1A, was launched atop a Delta rocket in April 1982 and, despite reaching its 35,600 km geosynchronous orbit, successfully deploying a jammed C-band antenna and returning valuable meteorological imagery, it inadvertently exhausted its attitude control propellant. The satellite was abandoned that September, far short of its advertised seven-year life span, but India's Department of Space received a $70 million insurance payout from the debacle.

Built by Ford Aerospace Corporation, the 1,150 kg cube shaped Insat-1B carried a dozen C-band and three S-band transponders for its communications and television

Glinting in the sunlight, Insat-1B and its attached Payload Assist Module booster drift away into the inky blackness.

services. Its meteorological payload consisted of a Very High Resolution Radiometer (VHRR), capable of acquiring visible and infrared images of Earth every 30 minutes, and a system of taking environmental data from unattended land-based and ocean-based stations. Between 1982 and 1990, four Insat-1s surveyed India's natural resources. Their data provided estimates of major crops, conducted drought monitoring, assessed the condition of vegetation, mapped areas at risk of flooding and identified new underground water supplies.

The deployment of Insat-1B was timed to occur during Challenger's 18th orbit, a little over a day into the mission, and, precisely on time at 7:48:54 am on August 31st, Gardner and Bluford flipped switches on the aft flight deck instrument panel to send the satellite on its way. Fifteen minutes later, Truly and Brandenstein performed a now-customary separation burn in readiness for the PAM-D ignition. Deployment from the Shuttle was so precise (within a tenth of a degree) that it saved Insat some 230 kg of station-keeping propellant which might otherwise have been needed had it been launched aboard an expendable rocket. At 8:34 am, the PAM-D fired to lift Insat to geosynchronous transfer orbit with a 35,600 km apogee. Later, ground controllers used the satellite's own hypergolic motor, which mixed nitrogen tetroxide and monomethyl hydrazine, to raise its perigee and circularise the orbit.

During its first few days of operations, under the direction of controllers at India's Department of Space, however, it came close to suffering the same fate as Insat-1A. Unconfirmed video recordings from the crew suggested that it may have been hit by space debris just 19.5 seconds after leaving the payload bay and, indeed, it was not until mid-September that ground operators at the Master Control Facility in Hassan succeeded in unfurling its single, five-panel planar solar array. Due to the presence of the radiometer on the opposite side, it was not practical to install two solar panels on Insat-1B. However, a 12.6 m solar 'sail' had been installed on its VHRR 'side' to provide passive compensation of the solar pressure torque about the satellite's body.

By this stage, however, Insat-1B was on station at 74 degrees East longitude – replacing its failed predecessor – and commenced full operations the following month. The debris, meanwhile, appeared to have originated from the orbiter's payload bay and a detailed, six-hour-long television scan was conducted after touchdown. Nothing on the satellite's sunshade or deployment mechanism appeared to be either missing or damaged and, upon inspecting still and video camera footage, no evidence of a direct strike on Insat was found. It seemed more likely, NASA's post-flight anomaly report concluded, that a stray particle had been spotted by the astronauts as it drifted between themselves and the satellite.

For almost seven years, Insat-1B provided satisfactory services, returning 36,000 images of Earth and providing communications and direct nationwide television services to thousands of remote Indian villages. On the ground, more than 5,000 Indian-built satellite dishes, some just three metres in diameter, were established to allow the satellite to broadcast social and educational programmes to rural communities. Insat-1B operated until July 1990, after which it served in a 'standby' capacity until it was replaced at 93.5 degrees East by Insat-2B in August 1993.

Despite the astronauts' intense focus on their mission, memories of simply being in space were aplenty. "The first impression," said Brandenstein, "is still the biggest. We were crossing Africa when I saw my first sunrise in orbit and, to this day, that is the 'wow' of my spaceflight career. Sunrises and sunsets from orbit are just phenomenal and the first one knocked my socks off! It happens relatively quickly because you're going so fast and you get this vivid spectrum forming at the horizon. When the Sun finally pops up, it's so bright; not attenuated by smog or clouds."

Throughout STS-8, they received daily updates from Mission Control on terrestrial events. "During our flight," said Bluford, "they kept me abreast of how Penn State [his alma mater] was doing in football and how the Philadelphia Phillies were doing in baseball. Each morning, we were awakened by a school song. We were informed about the shooting down of a Korean airliner, Dick Truly told me he was leaving the astronaut office to become Commander of the Naval Space Command and my wife sent me a message saying we had termites in our house!"

In addition to Insat-1B, the crew had a range of middeck investigations to tend. One was the venerable Continuous Flow Electrophoresis System (CFES), which had ridden on all three of Challenger's missions and, on STS-8, carried live human cells from a pancreas and a kidney, together with a rat pituitary gland. It was the first time that 'living' cells had been carried for electrophoretic separation in orbit. All of the samples were used to separate specific secretory cells with no apparent problems, although post-mission analysis revealed a larger residue of cells inside the spent CFES syringes than was anticipated.

Although not considered a problem with the machine itself, it was noted by NASA that it might represent a 'shortcoming' in the design of the CFES equipment to handle and separate living cells. Nonetheless, results from its three previous missions – one aboard Columbia – amply demonstrated its ability to separate 700 times more material in space than was achievable on Earth.

It was hoped that the pancreas cells in particular, which had been provided by McDonnell Douglas through an agreement with researchers at Washington University's School of Medicine, could be used in studies of purification techniques and, ultimately, new treatments for diabetes. The kidney cells, meanwhile, were supplied by NASA and the pituitary cells by Pennsylvania State University. In view of their 'living' status, one of Bluford and Gardner's key challenges on STS-8 was to keep them alive both before and after electrophoretic separation had taken place.

To accomplish this, the CFES hardware was fitted with a tray on which samples were carried aloft on a surface of micro-carrier beads in a fluid that was compatible with the living cells. Bluford, who tended the machine for several hours on August 30th, and Gardner, who monitored it the following day, transferred the cells to syringes before inserting into the separation chambers. Maintaining the cells and keeping them alive made it necessary to schedule CFES runs as soon as Challenger entered orbit. Hence, it was operated only on the first and second days of the flight.

Elsewhere, in addition to monitoring the astronauts' adaptation to microgravity, Bill Thornton kept a close watch on the behaviour of six male albino rats in an Animal Enclosure Module (AEM), housed in a middeck locker. One of the aims of the device, which would fly in support of a student experiment on Challenger's next mission in early 1984, was to assess how well the AEM contained micro-organisms and prevented 'leaks'. Apart from two micro-organisms, presumably introduced by the potatoes provided as a food and water source, the device maintained the rats' health satisfactorily during the mission.

Moreover, by posing no danger to the astronauts' own well being, it demonstrated the device's ability to maintain biological materials in full isolation. In fact, STS-8 marked the very first occasion on which a cage of animals had flown in the Shuttle's crew cabin. The rats consumed less food than predicted and did not gain weight at expected rates, compared with ground-based 'control' animals, although they returned to Earth in a healthy state. Their lower-than-expected food consumption level was attributed to the AEM's delivery system, which differed from ground-based units.

AN EYE ON THE FUTURE

By September 1st, with the Insat-1B deployment and completion of the lengthy electrophoresis experiment runs behind them, Challenger's astronauts set to work on their next major objective: testing the muscle of their ship's mechanical arm with the dumb bell-shaped PFTA. Although it would not be released into space, the device

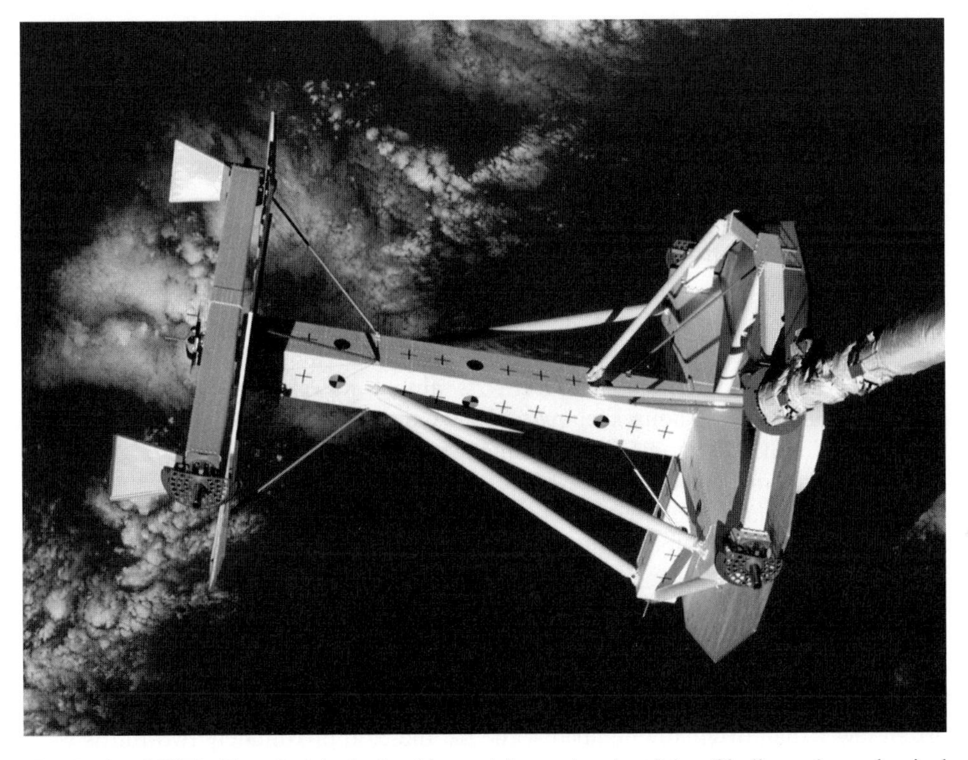

The Payload Flight Test Article during 'dynamic' exercises involving Challenger's mechanical arm.

was still the largest payload yet manipulated by the RMS – twice as heavy as the SPAS platform carried by Bob Crippen's crew – and, true to its nickname, was entirely passive, with no power, command or attitude control functions of its own.

Yet even PFTA was barely a third of the weight of the enormous Long Duration Exposure Facility, destined to be placed into orbit by another Shuttle crew in early 1984. Nonetheless, its forward and aft screens closely mimicked the visibility and manoeuvrability obstacles that future astronauts deploying large, cylindrical structures might face. In particular, PFTA became the first Shuttle-borne cargo with a 'five point' attachment to the payload bay – a keel and four longeron fittings – all of which were out of the direct view of the crew. Consequently, Gardner and Bluford would be reliant upon cameras fitted to the RMS.

With Dale Gardner at the controls, the dumb bell was first grappled by one of its two 'active' fixtures and subjected to a variety of tests, including evaluations of the mechanical arm's performance as Truly pulsed Challenger's Reaction Control System (RCS) thrusters. These tasks helped to satisfy a number of test objectives to verify ground-based simulations, assess visual cues for payload handling and demonstrate both hardware and computer software. During each activity, the RMS was employed in both 'manual' and 'automatic' modes.

The two grapple fixtures on the payload provided different geometries and mass properties for the mechanical arm and one of them – mounted in the centre of the PFTA's forward screen – offered a larger moment of inertia. The second active fixture was attached to the upper port side 'corner' of the aft screen. Much of the payload's weight was situated at its aft end, thanks to a quantity of lead ballast, and Gardner's evaluations helped to verify that the RMS could position a large structure within 50 mm and one degree of accuracy in respect to the Shuttle's axes.

Although the TDRS-B satellite had long since been deleted from the STS-8 roster, a number of important tests were performed during the mission to ensure that its doddery sibling, recently established in geosynchronous orbit at 67 degrees West longitude, would be able to support the Spacelab-1 flight, alone, later in 1983. Among these tests were evaluations of TDRS-1's ability to relay voice transmissions, commands and Shuttle housekeeping telemetry through its S-band communications channels, as well as demonstrating its high-quality Ku-band link.

This began only minutes after lift-off on August 30th, shortly after Challenger flew over Bermuda, and proved largely successful, although S-band telemetry was lost for a period of three hours at one point. However, crew voice communications were still available through other channels and the crew was asked to switch their data over to the S-band link. In addition to these problems, the White Sands Ground Terminal in New Mexico suffered a series of computers failures, which, in most cases, led to the loss of data.

In total, TDRS-1 supported Challenger during 65 orbital 'passes' – exactly two dozen fewer than originally planned – and, of these, approximately two-thirds were deemed fully successful. In particular, the performance of the TDRS-to-Shuttle S-band link was found to be highly dependent upon antenna 'look' angles, with instances in which the satellite was able to maintain return-link telemetry data, but forward-link lock could not be maintained. Still, TDRS-1's support of the Ku-band communications link proved excellent.

Meanwhile, despite the absence of Insat-1B, the payload bay was far from empty. Slightly forward of PFTA sat the U-shaped Development Flight Instrumentation (DFI) pallet, equipped with two scientific and engineering experiments, including a heat pipe that investigators hoped could provide a useful means of maintaining systems temperatures on future satellites. Dan Brandenstein monitored the performance of the pipe, activating its heater power switch and photographing temperature-sensitive tape through Challenger's aft flight deck windows.

The second experiment, known as the Evaluation of Oxygen Interaction with Materials (EOIM), consisted of a passive array of various samples – including coatings, composites and polymeric films – exposed to bombardment by molecular and atomic oxygen present in low-Earth orbit. Previous tests aboard Columbia had revealed that atomic oxygen in this environment was extremely reactive when in contact with solid surfaces; causing chemical changes, altering optical and electrical characteristics and even removing complete layers of material. This could, NASA feared, cause problems for valuable projects such as the Hubble Space Telescope or a future space station.

Admittedly, it was expected that Hubble's relatively high orbit would make atomic oxygen reaction rates fairly low, the long duration nature of other missions could lead significant erosion of solar arrays, optical coatings, light baffles and thermal control films. Among the materials flown on STS-8 were specimens of the Shuttle's new Advanced Flexible Reusable Surface Insulation (AFRSI) blanketing and thermal protection tiles to assess their degradation during orbital flight. As well as being mounted in trays and atop canisters on the DFI pallet, several samples were affixed to the RMS and exposed to space for a total of 40 hours.

The heat pipe investigation, performed early on August 31st, also proved highly successful, requiring 15 minutes to warm up and running at stable temperatures, which varied slowly in response to changes in the external environment. Although 36 photographs were taken, fewer than two dozen proved usable, due to a problem with the camera's film-advance mechanism. Still, about one and a half hours' worth of data was recorded and transcripts of the astronauts' visual observations were incorporated into the results.

Elsewhere in the payload bay were a record number of Getaway Special (GAS) canisters, four of which held scientific investigations and eight carried more than a quarter of a million first-day philatelic covers, intended to be sold by the US Postal Service after the mission. Each cover bore a recently released $9.35 postage stamp and featured the STS-8 crew's patch and logo to commemorate NASA's 25th anniversary that year. Unfortunately, the covers bore the mission's originally scheduled launch date of August 14th – which was also the stamp's release date – but this was rectified after landing.

Eclipsed by the philatelic covers, but no less important, were the other four GAS canisters, which included a Japanese effort to grow artificial snow crystals, a NASA-funded cosmic ray experiment, a test of the sensitivity of ultraviolet films in space and a contamination monitor to measure the impact of atomic oxygen particles on samples of carbon and osmium.

The Japanese study was actually a repeat of an experiment on Challenger's maiden mission, albeit with new and improved hardware. Sponsored by the newspaper 'Asahi Shimbun', its principal investigator, Shigeru Kimura, observed the growth of artificial snow particles in microgravity. Post-flight analysis after STS-6 had revealed that the temperature of the experiment's GAS end plate had fallen lower than expected, which led to a hardware redesign to warm its water in two tanks and thus provide sufficient vapour to generate crystals. On STS-8, it successfully produced the crystals and acquired high-quality video imagery.

Meanwhile, the Cosmic Ray Upset Experiment (CRUX), provided by NASA's Goddard Space Flight Center (GSFC) of Greenbelt, Maryland, helped to resolve long-standing questions about the probability of highly charged particles causing errors in memory-type integrated circuits. In some technologies, Principal Investigator John Adolphson explained, enough energy could be deposited to cause an effect known as 'latch-up', in which electronic devices literally destroyed themselves by drawing too much electrical current.

Also from Goddard was the ultraviolet-sensitive photographic emulsion experiment, whose results would pave the way for a major astrophysical instrument – the US Naval Research Laboratory's High-Resolution Telescope and Spectrograph

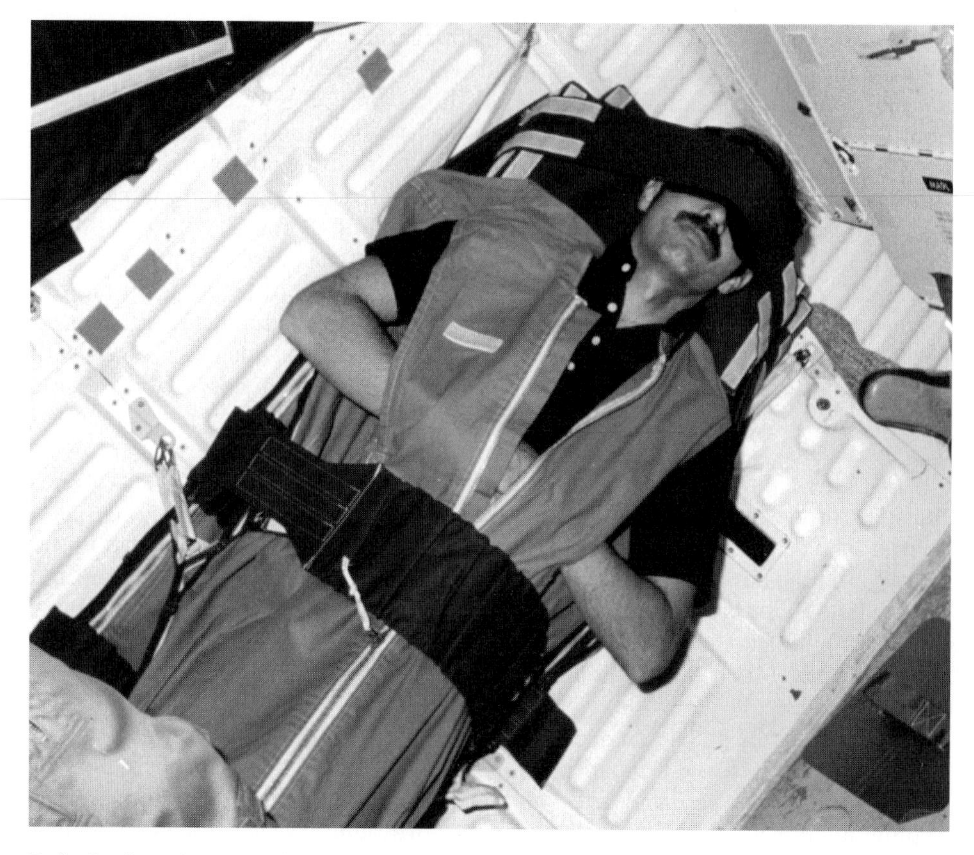

Dale Gardner sleeps on Challenger's tiny middeck. To his left a number of lockers are visible, together with floor-mounted foot loops.

(HRTS) – scheduled to be flown operationally aboard Spacelab-2 in the winter of 1984. In evaluating the effect of Challenger's gaseous environment, the emulsion experiment, led by Principal Investigator Werner Neupert, provided valuable insights into the impact of orbital hypersonic flight regimes on ultraviolet-sensitive films.

In fact, STS-8 was ideal for this kind of study, since the Shuttle's flight path was deliberately adjusted during several thermal tests. As a result, the six ultraviolet-sensitive films were oriented in the direction of travel – the 'velocity vector' – that produced a 'ram' effect, whilst the vehicle was in full sunlight. Laboratory experiments had already shown that charged particles, caused, perhaps, by clouds of ions produced in space through the action of solar ultraviolet radiation on residual gases from the orbiter, could cause chemical reactions and blacken emulsions. During STS-8, films were typically exposed for between three and 50 minutes.

Lastly, the Contamination Monitor Package, previously flown aboard Columbia on STS-3 in March 1982, was actually mounted outside of its GAS canister. Led by Principal Investigator Jack Triolo of Goddard, the experiment employed samples of carbon and osmium – two materials known to readily oxidise – to determine the detrimental effects of atomic oxygen flux in low-Earth orbit.

NIGHT FLYING

At the expense of being dubbed 'dull', Challenger's third mission had flowed exceptionally smoothly, with the only problems being a minor cabin pressure leak, later isolated to the toilet, and the presence of increasing amounts of floating dust. This became especially uncomfortable on September 4th – the night before landing – when the crew unstowed their clothes bags to prepare their flight suits for re-entry, stirring up the dust in the process. Cabin filters appeared to work properly, although Truly and Brandenstein were obliged to wipe dust from their flight deck computer displays before commencing re-entry preparations.

STS-8's nocturnal launch and Insat deployment requirement also meant that her touchdown, too, would occur in darkness; quite at odds with the seven previous Shuttle landings, which all occurred in daylight. To provide additional margins of safety, Challenger would return to Edwards Air Force Base in California, rather than attempting to land at KSC. "In other words," Brandenstein recalled, "if we had some problem and ran off the side of the runway, we wouldn't go into the moat!"

Additionally, the decision was taken to land on concrete Runway 22, rather than the dry lakebed, because "if we landed on the lakebed with the lights that we had devised to do the night landing, we'd kick up a cloud of dust, which attenuated the light," said Brandenstein. "We felt it was safer to take the approach to land on the concrete rather than the lakebed." The lights devised to support STS-8's homecoming were called Precision Approach Path Indicators (PAPI) and kept the pilots on their correct outer glide path of 19 degrees with a beam of half-white, half-red light.

The PAPI system was situated 2.3 km from the end of the runway and some three kilometres from the Shuttle's point of touchdown; the correct flight path was determined by Truly and Brandenstein by centring the white light onto the 'band' of red lights. Transition and area lighting, consisting of 800 million candlepower xenon floodlights, illuminated the whole area, with green marker lights indicating the 'end' of the runway.

Obviously, in view of the searing re-entry through the atmosphere, it was not possible to equip the orbiter with its own external lights, explained astronaut Loren Shriver, who worked on developing the night landing system. "The Shuttle is a hypersonic vehicle," he said, "and during re-entry, everything's got to be behind the tiles and inside the mould line. When the gear comes down, there are no landing lights on the Shuttle. A normal airplane has several lights that come down when the gear are extended or other lights that the crew can deploy or turn on. So, here we were, not wanting to not be able to land at night, because – to be a fully operational programme – we were going to eventually land at night somewhere. Without landing lights, we needed some kind of illumination on the runway, in addition to the normal runway lights. There were lots of cues for the pilots, but there was nothing illuminating the touchdown zone. We had to figure out a way to supply some of that lighting

Illuminated by powerful xenon floodlights, Challenger swoops into a darkened Edwards Air Force Base on September 5th 1983.

onto the touchdown zone and far enough ahead that the Commander could get the visual cues that he would normally have to fly in and land. We experimented with a number of methods to fly the glide slope on and then, after the pre-flare, to fly the shallow glide slope. We used various combinations of other high-powered lighting systems and ended up zeroing in on xenon lights. We found that certain arrangements of these lights in groups of two or four, and angled across the touchdown zone, not only headed the pilots in the right direction, but supplied the light. Then it became apparent that, once the pilots came in, if the light sources were behind them and they were trying to land on a lakebed, the wingtip vortexes and the Shuttle's rollout would produce a huge amount of dust, which would start to cut out the light in the rest of the touchdown zone. So it's maybe not a good thing to try to land on a lakebed at night, because the dust is soon going to block out all the light. Very soon after that, we put all that stuff on the concrete runways and decided if we were going to land at night, we wanted to land on a hard surface. It was an evolving process."

The mission had already been extended from five to six days, to incorporate an extra few hours' worth of TDRS-1 tests, by the time the astronauts fired Challenger's OMS engines for two and a half minutes, beginning at 6:47:30 am on September 5th, to start the glide to California.

"As we re-entered the Earth's atmosphere," remembered Bluford, "we began to feel the effects of gravity and saw the fiery plasma of hot air outside the front windows of the orbiter. Dale took pictures of the hot plasma as it enveloped us and he would occasionally hand me the camera. I could feel the camera getting heavier and heavier as we got closer to home."

For Truly, whose previous Shuttle landing aboard Columbia in November 1981 had been in daylight, STS-8 presented a whole series of novel challenges. "No engines. No moon. No correct dashboard info," he recalled years later. "The stars were blanked out because the window was frosted over. Then, finally, there were the lights of the California coast and Edwards. On the runway were the lines of red and white lights and that's what brought us in."

Returning to Earth after the first mission of what would ultimately become a four-flight astronaut career, Brandenstein explained how his brain reprogrammed itself each time to adapt to weightlessness and back to terrestrial gravity. "You just move around with a little push of the finger," he said, "and you don't do big pushes. You learn if you give a big push, you get clumsy. You have to be a little more patient and just go with small pushes and float small, slow and controlled. It takes about 30 minutes to reconfigure the Shuttle after you land. It was time to get out of the seat ... and nothing happened. To get out of the seat – because I now weighed my normal weight again – it took a conscious thought process to just get up! My natural instinct was for the brain to send little signals to the muscles, but the signals didn't do it once I was back on the ground. For the first couple of hours, a lot of things that used to come naturally were now conscious thought processes. Walking up steps was one. Naturally, if you just amble up a step, you wouldn't raise your foot high enough and you'd trip, so you'd stop and look down and make sure that your foot's a step high before you move forward with it. Your inner ear is a little bit desensitised and you watch people and they drift a bit. We were flying back from Edwards to Houston, and I felt like I had a cap on, and couldn't figure out what in the world it was. It was the weight of my hair, because for five days, the hair had been floating up and the weight of the hair made me feel that I was wearing a cap! It took about 24 hours after I got back to get back to operating on Earth. I didn't have to consciously concentrate anymore, but I had to be attentive or I'd get embarrassed, falling up the steps!"

Touchdown itself came at 7:40:43 am, completing a six-day journey which, although demonstrating that space sickness could not be effectively predicted, had helped immeasurably to further certify the Shuttle's mechanical arm for the Solar Max repair and prepare TDRS-1 for its vital role supporting Spacelab-1.

Before the Solar Max crew could fly, however, there was one more set of RMS handling tasks pencilled in for Challenger's next trip in January 1984. Moreover, the crew of that flight, the tenth Shuttle mission overall, would be required to conduct two ambitious spacewalks using another piece of equipment that was crucial for the Solar Max repair: a jet propelled backpack, which would turn astronauts Bruce McCandless and Bob Stewart into self-contained, free-flying spacecraft in their own right.

4

“At the peak of readiness”

A TOUCH OF SUPERSTITION

If Space Shuttle Challenger had been a sentient person, she would undoubtedly have been as confused as everyone else by the peculiar nomenclature with which her fourth mission was saddled. Following a nocturnal return to Edwards Air Force Base in California at the end of STS-8, she was slated to deploy two more communications satellites – one for the Indonesian government, a follow-on from Palapa-B1, and another for Western Union – on an eight-day flight in early 1984.

It would be the tenth Shuttle mission overall; yet, strangely, it was initially known as ‘STS-11’ and, ultimately, as ‘STS-41B’.

The reason can be traced to the fact that payloads for many flights were in constant flux during this time. Some were delayed or cancelled, higher priority ones were brought forward in the ‘pecking order’ and frustrating problems with satellite-delivery boosters – most notably the US Air Force’s problem-prone Inertial Upper Stage (IUS) – caused some to be dropped entirely. For instance, when astronauts Ken Mattingly, Loren Shriver, Ellison Onizuka and Jim Buchli were named as the STS-10 crew in October 1982, they confidently expected to launch aboard Challenger the following autumn on the Shuttle’s first top-secret Department of Defense assignment.

Unfortunately, an embarrassing debut performance by the IUS on STS-6, when it failed to properly deliver NASA’s first $100 million Tracking and Data Relay Satellite (TDRS) into geosynchronous transfer orbit, prompted a lengthy inquiry and the cancellation of several missions that relied upon the solid-fuelled booster. Mattingly’s STS-10 flight was supposed to carry a large intelligence-gathering satellite atop an IUS, but was postponed until the spring of 1985. The next flight on the Shuttle’s roster, STS-11, would thus become the reusable spacecraft’s ‘new’, tenth mission.

“As we started to do some of the background work and early training,” Shriver recalled years later, “it became apparent that STS-10 wasn’t going to go in sequence or on time. That was when we started to learn that the numerical sequence of the missions

Spacesuited Bob Stewart (left) and Bruce McCandless flank Ron McNair in the STS-41B crew's official portrait. Seated at the front are Vance Brand (left) and Robert 'Hoot' Gibson.

didn't mean a lot in those early days. As a matter of fact, that whole nomenclature went away!"

The plot thickened. For reasons best known within the higher echelons of NASA management, beginning with the tenth flight, the agency redesignated its missions with a cryptic and somewhat clumsy combination of numbers and letters. Consequently, when STS-11 Commander Vance Brand and his four crewmates boarded Challenger on the morning of February 3rd 1984, their flight would not be known as STS-10, or even STS-11, but rather as STS-41B.

The first number denoted the financial year under which Brand's mission was funded (1984 in this case), while the second digit ('1') pointed to the launch site of the Kennedy Space Center (KSC) in Florida. Another Shuttle complex at Vandenberg Air Force Base in California was expected to support a number of '2'-series flights from mid-1986 onwards. Lastly, the letter ('B') meant the second scheduled mission from KSC for 1984, although, due to changes in the manifest, the 'letters' for a given year were not always launched in sequence. Of course, as the year wore on and each new flight took to the skies, all the system really demonstrated was how best to perplex the public.

Columbia's STS-9 mission in late 1983 was also known internally as 'STS-41A', but due to delays the result was a roster that read as follows: STS-41B, 41C, 41D, 41G, 51A, 51C, 51D, 51B, 51G, 51F, 51I, 51J, 61A, 61B, 61C and 51L. The last of these – Challenger's final, tragic flight – seemed even more out of place because, originally funded under NASA's 1985 budget, it had been repeatedly postponed. "It was a neat new way of designating missions," said Vance Brand, "that confused everyone!" Although little official explanation for the change has surfaced, several intriguing theories have emerged.

One of them, said former flight director Jay Greene, was that the decision had been implemented in anticipation of the Shuttle's 13th mission – a tricky, six-day orbital ballet and spacewalking extravaganza to retrieve and repair the Solar Max satellite – whose launch had been set for Friday April 13th 1984. One of the five astronauts assigned to that flight was Terry Hart. "The Apollo 13 experience," he explained, "gave NASA a bad case of triskaidekaphobia, because there were a whole bunch of 'thirteens' in that and, of course, that mission had an oxygen tank explode."

Hart was referring to the ill-fated, Moon-bound journey of Jim Lovell, Jack Swigert and Fred Haise in the spring of 1970. Despite having often been labelled as NASA's finest hour, Apollo 13 infamously launched at 13:13 Houston time and suffered a major in-flight explosion two days later – on April 13th – from which the astronauts barely survived.

"We actually came out as 'STS-13' on the manifest," continued Hart. "Then, all of a sudden, three or four months after our assignment, there was an edict from NASA Headquarters that they were going to change the numbering system. No-one would say anything, but we were sure the reason that we were doing it was because they didn't want to fly an STS-13; so they went through this Byzantine structure and we became 'STS-41C'."

Jay Greene agreed that Hart's mission, which ultimately launched a week earlier on Friday April 6th 1984 and was actually the 11th Shuttle flight, was indeed the most likely catalyst for the bizarre numbering system. Indeed, when the STS-41B crew released their mission patch towards the end of 1983, it perfectly highlighted the suddenness with which the change took place. Around the edge of their patch were 11 stars, for the flight that should have been STS-11 ...

HUMAN SATELLITES

Also in pride of place on Vance Brand's patch was a snazzy, jet propelled spacesuit backpack, known as the Manned Manoeuvring Unit (MMU), together with the surname of an astronaut who had waited longer than most for his first orbital voyage. Mission Specialist Bruce McCandless joined NASA in April 1966, along with Brand, but his patient wait for space had exceeded by more than a decade that of many of the Thirty Five New Guys (TFNGs). Even old timers like Bo Bobko, Don Peterson and Story Musgrave had not waited quite as long as poor McCandless.

One of the reasons for his lengthy status as an astronaut-in-waiting was that he had been instrumental in the design and development of the MMU and had long been expected to test it on its first outing in space. His long wait – close to two full decades by the time Challenger lifted off at precisely 1:00 pm on February 3rd 1984 – would be worth it and his famous photograph, snapped by STS-41B Pilot Robert 'Hoot' Gibson, has since graced many a spaceflight book, magazine cover, wall poster and screensaver.

It also won NASA and the MMU's prime contractor, Martin Marietta of Denver, Colorado, the coveted Collier Trophy for 1984. Special recognition was also granted to McCandless, NASA's Charles Whitsett and Martin Marietta's Walter Bollendonk for a triumphant maiden mission.

In the wake of the Columbia tragedy in February 2003, it seems bitterly ironic that the original purpose of the MMU was to enable spacewalking astronauts to inspect and possibly repair damaged thermal protection materials on the Shuttle's wings and lower surfaces. Moreover, in the words of a NASA press release from October 1979, it would permit rescue operations and even the servicing and deployment of satellites. Despite the hype that encircled its first flight, however, it was actually the latest in a long line of manoeuvring packs whose heritage dated back to the early 1960s.

A decade before McCandless undertook his MMU sortie outside Challenger, astronauts had evaluated a similar device whilst wearing spacesuits and shirtsleeves inside the Skylab orbital workshop. Still earlier, in June 1965, the first American spacewalker, Ed White, had employed a hand-held pressurised gas 'gun' to move around the exterior of his Gemini capsule. Today, the MMU's own descendant – known as the Simplified Aid For Extravehicular Activity Rescue (SAFER) – is used routinely by astronauts working outside the International Space Station.

"In retrospect, I probably lavished too much attention on scientific and engineering interests, as opposed to the flying, flying and more flying," McCandless told me in a March 2006 email correspondence. "At any rate, I became interested in manoeuvring units shortly after the Gemini 9 fiasco in June 1966, in which spacewalker Gene Cernan was overwhelmed by immature pressure suit technology and was unable to fly the US Air Force's Astronaut Manoeuvring Unit [AMU]. This led to a retrenching to develop EVA technology and, unfortunately, removal of the AMU from Gemini 12, in order to guarantee ending the Gemini programme on a positive note. At about that point, I, together with a civil servant called David Shultz and Charles Whitsett became interested in showing that the concept of a manoeuvring unit was valid and that useful units could be built. We collaborated on the M-509

Stunning image of Bruce McCandless during his first flight with the Manned Manoeuvring Unit.

experiment – a multi-mode manoeuvring unit – to be demonstrated inside the Skylab workshop. I hoped to be the first to fly it, but that was not to be. I was named as backup Pilot for Skylab 2 and waved goodbye to being on the prime crew."

Unfortunately, as the first US space station was launched on May 14th 1973, a solar panel and micrometeoroid shield were torn off during its ascent to orbit. Temperatures inside the workshop soared and were only stabilised by the efforts of the first crew – that of Skylab 2, consisting of Charles 'Pete' Conrad, Joe Kerwin and Paul Weitz – in a series of complex spacewalks and emergency repair work.

"They, however, were prohibited from trying the manoeuvring unit out due to fears that its nickel–cadmium batteries had been damaged by the high temperatures inside the workshop following the loss of its micrometeoroid shield on launch," continued McCandless. "The two subsequent Skylab crews did use the M-509 and gave it glowing reports, thus enabling us to sell NASA management on building an MMU in connection with the Shuttle, initially planned for the conduct of tile inspection and repair. Ultimately, those tasks were scrapped and it was built and tested to support the Solar Max repair mission."

Although the need to potentially repair the Shuttle's heat-shielding tiles was one of the main reasons for the MMU, its development – which began in earnest in 1975 – was still hampered for some years by management disinterest and lack of firm funding. Then, in the spring of 1979, as Columbia was being moved from California to Florida, several tiles were lost and renewed vigour was injected into developing the backpack. By the time STS-1 took to the skies in April 1981, most of the tile problems, seemingly, had been solved and no MMU was aboard. Nevertheless, on opening the payload bay doors, the astronauts saw that some tiles were missing from one of the Orbital Manoeuvring System (OMS) pods, and, in response to concern that there might be tile damage to the belly of the vehicle, it was reportedly inspected by an imaging spy satellite.

It would instead be used, said NASA, for satellite repairs and maintenance and was rendered all the more useful by the provision of electrical sockets for tools, portable lights and cameras. The device was 1.2 m high, 81 cm wide and 66 cm deep and, according to astronaut Joe Allen, who flew it in November 1984, resembled "some kind of overstuffed rocket chair". On a typical mission, two MMUs were stored on opposing walls at the front end of the Shuttle's payload bay and spacesuited astronauts backed themselves into it and secured its two spring loaded latches into place.

After more than four years in the design definition stage, in February 1980 NASA awarded the $26.7 million MMU fabrication contract to Martin Marietta. The first two operational flight units, valued at around ten million dollars apiece, arrived at the Johnson Space Center (JSC) in Houston, Texas, in September 1983 to support astronaut training. Two months later, they were installed aboard Challenger. Each weighing 140 kg, they were painted white to achieve adequate thermal control in the harsh environment of low-Earth orbit and were fitted with electrical heaters to keep their components above minimum temperature levels.

Affixed to the back of each MMU were two propellant tanks, which supplied 24 tiny thrusters with six kilograms of high-pressure gaseous nitrogen. To operate them,

A view of the 'rear' of the Manned Manoeuvring Unit during vacuum chamber tests at the Johnson Space Center in March 1981.

the astronaut used hand controllers at the end of two armrests: one provided rotational acceleration for roll, pitch and yaw, while the other allowed him to move forward, backward, up, down and from left to right. Furthermore, by using both in unison, he could achieve very intricate movements. Particularly useful for repair missions, when a desired orientation had been reached, he could activate an automatic, 'attitude hold' function to free his hands for work.

Electrical power was provided by a pair of silver–zinc batteries, capable of supporting the unit for up to six hours of autonomous flight as far as 140 m from the Shuttle. In fact, one of the MMU's widely publicised features was that its wearer did not need to remain attached to the spacecraft by a safety tether. Of course, in the event of problems, most of its systems were redundant and neither McCandless, nor his spacewalking buddy, Mission Specialist Bob Stewart, ventured so far from Challenger that the pilots would not be able to rescue them if necessary.

"We didn't want to come back and face their wives if we lost either one of them up there," joked Brand.

Its controllability, though, was precise. "The minimal training and precision flying features," said one magazine editor, who flew a model of the MMU at Martin Marietta's Space Operations Simulator in Denver, "were demonstrated by my ability, with only a few minutes' practice, to manoeuvre safely in close proximity to fixed objects." Joe Allen, whose own MMU sortie in November 1984 salvaged an errant communications satellite, also remarked that, in space, it "glided" and displayed none of the idiosyncratic jerks, jolts, bumps and grinding sounds that were characteristic of Martin Marietta's simulator.

For Bruce McCandless, who backed himself into the device on February 7th 1984, securing himself with two mechanical latches and a lap belt, it represented "a heck of a big leap", in terms of spacewalking technology and his own personal odyssey. In a similar manner to the excursions conducted by Musgrave and Peterson almost a year earlier, preparations for the two STS-41B spacewalks began shortly after Challenger reached orbit, when her cabin pressure was lowered to 10.2 psi. This reduced McCandless and Stewart's 'pre-breathing' exercises from the three hours needed under 'normal', 14.7 psi conditions to less than an hour.

Another common thread between McCandless and Musgrave was that they were two of the most highly trained EVA specialists in the astronaut office at that time. "I am probably not a 'representative' EVA trainee," McCandless remembered years later. "I was grossly over-trained! I took every opportunity to get into a pressure suit, an altitude chamber or a water tank, commencing early in the Apollo programme. I helped design the Skylab M-509 experiment and made water tank runs on all of the Skylab and Hubble Space Telescope EVA tasks for development, validation and training. Concurrently, we used Martin Marietta's Space Operations Simulator (SOS) for manoeuvring unit development and, conversely, the manoeuvring units to drive enhancement and further improvement of the SOS. In discussing training for spaceflight, the first thing to recognise is there is currently no single, comprehensive device or system that gives a 'total' simulation. Eventually, training for spaceflight will consist of taking the trainees into space as passengers and conducting the training, *in situ*. Until that day arrives, however, training is accomplished on a 'part task' basis, leaving it to the individual to mentally integrate all of the pieces when the time comes. In the specific case of MMU training, the SOS allowed the pilots to 'fly' around inside a large 15 m long by 4.1 m wide by 4.5 m high room as though we were in space, as determined by computer software driving servo motors in all six axes in response to control inputs and the laws of physics. It was quite effective and could accommodate a fully suited astronaut and reasonable sized mock-ups of 'target' objects, such as the underside of the orbiter for tile repair. It also had the capability for introducing malfunctions for training purposes."

In spite of their complexity, McCandless and Stewart's excursions proved successful and the spacesuits performed admirably. The only 'nuisances' were static on the communication channels and difficulties attaching checklists to the suits' arms. "In spite of the 'sound-does-not-travel-through-a-vacuum' tenet of physics," McCandless told me, "it was noisy up there, thanks to two independent radio channels and plenty of people wanting to talk to me!" Then, just before leaving Challenger's airlock, Stewart reported a caution and warning alarm, which indicated the pressure of his suit's sublimator had risen to 4.0 psi. However, after being switched off and back on, it performed normally.

These subtle problems did not distract from the triumph of McCandless' Buck Rogers-style flight that day. Despite the sci-fi analogy, said Vance Brand, "it didn't have the person zooming real fast. It was a huge device that was very well-designed and redundant, so that it was very safe, but it moved along at about one to two miles per hour." At his furthest distance from the Shuttle, McCandless was 91 m away, politely offering to clean Challenger's cockpit windows as he floated over the flight deck. Watching intently from inside, an admiring Brand declined the offer.

"Having the opportunity to actually fly the MMU, the handling characteristics were exactly like those of the SOS," remembered McCandless, "with one, initially puzzling, exception. With the unit in 'attitude hold' mode, whenever I inputted a +/− X translation command, I heard and felt a chugging sound and vibration. On reflection, we collectively realised that this was caused by our bodies' centre of mass not being exactly co-aligned with the MMU centre of mass, thereby displacing that slightly. Consequently, a translational thrusting command tended to cause a slight pitching motion, which 'attitude hold' counteracted by modulating one or more active thrusters to 'off', as it had been designed to do, and counteracting the pitch moment, ultimately holding attitude right where it was supposed to. We ignored this effect and added a noise maker to the SOS to enhance the training for the STS-41C crew."

During their tethered work in the payload bay, McCandless and Stewart removed a failed television camera for replacement with an in-cabin unit and later installed it during the second spacewalk on February 9th. The MMU, too, performed admirably, but ironically, Brand undermined its *raison d'être*. The backpack was touted as being capable of achieving far more precise and intricate movements than the Shuttle, but on STS-41B and 41C the real value of Challenger's manoeuvrability and her Canadian-built mechanical arm were demonstrated – by retrieving one of McCandless' lost foot restraints.

"I scurried down the starboard handrail of the payload bay," McCandless remembered later, "and held up my right hand and Vance 'flew' my hand to the point where I could grasp, and retrieve the errant restraint."

"I don't recall now whether it was before or after he went out with the backpack," Brand added, "but he was trying to reposition his foot restraint, so that he could get into it to do work. Our EVA equipment was generally tethered, but it somehow got away from him. I looked back and saw it floating away. I thought about it for a second or two and decided that the ground wouldn't have time to come up with a decision whether we ought to chase it and go after it. It was going to get away from us very quickly, so I couldn't see anything wrong with going after it. We chased it, Bruce caught it and we didn't have to worry about encountering that as 'space junk' the next time we came around the world. I had one switch that was out of position when I first fired the Reaction Control System jets, which had the thrusters aligned to an axis system that was 90 degrees from what I needed at that time. After the first thrusting, I had to reposition the switch to get the proper orientation, so that the right thrusters would come on and I could accurately chase the restraint." By so doing, Brand showed that the Shuttle was capable of the same intricate motions as the MMU and, on the next flight, STS-41C in April, when a task involving the MMU was frustrated, the Remote Manipulator System (RMS) would prove itself equally capable. Despite the MMU's success during two satellite recoveries in November 1984, the manoeuvrability of the Shuttle contributed to its ultimate demise.

In fact, the year immortalised by George Orwell would be the only time the MMU was ever used in space. By the end of 1984, it had seen service on three Shuttle missions, flown by six astronauts – McCandless, Stewart, Allen, George 'Pinky' Nelson, James 'Ox' van Hoften and Dale Gardner – for a total of just ten and a half hours, spread across six spacewalks. Other assignments were expected but, in the wake of the Challenger disaster, safety upgrades imposed by the Rogers Commission proved costly and the units were mothballed to await further opportunities.

Sadly, as of 2002, by which time the smaller, backpack-mounted SAFER device had been operational for several years, no such opportunities had crystallised.

LOST, IMMOVABLE AND BURST IN SPACE

Despite the success of the backpack on its first excursion, two embarrassing failures characterised STS-41B, together with another problem that impacted part of McCandless and Stewart's second spacewalk on February 9th. Nestled inside Challenger's payload bay were the Indonesian government's Palapa-B2 and Western Union's Westar-6 communications satellites, together with the West German-built Shuttle Pallet Satellite (SPAS), which had previously flown aboard STS-7 in June 1983.

During Brand's mission, this 1,448 kg, free-flying platform would be equipped with the same experiments that it carried the previous summer, together with a dummy main electronics box, akin to that aboard Solar Max. The experiments themselves performed satisfactorily, with the only problem being a failed micro-switch on the SPAS' mass spectrometer. However, McCandless and Stewart adjusted this switch during their first spacewalk, achieving partial operating capability in the instrument.

Next, it was expected that on February 9th, Challenger's fifth crew member – Mission Specialist Ron McNair, the second African American astronaut to ride the Shuttle – would grapple SPAS with the Canadian-built mechanical arm. He would then raise it to a position three metres from the forward bulkhead, before rotating it at about one degree per second in order to simulate the attitude and dynamics of the slowly spinning Solar Max. The arm, whose wrist was capable of rolling to 'plus' or 'minus' 447 degrees, was expected to require about 15 minutes to reach the roll 'stop' points.

Meanwhile, McCandless and Stewart, spacesuited and ready to begin their second EVA day in the payload bay, would have approached the satellite and used a Trunnion Pin Attachment Device (TPAD) to duplicate 'docking' with SPAS. This

A simulated Main Electronics Box for Solar Max is installed onto SPAS-1A by technicians in the weeks preceding STS-41B's launch. It was intended that Bruce McCandless and Bob Stewart would use the box and satellite as part of ongoing demonstrations of spacewalk repair and servicing techniques.

would have provided a final dress rehearsal for a similar Solar Max docking procedure on STS-41C.

Preparations for SPAS operations with the mechanical arm duly began at 8:40 am when the satellite was transferred to internal battery power. However, as McNair worked through his procedures to check out the RMS, the arm's wrist yaw joint experienced a failure – refusing to move when commanded, even though it had worked perfectly during the first spacewalk – and the test was cancelled. Despite efforts to recover full capability in the RMS, including cycling power to the arm to clear the failure indication, it became obvious that SPAS would have to remain secured in the payload bay for STS-41B.

These troubles could be more serious during Challenger's next mission, because the Canadian-built device was to be instrumental in the planned repair of Solar Max. The problem McNair encountered could not be duplicated using engineering mock-ups at KSC and, after STS-41B's landing, the wrist joint and motor were shipped back

Bruce McCandless, carrying a TPAD device similar to the one earmarked for the Solar Max repair, prepares to dock onto SPAS-1A. Due to problems with the robot arm on STS-41B, all of McCandless' work had to be carried out whilst SPAS-1A was secured in Challenger's payload bay.

to their vendor for further investigation. Some minor corrosive effects were found, but thermal and vibration testing did not identify the problem. The cause remained unknown, although NASA decided to install a 'new' RMS for the Solar Max mission.

Fortunately, the mechanical arm difficulties did not significantly hamper the other tasks on February 9th, with MMU tests also being undertaken by Stewart, who became the US Army's first spacefarer and spacewalker. During the six hour and 17 minute excursion, they performed several successful TPAD 'docking' exercises with the berthed SPAS platform (minus the rotational aspect), retrieved McCandless' lost foot restraint, replaced the failed television camera and repaired a loose payload bay slidewire link. The failure of one of their spacesuit-mounted cameras also required them to make verbal comments to their colleagues inside Challenger for thruster firings in lieu of visual cues.

Both men found that conditions grew noticeably more frigid as they piloted themselves further from the Shuttle. "As I moved away from the payload bay, it got cold inside the suit," recalled McCandless. "The 'W' position on the temperature control turned out to actually stand for not 'warm', but 'minimum cooling'! As I left the relatively warm reflectivity of the payload bay and found my own physical activity limited to fingertip movements on the hand controllers, the heat balance shifted radically towards a cold equilibrium. This was ultimately solved by switching the sublimator feed water 'off' – a 'no-no' of the highest degree, since many thought it would not restart properly during an EVA. It did [restart] and I repeated this manoeuvre as required."

Both backpack evaluations, read Martin Marietta's post-mission review, "performed as expected and no anomalies were reported". Overall, McCandless flew the MMU for three and a half hours and Stewart for just under two hours. Yet, as has been noted, it was Challenger's own manoeuvrability, demonstrated by Brand, which rendered its future less certain. "We used the autopilot a lot," he said later. "We had the capability to manoeuvre the ship in rotation – roll, pitch, yaw – with a hand controller, but more often than not, we just punched something into the computer and set up the digital autopilot such that we got an automatic manoeuvre. That saved fuel, as we could move at very slow rates. We tested the RCS jets on orbit for translation up or down, sideways or forward and back. On the night side of the Earth, when we translated the ship down, the upward-firing RCS jets were used to do that. At night, it looked like a Fourth of July display because you could look out over the nose and you could see these tubes of fire going up. They were fantastic visual effects."

During neither excursion, which totalled 12 hours and 12 minutes, were McCandless or Stewart "lost in space", although both communications satellites were not so lucky. The first, the 580 kg Westar-6, was nearly identical to the two Hughes-built satellites placed into orbit by Bob Crippen's STS-7 crew the previous summer. Measuring 6.8 m tall and 2.1 m wide when fully deployed in geosynchronous orbit, and equipped with 24 C-band transponders, each facilitating either 2,400 telephone circuits or one colour television channel, it was twice as powerful as previous satellites in the Westar series.

Since the construction of the first American transcontinental telegraph system in the second half of the 19th century, the Western Union Telegraph Company had closely followed the development of new communications technologies, through the Morse key and sounder to the teletypewriter, microwave transmissions and message-switching computers. By the time that its sixth Westar satellite – destined for use exclusively by business customers – rode into orbit aboard STS-41B, the company was one of America's primary carriers of voice, data, video and fax telecommunications traffic.

During the early 1980s, Western Union contracted with both NASA and the European Arianespace concern for commercial launch services. However, when an

Bruce McCandless tests a mobile foot restraint, whilst attached to Challenger's Canadian-built mechanical arm.

Ariane rocket was lost in 1982, one Westar launch had to be rescheduled and the company began to reconsider its future dealings with the Europeans. Further, Western Union felt more confidence in the Shuttle, believing it to be less expensive and more reliable. By April 1983, they had opted in favour of the reusable spacecraft, rather than Ariane, to provide a workhorse to launch their satellites.

Already, efforts had been made to open negotiations with McDonnell Douglas, which had agreed to provide the Payload Assist Module (PAM)-D booster for Westar-6. Under the terms of an agreement signed in March 1983, Western Union would hold McDonnell Douglas "harmless" for any damage to their satellite and instead obtained insurance from Lloyds of London to cover potential losses.

After deployment from Challenger's payload bay, it was expected that Westar-6's attached PAM-D would duly insert it into geosynchronous transfer orbit. The satellite's own, Thiokol-built Star-30 apogee motor would then circularise its orbital path at an altitude of some 35,600 km. Supervised closely by the five STS-41B astronauts, Westar's sunshade was opened, it was spun-up to 50 revolutions per minute and ejected from the payload bay at 8:59 pm on February 3rd, almost exactly eight hours after launch.

Fifteen minutes later, as planned, Brand and Gibson pulsed Challenger's RCS thrusters to manoeuvre to a safe distance before the ignition of the PAM-D. "The impression," said Joe Allen, who watched the proceedings intently from Mission Control that day, "was that the rocket did indeed ignite. Then, somehow, they lost sight of the engine fire, but they weren't sure it was anything out of the normal. The ground controllers, however, detected that the rocket had ignited, the satellite had moved, but then the rocket had extinguished itself. Thus, it was in only a slightly higher orbit. It was a long way from geosynchronous orbit; a terrible disappointment." In fact, Westar was left in a lop-sided orbital path, with an apogee of barely 1,000 km and a perigee of around 250 km. The question now posed was whether Palapa-B2 – identical to Westar and mounted atop a similar PAM-D booster – might be subject to a similar failure. It was scheduled to be sprung free 27 hours into the STS-41B mission, but that was postponed by a day or so as troubleshooting of the Westar incident got underway.

The decision was taken, Allen said wryly, "probably with a vote that resembles our 2004 presidential elections, a hair's breadth of difference between those 'for' and 'against'." It fell in favour of taking the risk. The Indonesian government concurred with NASA and, at 3:13 pm on February 5th, Palapa popped into space. Surely, the Westar fault could not be common to both PAM-Ds, it was thought. Unfortunately, the lightning of bad luck, in this case, did indeed strike twice and Palapa's booster also fell silent and its motor nozzle went dark after just a few seconds.

"The deployments from the payload bay went flawlessly," Vance Brand remembered. "Everybody checked that backwards and forwards. The engine burns, which were solid rocket burns, each started and then after about 20 seconds, stopped. We had the underside of our vehicle pointed in the direction of the satellites, so any speeding particles from the burn would hit the underside and wouldn't do any harm. We were observing with a camera on the end of the RMS. It looked around the side of the ship to see what happened and that was recorded. I'm not sure, even today, that it's well understood why those rockets burned out prematurely, but each left its satellite stranded in an inappropriate orbit."

Consequently, Palapa, too, was virtually useless, with an apogee of just over 1,100 km and a perigee of 240 km. Its customers, not just from Indonesia, but also the member states of the Association of South-East Asian Nations (ASEAN), which included the Philippines, Thailand, Malaysia, Singapore and Papua New Guinea, would have to rely upon the services of Palapa-B1 alone. The owners of Westar and Palapa filed insurance claims of $180 million in total, although that of the former was later dismissed as Western Union had already signed a disclaimer with McDonnell Douglas to cover PAM-D failures.

The Indonesian government's case, on the other hand, went forward before a jury and the court ultimately agreed that an action in negligence would be allowed to go ahead. During the proceedings, the court heard extensive evidence of possible negligent design in the construction of the PAM-D booster and concluded that the only liable defendant was Thiokol for supplying a 'bad' rocket motor.

Efforts were underway, meanwhile, to retrieve both Westar and Palapa and return them to Earth. Within three weeks of Challenger's landing at the end of STS-41B, the satellites' manufacturer – Hughes Aircraft Corporation – had presented NASA with an option to attempt a salvage operation. By the first week of September 1984, following the spectacular repair of Solar Max (in April of that year), the increasingly confident space agency agreed with Hughes and the satellites' majority insurers to commit the Shuttle and MMU to a risky recovery mission.

In a complex, eight-day flight that November, the STS-51A crew, which featured former Challenger veterans Rick Hauck and Dale Gardner, together with Joe Allen, Dave Walker and Anna Fisher, successfully recovered both Palapa-B2 and Westar-6 and Hughes was contracted to refurbish them for subsequent reuse. Westar-6 was sold to the Asiasat consortium and blasted into geosynchronous orbit – successfully this time – aboard a Chinese Long March 3 rocket on April 7th 1990. When operational, its pair of nickel–cadmium batteries and solar cell coated body generated 935 watts of electricity to run a powerful communications payload.

The refurbished Palapa-B2 was also relaunched that same month, atop a Delta rocket. By that time, however, its operator, the Indonesian government-owned Perumtel concern, had already ordered a replacement satellite (known as 'Palapa-B2R'), which NASA had launched in March 1987, also by Delta. The 'original' Palapa-B2, meanwhile, was sold by its insurers to Sattel Technologies and eventually resold back to Perumtel under the new name of 'Palapa-B2P'. Perumtel maintained ownership of the satellite until 1993, when it passed to a private Indonesian company.

With two embarrassing satellite failures and an RMS problem that prevented operations with a third, it seemed that little else could go wrong on STS-41B. Sadly, however, it did! NASA's preparations for the vital resuscitation of Solar Max had shifted into high gear since the spring of 1983, with virtually every Shuttle flight conducting one test or another in direct support of the landmark STS-41C mission. In addition to the MMU and SPAS demonstrations, Brand's crew was assigned to deploy a large, inflatable balloon and conduct a simulated, 'closed loop' rendezvous with it.

Thermal vacuum testing of the IRT at NASA's Johnson Space Center in December 1983.

Known as the Integrated Rendezvous Target (IRT), the two-metre-diameter aluminised Mylar balloon was ejected, along with its Getaway Special (GAS) canister, from a longeron attached to Challenger's port side payload bay wall. Despite the Palapa deliberations, which had impacted several other mission objectives, the deployment of the 91 kg combination got underway as planned at 11:50 am on February 5th, drifting serenely away at 45 cm/sec. Very soon, however, it encountered difficulties.

The intention was that, a minute after leaving Challenger, the GAS canister would split in half and the balloon would inflate with nitrogen at a pressure of 0.3 psi. Rotating at three revolutions per minute, the balloon would then have been used by Brand and Gibson to firstly practice rendezvous manoeuvres from 9.2 to 14.8 km and, later, from a distance of more than 220 km, using the Shuttle's Ku-band antenna and other ranging systems to provide navigational data to the five General Purpose Computers (GPCs).

"One of the objectives," recalled Brand, "was to test out the rendezvous software in the computers for the first time. As I recall, the IRT was shot out of its canister by a spring. When it got out, it 'timed out' and filled with gas. We were watching it go and, all of a sudden, it exploded!" Post-flight analysis would confirm that a series of 'staves' enclosing the balloon failed to release correctly, due to a fabrication defect, although the IRT did start inflating. It ultimately burst when its progress was halted by the faulty staves.

Still, some success was achieved as the crew tracked several fragments of debris using the Ku-band antenna, their own eyes and Challenger's star trackers. The Ku-band dish, mounted on the starboard wall of the payload bay, had already caused minor headaches earlier in the mission when it failed to conduct a 'self test' and did not properly radar track McCandless as he flew the MMU. It also seemed to be susceptible to interference by external electromagnetic radiation. Nevertheless, it later tracked Stewart and post-flight analysis determined that a single self-test failure out of 19 attempts was "acceptable".

With the exception of the spectacular MMU excursions, Hoot Gibson remembered, years later, that the entire STS-41B crew felt "positively snakebit" at this point. Added Joe Allen: "They were now zero for three in satellite deployments! I've never asked him, but I wondered what Bruce was thinking at that point, because he was going to be the fourth satellite!"

McCandless, whose task had been to flip the two switches to deploy IRT, was far too occupied with his crewmates in seeking out the remnants of ballast from the burst balloon. However, he said years later, "we inherited the IRT from the cancelled STS-10 mission, so had not participated in any of the design qualification reviews of it."

A BITTERSWEET MISSION

Luckily for the spacewalkers, McCandless and Stewart's MMU sorties turned out to be the only fully successful 'satellite' deployments of STS-41B. Overall, 91 per cent of the mission's Detailed Test Objectives (DTOs) were satisfactorily accomplished, with the RMS fault preventing them from conducting MMU docking demonstrations with the deployed (and rotating) SPAS and the burst balloon providing only limited opportunities to evaluate Challenger's rendezvous and laser ranging gear. In spite of this, on the morning of February 11th, Brand completed one of the mission's most important test objectives by landing, for the first time, in Florida.

A KSC homecoming, within sight of the Shuttle's processing hangars and launch pads, had originally been planned for the end of the STS-7 flight in June 1983, but bad weather forced a 'wave-off' to California. Since STS-1, landings in Florida had been regarded as a key milestone in achieving truly 'routine' Shuttle missions, as well as saving an estimated million dollars and five days' worth of processing time. Unlike previous flights, Challenger would not be subject to this immense cost of being ferried from the West to the East Coasts atop the heavily modified Boeing 747 airliner.

For Vance Brand, the STS-41B touchdown provided him with the unique opportunity to have landed in three very different locations during his three-flight astronaut career. He had splashed down in the Pacific Ocean at the end of Apollo 18, the joint US–Soviet Apollo–Soyuz Test Project (ASTP) mission, in July 1975, made landfall at Edwards Air Force Base at the close of Columbia's STS-5 mission in November 1982 and now returned to the East Coast of the United States. However, despite the differences, he recalled similarities between all three. "On the return, of course, if you're landing near Hawaii or in the United States, whether it be Apollo or Shuttle," he said later, "you would do a de-orbit burn to slow down a little bit when you're over the Indian Ocean. The big difference comes after you're hitting the atmosphere. In each case, it takes about a half hour to coast before approaching Hawaii, where you hit the top of the atmosphere."

Following the completion of the 168-second de-orbit 'burn', executed at 11:16 am, Challenger's re-entry flight path took her across the Pacific Ocean to the Baja peninsula, over Mexico and southern Texas and towards Florida. Her final approach brought her squarely over the Titusville area and out over the Atlantic Ocean, where Brand and Gibson prepared for their landing from the north-west on Runway 15. Other missions have since also landed from the south-east on Runway 33. Touchdown at 12:15:55 pm, completing a mission just shy of eight full days – Challenger's longest to date – and covering five million kilometres, was perfect.

Surrounded by servicing vehicles, Challenger sits on the Shuttle Landing Facility (SLF), after becoming the first manned spacecraft to touch down back at her launch site.

For a flight with two failed satellite deployments, an RMS problem that prevented operations with a third and the burst rendezvous balloon, Challenger's fourth voyage had concluded triumphantly. In fact, in the official report, which referred specifically to Brand's landing, NASA remarked "the precision with which this objective was accomplished shows that all areas of the National Space Transportation System were at their peak of readiness". Indeed they were, for on Challenger's next mission, in barely eight weeks' time, she would be tested on her most ambitious assignment so far: the long awaited repair of Solar Max.

'LUCKY' THIRTEEN?

Despite NASA's seemingly ingrained case of triskaidekaphobia, which forced managers to impose the bizarre, '13-free' numbering system on its flights, the crew of perhaps the most important Shuttle mission to date clearly were unsure if STS-41C was supposed to be unlucky or not. Still internally dubbed 'STS-13', it would actually be the reusable spacecraft's 11th orbital journey overall; a decision had already been taken to cancel Hank Hartsfield's STS-12 mission and reassign his crew to the STS-16, or '41D', flight. The reason: Hartsfield's crew was supposed to deploy a Tracking and Data Relay Satellite, atop a still-grounded IUS.

Not until the troublesome, US Air Force-developed booster had successfully carried a top-secret intelligence gathering satellite aloft on Ken Mattingly's STS-51C flight (originally numbered 'STS-10') in the spring of 1985 would NASA, as second in the queue, regain confidence in its abilities. Instead, Hartsfield's crew eventually received a payload of three communications satellites (two of them utilising PAM-D boosters), a test of an experimental solar sail and the maiden voyage of the third operational Space Shuttle, named 'Discovery'.

By now rescheduled from Friday April 13th to Friday April 6th, perhaps to lessen the chances of ill fortune befalling STS-41C, this absurdity inspired a number of practical jokes from the crew. Pilot Dick Scobee designed the 'official' crew patch, although an 'unofficial' version lurked outside NASA Headquarters' approval: an insignia of a black cat, emblazoned with the number '13', surrounded by lightning bolts and a Shuttle hurtling from underneath its belly. Mission Specialist Terry Hart later admitted that they even had coffee mugs made, bearing the 'official' patch on one side and the 'unofficial' one on the other ...

When the crew was announced in mid-February 1983, one of them was unavailable to begin direct training until later that year. Commander Bob Crippen, who flew the first Shuttle mission, was preparing to lead STS-7 and his stint on the Solar Max repair would make him the first person to fly the reusable spacecraft three times. He would be joined by Scobee, together with Hart and Mission Specialists George 'Pinky' Nelson and James 'Ox' van Hoften. By early 1984, the mission was planned to last six days, launching on April 6th and landing well before the unlucky Friday 13th.

However, laughed Hart, "we had a problem during our mission that delayed us one day. So we ended up landing on Friday 13th after all. But we made it!"

Pinky Nelson's assignment to perform, with van Hoften, two spacewalks in support of the Solar Max repair had come at a restless time for himself and other members of the Thirty Five New Guys, as each waited impatiently for their first flight. "This was the mission I wanted," he said of STS-41C, "because it had EVAs. I remember meeting with Crippen shortly after that, in one of the little conference rooms at JSC, where he doled out the assignments and gave me the role of flying the MMU, which made my year! Here was a mission with four military pilots and they decided to let me fly the manoeuvring unit. Training for that mission was really fun. We were involved quite a bit with Vance Brand and Hoot Gibson. The mission before us was going to test out a lot of the equipment, so we worked closely with what they were doing and watched that flight closely. Ox and I were a great team. It was really the most complicated spacewalk that had ever been conceived and a real precursor to the much more complicated work they've done on the Hubble Space Telescope. We worked hard to choreograph this repair and we had it down to a dance. We knew all the steps and who was where when, what tools were needed and how we moved things."

Not only had they nailed down their mission to perfection but, in Crippen's case, even the pre-flight photographs turned into an art form. "I remember the day we posed for our crew picture," recalled Hart, "and all put our blue flight suits on and took maybe 20 pictures, trying to get the right expressions on our faces! Then, the tradition is that you bring them down to the astronaut office and ask the secretaries to pick which one is best. In one of them, one of us would be winking or our smile would be crooked or something like that. Every one of us had maybe a 50 per cent 'hit' rate on the pictures, having the right expression on our face. Then we looked at Crippen, who'd been in the public eye from STS-1 until this mission. Every photograph had the same expression on Bob Crippen's face! He had it down pat. He knew exactly how to smile!"

Like his STS-7 flight, Crippen almost gained an extra crew member for STS-41C. Although never 'officially' confirmed, the US Air Force briefly considered a 40-year-old naval engineer named Dave Vidrine for a 'Payload Specialist' seat aboard Challenger. For some time, efforts had been underway to train a cadre of Manned Spaceflight Engineers (MSEs) to accompany military payloads on the Shuttle. Although STS-41C was a 'civilian' mission, the rationale behind flying Vidrine seemed to be that observing the Solar Max repair with a 'satellite servicing specialist' could lead to opportunities for refurbishing important Department of Defense spacecraft in orbit.

In fact, by the time of the Challenger disaster, plans were afoot to launch Discovery, sometime in early 1987, to repair the military funded Landsat-4 Earth resources spacecraft. This mission would have flown from Vandenberg Air Force Base in California and, interestingly, Landsat incorporated a similar spacecraft 'bus' design to Solar Max. Dave Vidrine actually sat in with the STS-41C crew on several simulations, but in March 1984, his assignment was terminated by the head of the MSE group, Major-General Ralph Jacobson, as having "no value" to the Air Force.

As Nelson and 'Ox' van Hoften – the nickname came from his status as NASA's biggest astronaut – worked in the Weightless Environment Training Facility (WETF) to perfect their orbital 'dance', Crippen, Scobee and Hart busied themselves with

On March 29th 1984, after depositing the STS-41C stack on Pad 39A, the gigantic 'crawler' inches its way back to the Vehicle Assembly Building.

rendezvous procedures in the Shuttle simulator. In Hart's case, the RMS was another of his responsibilities. This had given trouble on its previous mission, when the wrist yaw joint failed. Although the cause of that failure was still unknown when STS-41C lifted off, the faulty arm (serial number 201) had been replaced by another (serial number 302) and verified on the ground.

Other work performed on Challenger between her missions included replacing her left-hand Orbital Manoeuvring System (OMS) pod with one from sister ship Discovery. Significant damage had been identified during inspections after STS-41B. This was caused, apparently, by ice from the potable and waste water dump nozzles. Although these nozzles are situated close to the side hatch in the Shuttle's forward fuselage, the ice apparently detached 22 minutes after re-entry interface and hit the OMS pod as Challenger flew at Mach 4.5 – some 5,500 km/h! Tile damage had also been caused by debris falling from the External Tank during ascent.

Further, albeit less serious, problems were encountered with the brakes during the STS-41B touchdown, chipping carbon liner edges and causing retaining washers to fail. Following the meeting of an industry-wide committee at JSC in January 1984, it was concluded that, in view of stresses imposed on the orbiter's brakes, such problems were "normal" and not safety-of-flight issues. 'Hard' braking on the runway had been demonstrated safely by Paul Weitz at the end of STS-6. Nonetheless, NASA opted to install extra instrumentation aboard Challenger for her STS-41G flight to better understand the dynamic interaction between the brakes and hydraulic systems.

DEPLOYMENT OF LDEF

The satisfactory performance of the RMS was vital on STS-41C, not only for the retrieval and repair of Solar Max, but also for the deployment of the crew's own payload: a 12-sided structure called the Long Duration Exposure Facility (LDEF). As its name implied, it was intended to accommodate experiments that required exposure to the hostile environment of low-Earth orbit for protracted periods. No-one could possibly have foreseen, at the time of LDEF's launch, exactly how long it would remain in space before being retrieved by another Shuttle mission and returned to Earth.

NASA intended to collect the satellite during Commander Brewster Shaw's STS-51D mission in February 1985, but that was repeatedly delayed. By the time Challenger exploded, the retrieval had been rescheduled for Commander Don Williams' STS-61I flight in the autumn of 1986. In fact, it would not be recovered until January 1990, by which point it was only weeks away from an uncontrolled and fiery re-entry.

It was a peculiar object, measuring 9.1 m long by 4.2 m wide and weighed just over 9,750 kg. At its most basic, it consisted of a bus-sized contraption made from aluminium rings and longerons, loaded with trays for 57 scientific experiments (some of which occupied more than one tray). Shortly after the formation of NASA in October 1958, researchers began to seriously consider building a satellite that could

The Long Duration Exposure Facility in March 1984, shortly before installation aboard Challenger. After launch, it would not return to Earth for almost six years.

carry material samples and assess how the harsh environment of low-Earth orbit caused them to degrade over time.

By the early 1970s, these ideas had acquired a name: the Meteoroid and Exposure Module (MEM), which, it was proposed, would be carried aloft by the Shuttle – then scheduled to make its first flight sometime in 1978 – and picked up a few months later. As the name implied, its primary focus was the impact of micrometeoroids on satellites and how best to protect them. Subsequently renamed LDEF, contracts for its design and development were granted to NASA's Langley Research Center in Hampton, Virginia.

The structure was complete by 1978 and, after tests, was kept at Langley until a Shuttle flight became available. By this point, its objectives had expanded from micrometeoroid research to studies of changes in material properties over time, performance tests of new spacecraft systems, evaluations of power sources and conducting crystal growth and space physics investigations.

The satellite was designed to be reusable and adaptable for differing lengths, if desired, although ultimately it would fly only once. Its length was divided equally between six bays for the experiment trays, with a central 'ring' at the midpoint connected by longerons to the end frames. Aluminium 'intercostals' linked each longeron to adjacent rows of longerons on each side and removable bolts joined the longerons to the end frames and intercostals. This meant LDEF could be made 'shorter' or 'longer' if a mission required it. Experiment trays were then clipped into the rectangular openings between the longerons and intercostals.

Two RMS grapple fixtures were provided on the satellite: one to allow it to be picked up by the robot arm for deployment and subsequent retrieval and a second to send signals to initiate the experiments. It had no attitude control system and, said one engineer, "what you saw was what you got": a passive container with no manoeuvring capabilities. It was designed to remain in orbit by being placed into a 'gravity gradient' attitude, with one end facing Earth, which made an onboard propulsion system unnecessary. This also freed it from acceleration forces or contamination caused by thruster firings.

The orientation of LDEF also meant that the two 'ends' would be subjected to a unique thermal environment, although all parts of the satellite were subjected to daily temperature changes as the Sun 'rose' and 'set' every 90 minutes and solar angles changed annually. Heat management was accomplished by coating the interior surfaces with high-emissivity black paint, which kept thermal gradients across the structure to a minimum and maximised heat transfer across LDEF's body. The experiments were also spread evenly to equalize thermal properties across the satellite.

Eighty-six trays – 72 around the circumference, six on the Earth-facing end and eight on the space-facing end – accommodated 57 investigations. The 1.3 m × 86 cm trays came in several different depths and housed experiments weighing up to 90 kg. These covered four disciplines: materials and structures, power and propulsion, science and electronics and optics. They captured interstellar gas atoms to better understand the Milky Way galaxy's formation, observed cosmic rays and micrometeoroids, studied shrimp eggs and tomato seeds and investigated the impact of atomic oxygen on different materials, including solar cells.

Originally scheduled for launch in December 1983, delays to several Shuttle flights that year pushed the satellite into the following spring. In June 1983, encased in a specially constructed LDEF Assembly and Transportation System (LATS) crate, it was transferred aboard a Second World War-era landing craft from Langley down the coast to KSC in Florida. Upon arrival, it was ensconced in the Spacecraft Assembly and Encapsulation Facility (SAEF), during which time its experiments were prepared and integrated. Eventually, it moved to the Operations and Checkout Building for final processing, before being loaded aboard Challenger on March 20th 1984.

Deployment occurred 24 hours into the STS-41C mission, on April 7th, and, although Hart admitted "that was exciting", it hardly compared with his first and only Shuttle launch a day earlier. "It was a clear, cool morning," he said of that Friday, "and we went through the traditions of having breakfast together and there was always a cake there for the crew before they went out. Then, going into the van and realising that all the Mercury astronauts went on that van made it a very heady experience. Next, we went out to the launch pad and up the elevator. As usual, people don't say much in elevators – whether you're in a hotel or on the launch pad – and you watch the numbers tick by and, instead of floors, they do everything in feet in the launch pad elevators, so you're so many feet above sea level. When you walk across the gantry to board the Shuttle, you can look down into the flame trench. The obvious thing that's striking you is that this is for real: we're going to go! Everything was pretty smooth on our launch countdown. We got strapped in and, again, the guys strapping us in were a lot of the same guys that strapped in Al Shepard on his Mercury flight [in May 1961]."

One of Hart's main responsibilities during ascent was to act as a 'second flight engineer'; seated behind Scobee, he assisted van Hoften with checking off the milestones and monitoring the procedures needed in the event of contingencies. There were none. "Off we went," he said of the 1:58 pm lift-off, "right on time on a perfectly clear day. I had a couple of surprises: the shake, rattle and roll of the Solid Rocket Boosters for the first two minutes is a very low-frequency rumble; just a tremendous sense of power. You can look back over your shoulder or look out the top window when you're in the flight deck and watch the world disappearing behind you. Very quickly, the SRBs taper off and separate and that was the surprise I had, because your g-loading builds up close to two and a half gs as the boosters reach their peak thrust. As the solid rockets burn off and separate, the sensation that you have at that point I wasn't quite prepared for, because you go from two and a half gs back to about one and a half. The sensation you have is that you're losing out, that you're falling back into the water! You don't think you're accelerating as much as you should be to get going and, of course, I'd worked on the main engine programme anyway, so I was very familiar with what the engines could do or not do. I think in the next minute I must have checked the main engines to make sure they were running, because I swear we only had two working: it just didn't feel like we had enough thrust to make it to orbit! Then, gradually, the External Tank gets lighter and as it does, of course, with the same thrust on the engines, you begin to accelerate faster and faster. After a couple of minutes, I felt like – yes – I guess they're all working."

Indeed, Challenger's fifth launch had proceeded without incident. The External Tank behaved superbly and the performance of the main engines, read NASA's post-mission report, "appeared to be normal". The only minor deviation was when the engines throttled down to 67 per cent, rather than the predicted 71 per cent, as Challenger passed through maximum aerodynamic pressure a minute into the flight; this lower level was later attributed to a higher than anticipated SRB impulse during the first 20 seconds. Chase aircraft also revealed that one of the main parachutes on the right-hand booster failed to inflate, although both were recovered successfully.

Watching the ascent from the roof of the launch control centre at KSC was McDonnell Douglas engineer Charlie Walker, who had spent almost a year in training as part of Hank Hartsfield's STS-41D crew, due to blast off in late June. "To watch the launch of that vehicle, knowing that you are going to be the next crew that will do that," Walker said later, "is an exciting experience; an emotional experience. As soon as the shockwaves of the SRB ignition and the lift-off struck us, the panels on the doors and walls of the Vehicle Assembly Building started rattling. Now, here are several tens of acres of aluminium and steel, rattling, and it's like a thunderstorm going on behind you at the same time the sky is burning with millions of pounds of propellant that's pushing this rocket into space in front of you. It was a mighty experience!"

Experiencing the launch from a somewhat different perspective, seated 'downstairs' on Challenger's middeck, next to the side hatch, Pinky Nelson did not have the luxury of viewing the ascent from Walker's point of view, nor through the wrap-around windows of the cockpit as Crippen, Scobee, Hart and van Hoften could. Nevertheless, he recalled the rapidity – and loneliness – of his first ride into orbit. He was also able to peer through a tiny circular window in the hatch and catch a fleeting glimpse of the enormous, controlled explosion that was underway outside. "I could see the tower go by and the sky and horizon as we ascended," he recalled. "It was a bit lonely down there, but Crip kept a running commentary on how the launch was going, since we were all rookies, but him. That helped keeping up with the events. My first experience with weightlessness was problematic. I'd had many flights on the KC-135 aircraft and hundreds of hours in the water tank, so was familiar with the sensations of weightlessness. I remember how pleasant a sensation it was and how surprised I was that I didn't get sick!"

Nelson's rapid and comfortable adaptation to microgravity was not shared by crewmate Terry Hart.

STS-41C marked the Shuttle's first 'direct insertion' ascent. In other words, only one OMS firing – rather than two – was needed to circularise Challenger's orbit at an altitude of around 530 km. Previously, when less performance data was available for the main engines and some targeting precision was lacking, two OMS burns raised the apogee and finally boosted the perigee to circularise the orbit. On STS-41C, however, the first of these two firings was omitted and the 'OMS-2' burn not only achieved orbital insertion, but also enabled the engines to provide more energy and permit the easier use of onboard software. The high orbit was needed for the rendezvous with Solar Max.

As Crippen, Scobee and van Hoften busied themselves on the forward flight deck with readying their ship for orbital operations, Hart was granted the opportunity to unstrap and leave his seat to photograph the jettisoned External Tank as it tumbled Earthward. It was perhaps fortuitous that the LDEF deployment was still a day away, because Hart's initial euphoria turned rapidly into a severe dose of space sickness.

"I had never had any motion sickness," he recalled years later. "I was a fighter pilot and could do anything in an airplane. I had a light airplane I used to do aerobatics in and nothing ever bothered me in terms of flying or riding a boat or a train or a car or whatever. I wasn't weightless for more than three minutes and I knew I was in trouble! I could just tell my whole gastro-intestinal system was going into high-speed reverse and I didn't understand it because, psychologically, I was elated. Maybe I got up too quick and started moving around or started looking out the window too soon, but for the whole first day I was really out of it. I felt awful and was throwing up every 30 minutes or so for a day, but we suppressed that! I got on camera once during the day just so they knew I was there. My wife saw me and said 'He's sick' and everyone replied 'Naw, he's fine, he's fine'. But I could barely force myself to get out of the corner of the cabin and on camera.

"There were some things I had to do that first day, but they were minimal," Hart continued. "I had to unstow the RMS and barely made it through that. I really was totally incapacitated for the first day and I tried the usual drugs that they give you to

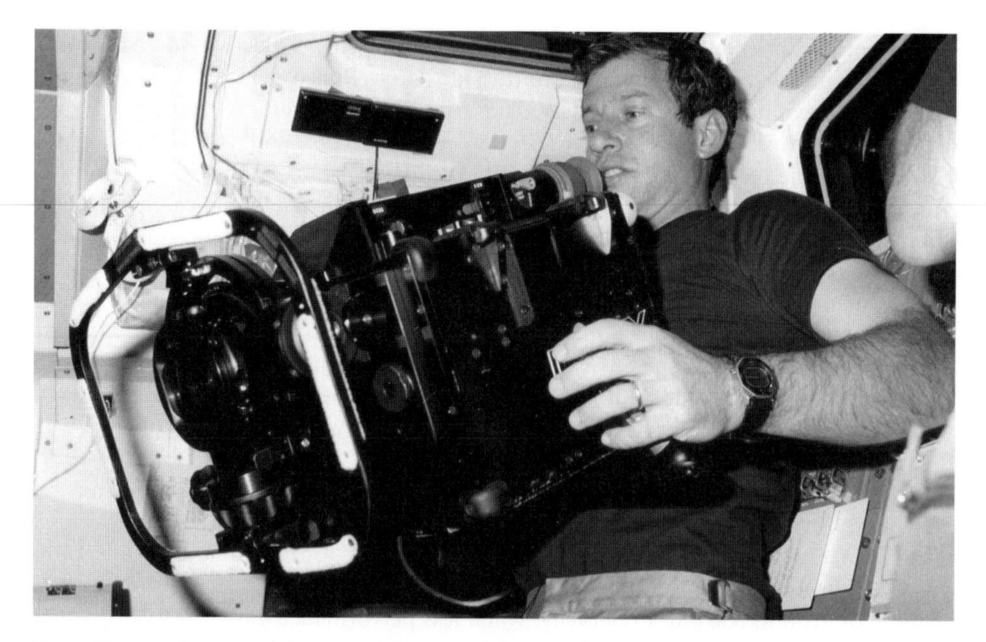

Terry Hart – after recovering from his space sickness episode – prepares the IMAX camera to film aspects of the Solar Max repair effort.

help, but I had it so bad that nothing helped at all. That night, when we got ready to go to sleep, I was exhausted, really depleted. I remember falling asleep and was asleep for maybe a half hour, when I dreamt that I was falling – I had a visceral reaction to a fear of falling – and I remember reaching out to grab something and I did it with such force that I ripped my sleeping bag. I don't think the other guys were asleep yet, but if they were, I woke them up when I yelled out. That was kind of a low spot and, after that, I acclimatised. I had some kind of fundamental neurological brainstem reaction – totally subconscious – to a fear of falling. I think my initial sickness, after three or four minutes of weightlessness, was something that triggered my basic instincts of falling, even though it wasn't conscious. I couldn't detect it consciously and I think it stayed with me for that first night. The next day, I was able to do all my duties, but it was just a terrible experience. I never heard anyone else relate such a bad experience."

Fortunately, by Day Two, Hart had recovered sufficiently to take the lead in the LDEF deployment, successfully releasing it into space at 5:19:27 pm, as Challenger travelled 'upside down', her open payload bay facing Earthward. To activate its many investigations, he firstly grappled the satellite by its so-called Experiment Initiation System (EIS) fixture and then regrappled the second fixture to actually pick it up and deploy it. "The concern," he remembered, "was that I was going to get it stuck, then we couldn't close the payload bay doors and couldn't come home. Crippen and I were trained on the RMS, with him watching and making sure everything was going well. First, I had to lift it out 'straight' and then the arm did everything it was supposed to do. I put it back in the payload bay, just to make sure it would go back in before I lifted it out one more time to deploy it. We left it out on the arm and did some slow manoeuvres to verify all the dynamics that the engineers wanted to understand about lifting heavy objects out of the Shuttle. Then we very carefully deployed it. It wasn't detectable at all when I released it; totally steady and we very carefully backed away and got some great photographs."

As LDEF drifted serenely into the inky blackness, Crippen and Scobee pulsed Challenger's RCS thrusters to increase their distance from the satellite, confirming the separation rates using the Ku-band radar. One problem they highlighted was that their view of LDEF's trunnion pins and berthing guides using the television system was not satisfactory and they expressed concern about its effectiveness during the retrieval, which, at the time, was planned for the following spring.

REPAIRING SOLAR MAX

Despite the importance of LDEF, it was overshadowed by the repair of NASA's malfunctioning Solar Max satellite. In fact, virtually every Shuttle flight since November 1982 had helped lay the groundwork for the reusable spacecraft's most ambitious mission so far. Extensive tests had been undertaken to validate the Canadian-built mechanical arm, requiring it to manipulate larger and more bulky payloads, and three spacewalks had verified the performance of the suits, tools and MMUs, together with the ability of astronauts to work effectively with them.

However, the uncertainties, said Terry Hart, remained. "Sally Ride had used it for the SPAS payload, but the engineers wanted to understand the full capabilities of the arm to move very large payloads and the largest one flown to date was LDEF. The arm engineers wanted to make sure we properly tested the arm moving such a large object, so they could understand that it was going to be able to do what it was designed to do. The bulk of my training was done in the facilities at JSC, but the best simulator in terms of the dynamics of the arm was actually at the manufacturer up in Toronto. I spent several weeks there to simulate lifting up the LDEF to test the flex of the arm with a heavy payload and using it to capture Solar Max in a rotating mode. Their concern was whether it could track something that was moving and, when you snared it, whether it would it cause stresses that were undesirable in the arm."

Elsewhere, Nelson and van Hoften paid a great deal of attention to the MMU assisted spacewalks undertaken on Challenger's previous mission in February 1984. Crippen and Scobee, meanwhile, would have the opportunity to put their rendezvous training to the test. Their target – Solar Max – had launched atop a Thor-Delta rocket from Cape Canaveral Air Force Station in Florida in February 1980, with the intention of spending a decade aloft (essentially a full 'solar cycle') to provide broad spectral coverage of the underlying mechanisms responsible for causing solar flares. To do so, it was equipped with a battery of gamma ray, X-ray, ultraviolet and other instruments.

It is perhaps ironic that, only months after the MMU fabrication contract had been awarded to Martin Marietta, an unfortunate series of circumstances conspired to lead to the jet backpack's first operational use. One of Solar Max's instruments was a white light coronagraph and polarimeter, provided by the High-Altitude Observatory of Boulder, Colorado, which operated satisfactorily from March to September 1980, before suffering an electronics failure that left it inoperative. Then, in December, a fuse blew in Solar Max's attitude control system, causing it to 'wobble' and rendering it incapable of pointing precisely towards the Sun.

All was not lost, however, because it had been designed as one of several Multi-Mission Modular Spacecraft (MMS), part of NASA's vision to permit certain satellites to be serviced by the Shuttle. Measuring four metres long and fitting into a circular envelope some 2.3 m in diameter, the 2,315 kg Solar Max had two sections: a payload module, laden with eight powerful solar instruments, and the MMS for attitude control, power, communications and data handling functions. Connecting the sections was a transition adaptor, which supported two fixed solar array paddles to provide between 1,500 and 3,000 watts of electrical power.

This modular 'bus' was also being used for a number of other spacecraft, including the Upper Atmosphere Research Satellite (UARS), the Extreme Ultraviolet Explorer (EUVE) and the Landsat-4 Earth resources platform. Interestingly, at the time of the Challenger disaster, a repair and servicing mission for Landsat-4 – which had suffered problems with power cabling to its solar panels – had been provisionally booked for early 1987 on one of the Shuttle's operational launches from Vandenberg Air Force Base. Had it flown, the procedures involved would probably have closely mirrored those followed by the STS-41C crew.

In view of its problems, Solar Max was placed into a slowly spinning 'safe' mode, which it maintained for three years, and although three instruments returned valuable data, the primary mission was effectively suspended. Its Hard X-ray Imaging Spectrometer (HXIS) – essentially a 'flare alarm' to alert the other instruments electronically to major solar events – later malfunctioned in June 1981 and was left useless. However, by keeping the spacecraft rotating at one degree per second and aiming its solar panels constantly in the direction of the Sun, NASA engineers kept it feebly alive.

In the meantime, as efforts got underway to build a replacement electronics box for the white light coronagraph and polarimeter, it was decided to incorporate changes into the 'new' device to improve the instrument's imaging resolution – which had already begun to degrade as early as July 1980 – and permit space-to-ground communications through the Tracking and Data Relay Satellite network. Construction of the new box got underway in November 1982 and, after extensive tests, was complete by the following October. Shortly before Christmas 1983, the box was declared 'flight ready' and transferred to KSC for final checkout.

By the morning of April 8th 1984, after executing a series of thruster firings to gradually align their orbital path with that of the satellite, the STS-41C crew glimpsed their quarry as a steadily brightening star. Crippen gingerly manoeuvred Challenger into position, about 70 m from the slowly spinning Solar Max as Nelson and van Hoften completed donning their spacesuits and entered the payload bay at 2:18 pm. The plan was for Nelson, designated 'EV1', to fly the MMU out to the satellite and dock himself to its midsection using a specially designed Trunnion Pin Attachment Device (TPAD), which had been tested on an earlier mission.

George 'Pinky' Nelson, equipped with a Manned Manoeuvring Unit, makes his first unsuccessful attempt to dock onto the slowly spinning Solar Max.

Although Solar Max was not spinning too fast for Hart to grapple it with the RMS, "we felt it was more prudent to have Pinky fly over with a backpack, dock himself to the satellite, stabilise it and then I could grab it with the arm".

"Donning the MMU went very smoothly," Nelson recalled two decades later, "just like training. Ox and I had practiced so intensely that it was more like a well-choreographed dance than anything else. Once I left the docking station in the payload bay, the MMU flew just like the simulator at Martin Marietta in Denver, where we trained. I had been very well debriefed by Bruce McCandless about the few differences between the simulator and the real unit, such as 'chatter' when accelerating, so I didn't experience anything unexpected."

Precisely on time, after a ten-minute solo flight, Nelson arrived in Solar Max's vicinity, using the MMU thrusters to gently match its rotation. However, when he moved in to dock his TPAD onto the satellite, it did not clamp properly into place.

"We didn't know what was wrong," explained Hart, "but, being mechanical engineers, we said 'If a small hammer doesn't work, use a bigger hammer!' So Pinky went in twice as fast the next time and he hit again and bounced right off again." A third try, which imparted yet more force, also failed. Had the TPAD been affected by the cold of orbital 'night-time', Mission Control wondered? Its temperature after removal from the payload bay storage locker had not been maintained, but pre-flight tests – and experience on STS-41B – determined that it was capable of withstanding at least a few hours in the frigid darkness.

In fact, on the ground, it had shown that it could operate satisfactorily for up to six hours at temperatures as low as minus 40 degrees Celsius. So far, in 'real' space, it had been outside for less than two hours and subjected to a relatively balmy minus 12 degrees Celsius ...

Low temperatures, therefore, did not seem to be a contributory factor. Furthermore, when Nelson pushed the TPAD against Solar Max's midsection, its 'trigger' activated and released a pair of 'jaws' in an attempt to grab onto its quarry. This ruled out any kind of malfunction in the docking hardware.

However, as STS-41C's first spacewalk continued, the crew saw another problem brewing: Nelson's efforts had 'jostled' Solar Max out of its previously slow spin and Crippen asked him to grab a solar panel to steady it. The gyroscopic effect of this action worsened matters and, with his MMU's nitrogen supply running short, Nelson returned to Challenger. Instead of revolving gently, like a top, Solar Max was now tumbling unpredictably around all three axes. Four tries by Hart to grapple it with the RMS proved fruitless and Crippen opted to withdraw to a distance of about 160 km until a new strategy could be thrashed out.

"The grappling pin I had to grab was underneath one of the large solar panels, so I could only get there under certain conditions," recalled Hart, "and it was very hard to predict how it was doing. I got close to it and I was maybe a foot away from getting it, but I'd reach some limit on the elbow or the wrist. I couldn't go far enough or fast enough to get it. It may be a good thing, because the satellite was tumbling so much that if I had gotten it, it may have actually broken the arm! Crippen, rightfully, said 'King's X. Let's go back'. We got the Shuttle back in position in front of the satellite and then we stabilised everything. We had fuel left, but not enough to do what we were doing anymore."

Overnight, as the astronauts slept, Goddard Space Flight Center (GSFC) engineers in Greenbelt, Maryland, battled to regain control, but since its solar panels were no longer pointing towards the Sun, battery power was gradually dwindling away without recharging. The engineers switched off as many systems as possible, including heaters, but still had only six to eight hours of battery life left. When it became clear that Solar Max's magnetic torquer bars were not showing the rotation, Goddard implemented a new technique, using a different method of sensing its position. This made the bars more effective in 'pushing off' against Earth's magnetic field and the satellite quickly stabilised itself. Then, just as battery life was running out, it came around in its orbit in such a way that the electricity-generating panels faced sunward once more and began to recharge. When Crippen's crew awoke on April 9th, the batteries were powered and it was rotating serenely at half a degree per second.

"Then we talked about what we had to do and Mission Control worked out the available fuel," said Hart, "but we took an extra day and decided we would do a second rendezvous. This time, Pinky and Ox would stay inside the orbiter and I would try to capture it with the arm." It also became clear during the second spacewalk precisely why Nelson's attempts to capture Solar Max had been thrice frustrated: a small grommet, just 20 mm high and 6.4 mm thick, had obstructed the full penetration of the TPAD onto the satellite's trunnion pin.

The grommet, which was installed near the pin, helped to hold part of Solar Max's gold-coloured thermal insulation blanketing in place. "What no-one noticed," explained Hart, "is that one of the blankets had been put on with a little fibreglass standoff that the grommets would fit over. The engineering drawings didn't specify where those standoffs could be, so when they assembed the satellite, the technicians just put one wherever the grommet was. They glued it onto the metal frame, then stuck the blanket on. That was the correct thing to do, because no-one envisioned using that pin for anything."

A use for the pin did emerge, however, a year after Solar Max's launch, when the option of a Shuttle repair was first explored in depth, "but when they were designing the TPAD," Hart continued, "no-one noticed that there was a grommet there. When Pinky went to dock, it interfered with the docking adaptor." It was later determined that, if Nelson had come within a very narrow pitch angle 'corridor' to the pin, he might still have succeeded and captured Solar Max. However, during his second spacewalk, he took measurements of where the grommet was and the obstruction it posed.

Upon investigation, Nelson revealed that the grommet stuck out barely one and a half centimetres too far ...

The TPAD, clearly, would not work. Either way, Challenger's onboard fuel was now too low (at just 22 per cent) to support a rescue if Nelson's MMU happened to fail. Instead, Crippen would fly close enough to Solar Max for Hart to grapple it with the mechanical arm. As the pilots manoeuvred to re-rendezvous with the satellite, the off duty EVA crewmen tended a couple of experiments in the middeck, including a

James 'Ox' van Hoften (left) and George 'Pinky' Nelson pre-breathe pure oxygen using their launch and entry helmets before one of their two spacewalks.

student investigation into how well a colony of 3,300 bees made honeycomb cells in space, which Nelson later called, somewhat half-heartedly, "goofy science".

Devised by student Dan Poskevich of Tennessee Technological Institute, the experiment theorised that by comparing bee-built structures on Earth and in space, generalisations may be formed for studies of other populations of the order hymenoptra, including wasps and ants. For Poskevich's investigation, two frames were enclosed in an environmentally controlled box, which provided lighting and temperature to simulate terrestrial conditions. Despite noting some disorientation in the bees, they ultimately proved that they could walk, fly and float without difficulty.

Moreover, they built a sizeable, structurally 'normal' honeycomb cell and the queen bee laid around 35 eggs. Only around 120 of the insects died during Challenger's flight, representing a little over three per cent of the population and significantly fewer than anticipated by Poskevich.

Elsewhere, the Solar Max retrieval was being pursued with renewed vigour. Early on April 10th – on his first attempt – Hart successfully grappled the satellite with the mechanical arm and anchored it onto a Flight Support Structure (FSS) platform at the rear end of the payload bay. "It was a dramatic moment for Mission Control," he remembered later. "We were euphoric when we succeeded. We really felt that the mission was at risk, which it was, and we were really on a mission that was demonstrating the flexibility and usefulness of the Shuttle to do things like repair."

The spectacular success, sadly, would prove to be the MMU's death knell.

An umbilical line was connected to the satellite to feed it with power from the orbiter and it was pivoted around so that Nelson and van Hoften, during their second spacewalk, which began at 8:58 am on April 11th, could reach and fix its broken attitude control system and the main electronics box of the disabled coronagraph and polarimeter. These repairs were originally scheduled to occupy one EVA apiece, but with the condensed and re-timetabled flight plan, it was decided to attempt both during the same excursion.

Replacement of the attitude control box – responsible for crippling the $240 million project more than three years earlier – took the spacewalkers barely 45 minutes to complete. Standing on the end of the RMS, his feet anchored in restraints, van Hoften removed a pair of screws, pulled the box out smoothly and plugged in a new unit. The second procedure of fixing the main electronics box to the coronagraph and polarimeter, which was not designed for replacement in orbit, was expected to be a longer and trickier task.

Nonetheless, with surprising dexterity and outstanding skill, van Hoften pulled back a panel covering the box, cut and taped back a layer of insulation, removed two dozen screws and cut several wires; all done whilst encased in his bulky spacesuit. Nelson then took over, installing the new electronics box using large, gold-plated beryllium clips, instead of tiny screws, for the connectors. An hour after their second task had begun, the two men were finished and were able to place a 'baffle cover' over the X-ray polychromator to vent its exhaust gases away from Solar Max's other instruments.

The second excursion had lasted six hours and 44 minutes, which, in addition to their two and a half hour outing on April 8th, brought Nelson and van Hoften's spacewalking time to more than nine hours.

"The repair itself was a kick," Pinky Nelson recalled years later. "It was so much easier to work in space than it is on the ground. Ox and I, and TJ Hart running the arm, just kind of 'did' this repair. It was a piece of cake! It was so much fun riding on the end of the arm and much easier than working underwater." The pilots, too, were just as excited, particularly Scobee, who persuaded his crewmates to don T-shirts for the space-to-ground press conferences, emblazoned with the legend, 'Ace Satellite Repair Company'.

By this time, of course, Hart and his crewmates had long since found their 'space legs' and had adapted exceptionally well to the peculiar microgravity environment. "The first day or two," said Hart, "you tend to 'over control' your body a little bit and you tend to use your feet too much, so you flail and bounce into things. By the third day, you really get the hang of it, so you just use your fingertips to pull yourself around. It's almost like swimming underwater ... a graceful motion."

Finally, after a day of checkout in the payload bay, on April 12th Hart regrappled the satellite and deployed it back into space. By this time, in view of the additional day of planning needed to retrieve Solar Max, the mission had been extended by 24 hours and rescheduled to land on 'unlucky' Friday 13th after all. Yet, despite the huge success of the repair, bad luck had one more card to play.

Much to Crippen's chagrin, the planned landing at KSC was postponed and finally cancelled due to showers in Florida, obliging the crew to land at Edwards Air Force Base. Shortly after Challenger's loss, the Rogers Commission would hear evidence that eight KSC homecomings had been planned between June 1983 and January 1986, with three having been diverted to California at short notice, entirely due to Florida's unpredictable weather. In fact, Crippen himself would testify to the

Crowded into Challenger's aft flight deck and framed by her overhead and payload bay-facing windows, the STS-41C crew celebrates their success. From left to right, holding 'Ace Satellite Repair Company' cards, are Dick Scobee, George 'Pinky' Nelson, James 'Ox' van Hoften, Terry Hart and Bob Crippen.

commission on April 3rd 1986 that, despite his eagerness to land at KSC, he felt convinced "that you are much safer landing at Edwards".

The predictability of 'favourable' weather conditions was essential for returning Shuttle missions because, after performing the 'de-orbit' OMS burn, the crew are committed to touch down approximately an hour later, with no option of returning to space or choosing an alternate landing site. Consequently, weather officers had to be certain, nearly one and a half hours before the burn, that conditions would be acceptable for the Shuttle to land. Thunderstorms in Florida, which build and dissipate quickly in the summer months, together with early morning fog, have made Cape Canaveral a notoriously difficult region to forecast.

Edwards, on the other hand, has much more stable weather and, in the case of the STS-41C return, proved the more reliable and safest option. For Terry Hart, who had already decided before the mission that this would be his only spaceflight – despite having been offered, by George Abbey, the chance to fly on a West German-dedicated Spacelab mission in the autumn of 1985 – the re-entry came after a sleepless last night in orbit, soaking up as much of the experience as possible. Pinky Nelson, who had ridden into orbit on the middeck, changed places with Hart on the flight deck for the return to Earth.

"I didn't have a lot of time, since we were busy on the flight, but the last night on-orbit, I had no duties at all," recalled Hart. "I just figured that I wasn't going to sleep at all. I'd turned down a second mission and was going back to AT&T Corporation, so I was damned if I was going to sleep. I stayed up all night and looked out the window while the rest of the crew was sleeping and watched the Himalayas go by and other parts of the world that I didn't see during the regular shifts."

Re-entry the following morning, he remembered, "was a wonderful thing. Watching the fireball around the vehicle was breathtaking. The 'engineer' side of me wanted to see the 'g' buildup, but I had a camera and remember just letting it go and it would sit there, of course, when we were weightless. As we started to hit the upper parts of the atmosphere, I watched the camera accelerate forward as I let go, because the vehicle was decelerating. I was downstairs, but I was able to stick my head up every once in a while before I strapped in and looked out and could see the fireball overhead 'flickering' – a very impressive experience coming through that, but very smooth and quiet all the way down."

True to the weather forecasters' predictions, an ominous thunderstorm arrived over KSC's Shuttle runway at precisely the time Crippen might otherwise have been landing there.

"My family were in Cape Canaveral," continued Hart, "and we were landing in California, but it was beautiful. When I got out of my seat, I felt like I was using almost all of my strength just to get up! I was used to moving my body around with just my fingertips and now, all of a sudden, I had to exert all this force to get up. We didn't want to fall down the stairs on national television, so we were all doing deep knee bends to make sure we get our blood flowing again."

Pinky Nelson, having exchanged seats with Hart on Challenger's flight deck, now sat shoulder to shoulder with his spacewalking buddy, van Hoften, during the descent to Earth. "I got some great movies of the shock that hangs over the tail," he remembered years later. "My main memory is just how strong gravity felt after a week of weightlessness! I'd done so many approaches and landings in the simulator, the Shuttle Training Aircraft and the T-38 jets, that the landing felt very normal. It was disappointing that my family could not be there."

Unfortunately, Nelson's next mission, aboard Columbia in January 1986, would also suffer from having its landing site diverted; his family would thus 'miss' seeing two of his three Shuttle touchdowns.

Challenger made landfall at 1:38:06 pm on Edwards' Runway 17, completing a mission of almost seven days. As STS-41C's epic voyage was ending, Solar Max's rejuvenated voyage of exploration was scarcely beginning. After a four-week check-out, it set to work on what would turn into five and a half years of observations of changes in the Sun's energy output. Its coronagraph and polarimeter, repaired by Nelson and van Hoften, resumed work in June 1984 and, despite a few interruptions, continued to capture images of the solar corona during 'daytime' portions of its orbit until the end of the mission.

Minor problems arose for most of January 1986, ending just two days before Challenger's untimely destruction, when Solar Max suffered a loss of memory in its onboard computer. Observations were again interrupted in December of that year, when the coronagraph's dedicated tape recorder failed, only coming back online in March 1987, thanks to the use of a backup device. Nevertheless, the satellite continued operating – in spite of atmospheric friction gradually dragging its orbit downwards – almost until the end of the decade.

Shot through the small circular window in the middeck side hatch, this view of the Kennedy Space Center was taken by Kathy Sullivan moments before touchdown at the end of STS-41G.

Other scientific results included the discovery that the Sun is much brighter during periods at which 'sunspot' activity on its surface reaches its peak. Solar Max's instruments confirmed that, although the sunspots themselves are dark, they are surrounded by bright 'faculae', which more than offset the dimming effects of these Earth-sized blotches. By the time the satellite's mission finally ended in mid-November 1989 – to re-enter the atmosphere two weeks later – it had chalked up an impressive tally of 15 observed deep-space gamma ray bursts, a quarter of a million images of the Sun's corona and over 12,000 recorded solar flares.

Additionally, its onboard gamma ray spectrograph made important contributions to the international study of Supernova 1987A, which had provided astronomers with their first 'local' opportunity to examine such a major stellar event since 1604. Solar Max's ultraviolet spectrometer and polarimeter, despite suffering from a jammed grating drive mechanism in April 1985, managed to provide pointing and timing information for the other instruments and even conducted four years' worth of ozone-concentration measurements in Earth's atmosphere.

Original plans, nurtured both prior to and in the wake of the STS-51L, to retrieve Solar Max once more, bring it back to Earth and refit it as the Extreme Ultraviolet Explorer came to nothing. Pre-Challenger plans from as late as December 1985 called for a retrieval of Solar Max in 1987 or 1988 for a series of EUVE upgrades and a re-launch aboard the Shuttle in 1989 or 1990. However, another MMS bus was ultimately built for EUVE, which reached space thanks to an expendable rocket in mid-1992. Interestingly, however, part of the original Solar Max is still in orbit. When the broken attitude control system was returned to Earth by Crippen's crew, it was refurbished and placed aboard NASA's Upper Atmosphere Research Satellite. Since September 1991, that mission has closely monitored the chemistry of Earth's middle and upper atmosphere.

Despite the immense success of the Solar Max repair, plans were already afoot to improve future servicing missions. In November 1985, a satellite services workshop held at JSC, part of which was chaired by STS-41B's Bruce McCandless, heard proposals to install not one, but two, RMS mechanical arms aboard future Solar Max-type repair missions. Although the instrumentation on the Shuttle's aft flight deck was not capable of operating both arms simultaneously, it was suggested that one could be employed 'passively' to hold a target satellite steady whilst its 'active' sibling manoeuvred tools, replacement units or spacewalking astronauts into place.

The proposal noted that, if two RMS devices had been available to Crippen's crew, the need to fully berth and latch Solar Max into the payload bay might have been unnecessary. Moreover, the potential risk of damaging the satellite's solar panels or the orbiter itself during berthing or redeployment would be neatly sidestepped and the entire servicing could have been undertaken 'outside' the bay. An orbiter thus equipped would, the paper's author pointed out, have greater ability to reposition the satellite throughout its repair than the more limited 'turntable' option offered by the Flight Support Structure.

If STS-41C was stricken with bad luck at all, the greatest victim must have been the MMU jet backpack first tested by Bruce McCandless two months earlier. Admittedly, it had performed admirably under the control of McCandless, Stewart, Nelson and van Hoften. It could hardly be blamed directly for the failure of the TPAD. Indeed, it would prove its worth in November 1984 when, flown by Joe Allen and Dale Gardner, it was instrumental in the retrieval of the errant Palapa-B2 and Westar-6 communications satellites.

However, what STS-41C did prove was the crisp manoeuvrability of the Shuttle itself and the precise handling characteristics of the RMS. It was, observers said later, "Hart's small grab", rather than "Nelson's free flight", which had pulled success from the jaws of what might have been an ignominious defeat.

It is perhaps significant that on the reusable spacecraft's next satellite rescue (that of the crippled Leasat-3 communications satellite in August 1985), astronauts relied on their own spacesuits – *sans* MMU – and a mechanical arm deftly manipulated by colleague Mike Lounge to perform a breathtaking repair. Missions subsequent to the Challenger disaster have also demonstrated that other equipment such as tethers, safety grips, hand bars and foot restraints can allow astronauts to conduct a multitude of tasks without the need of a bulky, jet propelled backpack.

Similarly, Vance Brand's ability on STS-41B to fly the Shuttle with pinpoint accuracy to collect Bruce McCandless' lost foot restraint and the use of the RMS on a subsequent mission to knock a chunk of ice off a waste water port removed the need for additional risk. Despite much criticism of NASA's cavalier attitude towards Shuttle operations before the Challenger disaster, there was also, said McCandless, "a sort of creeping conservatism and EVAs came to be regarded as hazardous, to be scheduled only if absolutely required. In the same timespan, new techniques – such as 'Low-Z' translation, in which both the Shuttle's +/−X thrusters were fired simultaneously, almost cancelling each other out, but yielding a small, ten per cent +Z braking component for approaching a satellite without 'blasting' it – had been developed. This permitted the Shuttle to fly right up to a satellite to the point where an astronaut could reach out and grab it without needing an MMU."

More tellingly, Pinky Nelson doubted, even in the heady days before January 1986, that the jet backpack would have flown again, except "maybe for a vehicle-to-vehicle rescue in a Columbia-like scenario, but not for any operations that were envisaged in the pre-Challenger programme". Like McCandless, he stressed to me in a March 2006 email correspondence that it was "well conceived and engineered but, unfortunately, the planned uses of the MMU were superseded by other capabilities that we developed, but couldn't anticipate".

Finally, when the Rogers presidential inquiry into Challenger's loss presented its findings, renewed emphasis was imposed on increasing the safety of other Shuttle components. "The cost of recertification, coupled with the 'Low-Z' orbiter manoeuvring enhancements," recalled McCandless, "eventually killed it off. In the fall of 1989, there was an effort to fly them again. A proposal was solicited from Martin Marietta for recertification and refurbishment for one Shuttle mission. It came in at $6.1 million, which was deemed too expensive, and despite some small sums for clean room environmental storage, 'just in case', they were eventually retired."

Today, the MMU flight unit first used by McCandless hangs in the Udvar-Hazy annex of the National Air and Space Museum at Dulles International Airport in Washington, DC. The second jet backpack was loaned to NASA's Marshall Space Flight Center (MSFC) in Huntsville, Alabama, for use as a possible 'flying testbed' for autonomous rendezvous and docking systems – "subject," said McCandless, "to the constraint that it be maintained in a condition that could be restored to flight configuration." Both units, therefore, were mothballed until such time as their unique capabilities were needed again.

They never were.

Perhaps, through the drawbacks it uncovered with the multi-million-dollar backpack, STS-41C proved to be unlucky after all.

FIRE ON THE PAD

By the beginning of June 1984, the Shuttle seemed to be prospering. Eight more missions were scheduled before year's end, beginning with the maiden voyage of the orbiter Discovery. On that flight, designated 'STS-41D', Commander Hank Hartsfield would lead Pilot Mike Coats, Mission Specialists Mike Mullane, Steve Hawley and Judy Resnik and McDonnell Douglas Payload Specialist Charlie Walker to deploy two communications satellites and activate an experimental solar 'sail' in the payload bay. Next, in July, would come Ken Mattingly's top-secret Department of Defense assignment, originally labelled 'STS-10' but now redesignated 'STS-41E'.

The 'original' STS-41E crew – Commander Karol 'Bo' Bobko, Pilot Don Williams and Mission Specialists Jeff Hoffman, Dave Griggs and Rhea Seddon – should have deployed two communications satellites and operated the solar sail, which was funded by NASA's Office of Aeronautics and Space Technology. However, in a September 1983 press release, the agency announced that future crews would be detailed by payload, rather than flight number, and Bobko's team found themselves rapidly reassigned to STS-41F, scheduled for August 9th. Finally, two other 1984-funded missions – STS-41G and STS-41H – would launch in late August and September, respectively.

For poor Bobko, who had flown as Pilot on STS-6, by the time he next rocketed into space, in April 1985, he and his crew would have designed no fewer than four different mission patches! "Mary Lee used to be the lady that arranged the patches," he said, "and along the top of her office she had different plaques with all the different patches, and then you got to a corner and there were four of them, which were all for our mission or its derivatives."

On the STS-41G mission, Bob Crippen would become the first person to complete four flights aboard the reusable spacecraft, earning him the nickname 'Mr Shuttle' and leading Pilot Jon McBride and Mission Specialists Kathy Sullivan, Sally Ride and Dave Leestma to operate a battery of Earth resources instruments. Three weeks later, STS-41H Commander Rick Hauck, Pilot Dave Walker and Mission Specialists Joe Allen, Anna Fisher and Dale Gardner would undertake a top-secret Department of Defense mission or, if IUS woes were rectified, deploy a second Tracking and Data Relay Satellite. Thus, in a situation that would become increasingly common before STS-51L, Hauck's crew was not tied one specific payload.

Since NASA's financial year ended on September 30th, the other three Shuttle flights planned for 1984 – in October, November and December – were labelled as '51-series' missions. They would, respectively, have conducted materials processing research on Commander Dan Brandenstein's STS-51A mission, operated the Spacelab-3 facility on STS-51B (led by Bob Overmyer) and, perhaps, inserted a third TDRS into orbit on Joe Engle's STS-51C flight. The peculiar numbering system had already caused much confusion, despite the fact that, so far, it had followed a more or less logical sequence. Then, on June 26th 1984, during her second attempt to begin STS-41D, the new orbiter Discovery suffered a dramatic main engine shutdown, seconds before lift-off.

The STS-41G crew at a pre-flight press conference. Bottom row (left to right) are Marc Garneau, Paul Scully-Power and Bob Crippen. Middle row (left to right) are Jon McBride, Dave Leestma and Sally Ride, with Kathy Sullivan at the 'top' of the pyramid.

Fortunately, as it turned out, the crew was kept aboard throughout the crisis, as their vehicle was 'safed'. Later inspections revealed scorched paint on the launch pad, close to where the astronauts would have evacuated the orbiter. It was caused by hydrogen and oxygen propellants that had gone unignited, down into the flame trench, then been ignited by one of the engines. "There was literally an invisible hydrogen fire burning up and around the outside of the Shuttle," said Charlie Walker. "It would have been a bad day for us if we'd tried to get out of there!"

Repairs were implemented on a failed actuator in Discovery's main fuel valve, but Hank Hartsfield's crew would not reach orbit until August 30th – the very day that Bob Crippen's STS-41G crew should originally have launched. In the meantime, on August 3rd 1984, a jittery NASA opted to cancel the two remaining flights for that financial year, transferring some of Bobko's STS-41F payload onto Hartsfield's mission. The decision was met with disappointment by members of his crew.

"The end result was the 41D crew ended up taking the payloads that we were supposed to fly," remembered Don Williams. "It was very disappointing, because to go that far and be within three or four months of flying and then go back to square one was tough. When it was sorted out, Hank and his crew went off to do that mission and we were assigned to do a Tracking and Data Relay Satellite deployment. We jumped into that and went back to work again."

Ken Mattingly's STS-41E assignment, meanwhile, slipped until January 1985 and was redesignated 'STS-51C', while Bobko and his team were pencilled in for the STS-51E flight a month later in February. Subsequent missions were also renumbered, with Hauck's crew now known as 'STS-51A' and the last three flights of 1984 quietly pushed into the following spring. Of these, only Overmyer's Spacelab-3 mission kept its original designation. In fact, STS-41D's Steve Hawley would recall that, in those days, "it wasn't very unusual to change flights several times. Many of the flights were similar: launching satellites or running experiments that could be quickly learned. It wasn't as important to stick with a payload."

Consequently, when STS-41D landed, only one '41-series' mission remained in place and yet, with its new launch date of October 5th, would actually fly at the beginning of NASA's 1985 manifest. Crippen's crew also picked up two Payload Specialists: Paul Scully-Power, an Australian-born oceanographer working for the US Navy, and Marc Garneau, who would become the first Canadian spacefarer. They were non-career astronauts selected for their expertise with particular onboard experiments. Making history as the first seven-person astronaut crew, they would deploy the Earth Radiation Budget Satellite (ERBS) and operate NASA's third Office of Space and Terrestrial Applications (OSTA-3) payload.

AN EXPERIMENTAL CREW

Payload Specialist training was much shorter and less intense than that of the career astronauts. "It's basically all the categories, except for maybe the big expense ones," said Charlie Walker, who flew three times as a Payload Specialist, more than anyone else. "I didn't get emergency water training or survival training in the deserts or jungles, but I did go through all the systems; both briefings as well as stand-alone simulator training. I knew what the electrical and environmental systems did and I knew the computer interfaces. I was there as a working passenger. I wasn't a fully-fledged crew member and I took no real exception to that. Occasionally there were circumstances in which it was made clear that 'You're not one of us. You're along for the ride and you've got a job to do', but it was only a few individuals – some in the astronaut office – from whom I got that impression. There was no belligerence, really, expressed openly, and no offence on my part taken."

The STS-41G crew was an unusual one. Although Crippen was responsible for the success of the mission, his Solar Max commitments during the first half of 1984 meant the remainder of the crew had completed a sizeable portion of their training without him. As the only experienced astronaut, therefore, Sally Ride took the mantle of 'surrogate Commander' to nurse three rookies and two Payload Specialists through their preparations. Only at the end of April, by which time the crew had been training for five and a half months, did Crippen join them on a permanent basis.

"I was the only one on the crew who had flown before," Ride remembered years later of what, unofficially at least, and only on the ground, made her the 'first' woman to lead a mission. "I had also flown with Crip before, so I knew how he liked things done and I knew what his habits were. On launch and re-entry, I knew what he wanted to do and what he wanted the Pilot and the flight engineer to do, so our crew started launch and re-entry simulations without Crip. During those simulations, I was the flight engineer, Jon McBride was the Pilot and then one of the other Mission Specialists sometimes played 'Commander'. We were basically in there to train Jon and me: part of my job was to say 'This is the way Crip likes to handle this situation or this sort of problem and this is how he would want us to work'. I tried to give the rest of the crew some indication of the way that Crip liked to run a flight and run a crew. Then, thankfully, he landed from STS-41C and joined us. He's very easygoing, so I don't think our group dynamics changed much when he joined the crew. Everyone knew him really well and had worked with him in the astronaut office for years, so he wasn't an unknown presence joining us." Ride recalled that, since the STS-41G crew had a lengthy training regime ahead of them, her personal burden was not excessive. "The role I was playing," she said, "was really to talk about the basics of being in space and giving them some familiarity with the space environment."

Crippen's assignment to command yet another (almost)-all-rookie team was, said oceanographer Bob Stevenson, who worked with him during the build-up to STS-41G, possibly the deliberate intention of NASA's director of flight operations, George Abbey. "Bob is an exceptionally able astronaut," Stevenson, who died in 2001, explained in a January 2000 interview, "but also an exceptionally able leader of people and I think George, at the time, wanted to use Crip as much as he could to help train and pass on what he could to the new crews. Crippen had a lot of duty in terms of flights!"

Indeed, had Challenger not been lost in January 1986, he would have journeyed into orbit yet again: this time in command of STS-62A, the Shuttle's first launch from Vandenberg Air Force Base in California.

Among his multi-faceted STS-41G crew was the first American woman to make a spacewalk (Sullivan) and the first member of NASA's 1980 astronaut intake to fly into orbit (Leestma). Years later, Leestma remembered fondly the arrival of his astronaut group, who wryly dubbed themselves the 'Needless Nineteen'. "We earned that nickname because there were 35 ahead of us, plus all the other astronauts that were there from the Apollo days that either hadn't flown or were still there, and they were all waiting to fly the Shuttle, so we were way down the line!"

One of Leestma's first assignments in the astronaut office was to devise a checklist for operating the orbiter's Auxiliary Power Units and journalist Henry S.F. Cooper, who wrote a book about the STS-41G crew's training regime, believed it was this work that assured him a seat on Crippen's mission. Leestma recalled the training well, as Ride, or perhaps an invited 'guest', acted as surrogate skipper for each simulation run. "We asked Sally to be our training co-ordinator," he said, "and she became the *de facto* Commander, at least for organising our training and assignments and making sure that we were progressing. We trained as a crew of four for a long time. Under those circumstances, there was a lot of pressure on us to know what we were doing and not screw up, because Mission Operations were looking at us carefully to see if this was something that could be done or not. Can you train without one of the crew members, who is doing another flight?" Even at this early stage, NASA was looking at flying astronauts more rapidly than ever before – two or three times per year – and STS-41G's training cycle gathered valuable, 'real world' data to support this.

In his 1987 book, Cooper highlighted using Crippen so often "went against NASA's policy of building up a pool of experienced astronauts – essential if it [the agency] is ever to achieve a rate of one flight a month – and using him again seemed even more unusual in light of his late arrival". However, in the days before the Challenger disaster, the intention was to fly 14 times in 1986 and a marathon two dozen times the year after. Clearly, with a hundred astronauts and between five and seven seats available on each mission, people would indeed be flying relatively routinely.

Since the crew included Sally Ride, who made history as the first American woman to reach orbit in June 1983, and Sullivan, who would make a spacewalk, the two female astronauts became unofficial 'spokespeople' for the mission. "Jon and I could easily just stand in the background," chuckled Leestma, "and be 'just one of the crew'. It took a lot of the spotlight off us, which was fine." As training reached its peak, in June 1984, with Crippen finally done with his post-Solar Max duties and dedicated to STS-41G full-time, the astronauts averaged 80–90-hour work weeks in the JSC simulators.

During the course of their training, the simulators were of pivotal importance. Not only did they enable the astronauts to hone their skills by responding to literally hundreds of abort scenarios, but they also allowed them to begin working together as a cohesive unit. Crippen would comment later that it was vital he understood exactly how each of his crewmates would respond to certain malfunctions and under specific conditions; that a bond of mutual trust should develop between them. This was made more difficult by his absence until the early summer of 1984.

In the wake of the STS-41D main engine shutdown on June 26th and the cancellation of the two missions scheduled to follow it, Crippen's mission was now at the head of the queue, waiting for launch. "We knew that our training was going to be hot and heavy that summer," grinned Leestma.

EARTH WATCHING

When Crippen, McBride, Sullivan, Ride and Leestma were named to STS-41G in mid-November 1983, this mission was scheduled to begin on the penultimate day of the following August. However, for a time, the identity of their orbiter was still in question. "It was designated STS-17 at the time," said Leestma, "and, when we first got assigned, it was on Columbia. We ended up flying on Challenger. NASA always told us when we got selected for a spaceflight, not to fall in love with our orbiter or our payload, because they were liable to change!"

Columbia, the first-flown vehicle of the reusable fleet, had already journeyed into space six times, including the first Spacelab mission in the winter of 1983; in fact, the inclusion of additional cryogenic oxygen and hydrogen tanks under her payload bay floor in support of that flight enabled her to remain comfortably aloft for ten days. Had she flown STS-41G, the mission was expected to last around that length of time, but problems had arisen. Following her Spacelab flight, Columbia was transported to Rockwell's Palmdale plant in California in January 1984 for a number of modifications. These included the removal of a pair of ejection seats – fitted to support two-man crews on her early test flights – and the installation of new equipment to gather aerodynamic data during future hypersonic re-entries, together with a Heads-Up Display (HUD) to rival those aboard Challenger and Discovery. Unfortunately, these upgrades, coupled with the need to attend to wear and tear from her six previous missions, meant she spent longer in California than timetabled. Dave Leestma had been detailed by Crippen to follow Columbia's progress, but he quickly became aware that the flagship would not be ready in time for STS-41G.

NASA management was already considering the roomier Challenger as a more attractive alternative to Columbia, particularly in view of Crippen's large crew. For the astronauts, however, the 'old girl' was preferable in terms of her ability to stay longer in space, but without a HUD, she could not yet confidently support a precision landing on the swamp-fringed, relatively narrow KSC runway. Not only were touchdowns at the Florida spaceport highly desirable as NASA sought to make Shuttle missions 'routine', but Crippen himself wanted a shot at landing there, having been thwarted twice by bad weather on STS-7 and STS-41C.

Still, even the HUD-equipped Challenger needed lengthy refurbishment before she could fly again and was not in a position to conduct STS-41G until the beginning of October 1984. During her touchdown at Edwards, following the Solar Max repair, all of her brakes had suffered varying degrees of damage: cracked rotors, chipped carbon edges, missing washers and contamination by surface debris. Tile and thermal blanketing discolouration also demanded attention.

By mid-June, the word from NASA Headquarters was official: Crippen's crew would use Challenger, allowing Columbia to spend more time in California having her ejection seats removed. However, STS-41G's duration would be cut from ten to eight days, due to fewer cryogenic tanks aboard Challenger. Lift-off was pushed back slightly to October 5th, making it the first flight in NASA's 1985 financial year, but still laden with its '41-series' designation. To demonstrate the readiness of himself and his crew, one of Crippen's first actions had been to ask the instructors for a 'fully integrated' simulation on May 8th.

Normally, such simulations – which involve not only the astronauts, but also the entire flight control team for the 'real' mission – were undertaken only in the last eight weeks or so before launch. Despite reservations on the part of some STS-41G trainers, the integrated 'sim' went ahead and proceeded perfectly. By this point, Crippen had three Shuttle missions under his belt and was easily the most experienced, 'in-training' astronaut at the time. "With Crippen there, we had a harder time fooling the crew," lead instructor Ted Browder told Henry Cooper. "Crippen has seen about every training scenario there is!"

Training was complicated yet further by the addition of Garneau and Scully-Power as Payload Specialists in May 1984. Although the astronauts welcomed the newcomers, with Bob Crippen proudly telling a press conference that – excepting Ride – his entire crew had a naval, or at least 'nautical', background, there was some concern at the sheer number of bodies and available room aboard Challenger. Window space, explained Dave Leestma, was a precious resource, together with very real concerns about whether the Shuttle's trouble-prone multi-million-dollar toilet could handle the additional stress.

Yet, Leestma was happy with his lot. "When we finally did fly, we moved up through a lot of flights, because they were having trouble with the IUS and those flights slipped behind us," he said, "but we stayed with our particular flight. I ended up flying earlier than the other people in my class." In fact, the next 1980 arrival would not reach orbit until July of the following year and almost half of the Thirty Five New Guys were still awaiting their first missions. Clearly, Leestma's chance to fly earlier than expected was a fortuitous one.

Paul Scully-Power, too, had been bitten by exceptional good fortune. His seat on STS-41G, originally, was assigned to another scientist, Bob Stevenson, of the Institute of Oceanography at the University of California in San Diego. The latter's involvement with the Shuttle effort had begun in the summer of 1978, when he was asked by geophysicist Kathy Sullivan to give oceanography classes to the Thirty Five New Guys. Later, during the early Shuttle missions, he and Scully-Power supported crews with their observations of sun-glint on the oceans, sea water temperatures, photography, analysis of ship wakes and elaborate spiral eddies.

During STS-8, Dick Truly had expressed astonishment at spotting spiral eddies "as far as he could see," recalled Stevenson, "either side of the flight path, from the western–southern part of the Indian Ocean, all the way past New Zealand", for five continuous days. Truly's interest provided the oceanographers with an astronaut ally and he pushed for the inclusion of Stevenson or Scully-Power on a Shuttle flight. In fact, George Abbey had considered such a move in 1982, but the STS-5 space sickness episodes obliged NASA to add physicians Norm Thagard and Bill Thornton to two missions instead.

Then, in March 1984, Stevenson received a phone call from Scully-Power, informing him that Abbey had proposed him as a Payload Specialist on STS-41G. However, Stevenson's wife was battling breast cancer at the time and he declined the offer, suggesting that Scully-Power go in his stead. Jovially, he offered to take over if his wife recovered in time or if Scully-Power happened to fall into a hole the week before Challenger's launch. As circumstances transpired, Scully-Power did not break his leg but, tragically, Stevenson's wife passed away a few days before STS-41G lifted off. "The funeral was the day of the launch," Bob Stevenson remembered. "The whole crew called me from the crew quarters that morning. They sent a beautiful arrangement; all orchids. Kathy chose those."

It was the nature of the onboard Earth resources instrumentation, rather than a question of good or cruel luck, that kept Crippen's crew close to their original schedule. Henry Cooper has also hinted that STS-41G's planned tests of an orbital refuelling system, which NASA hoped to use on several lucrative satellite servicing missions for the Department of Defense – Landsat-4 being the first – was also a key player in averting its cancellation or postponement. "Due to the requirements of our payload," explained Leestma, "we had to fly at a certain time of year and a certain inclination and so they held us in our place."

Not only the high inclination – 57 degrees to the equator, enabling the crew to 'see' much of Earth's surface – but also the altitude would be unusual, for the astronauts would adjust Challenger's orbit in order to support two very different payloads. The first, the Earth Radiation Budget Satellite, was first in a series of three platforms to explore the impact of solar radiation on our planet, including its absorption and re-emission by the atmosphere. When NASA began developing ERBS in 1978, the agency hoped it would provide a clearer understanding of the radiation 'balance' between the Sun, Earth, atmosphere and space.

This, in turn, could expand knowledge of the mechanisms responsible for terrestrial weather and climatic change. Following the announcement, two remote-sensing instruments were identified: an active 'scanner' and a passive 'non-scanner', which employed seven radiometers to measure energy intensities in the atmosphere and one sensor to determine solar intensity. Both were fabricated and calibrated by TRW at its Redondo Beach facility in California and flew not only aboard ERBS, but also two other Earth resources missions – the National Oceanic and Atmospheric Administration's NOAA-9 and NOAA-10 satellites, launched in December 1984 and September 1986 – to permit continuous climate change analysis.

Collectively, the triad was dubbed the 'Earth Radiation Budget Experiment' (ERBE) and its data has improved scientists' comprehension of how clouds, aerosols and 'greenhouse gases' contribute to the planet's daily and long-term climate. In particular, their results have highlighted how clouds formed over water differ from those created above land, which, consequently, affects their ability to reflect sunlight back into space. New insights have been provided into Earth's upper atmospheric radiation levels and, in fact, led directly to follow-on satellite projects, including the 1997-launched Tropical Rainfall Measuring Mission and subsequent Earth Observation System.

In addition, data from ERBS and the two NOAA satellites has provided a better awareness of how the amount of energy emitted by our planet varies between 'day-time' and 'night-time'. Such 'diurnal' changes are known to be important factors in our daily weather and climate. Furthermore, the amount of radiation emitted by the Sun has increased by nearly 0.05 per cent per decade since the late 1970s, which, "if sustained over many decades, could cause significant climate change," said Richard Willson, a researcher affiliated with NASA's Goddard Space Flight Center and Columbia University's Earth Institute in New York.

Also aboard ERBS, in addition to its scanner and non-scanner, was the 29 kg Stratospheric Aerosol Gas Experiment (SAGE)-2, designed to assess the effects of human and 'natural' activities – ranging from the burning of 'fossil fuels' and use of chloroflurocarbons (CFCs) to volcanic eruptions – on Earth's radiation balance. During every orbital sunrise and sunset, SAGE-2 used a technique called 'occultation' to measure attenuated solar radiation through our planet's limb and produce a spectrum that would enable the chemical species along the line of sight to be determined. These measurements focused specifically on the lower and middle stratosphere, some 15–25 km above the surface, although retrieved aerosol, water vapour and ozone profiles often extended much lower into the troposphere.

Significantly, the role of nitrogen dioxide, whose levels in the atmosphere have grown substantially due to increased industrialisation in the 1980s, as a major player in the destruction of stratospheric ozone was first traced by SAGE-2. It has also measured the decline in stratospheric ozone quantities over the Antarctic region since the much-publicised 'hole' was first identified in 1985. Additionally, the instrument highlighted natural contributors, including the Mount Pinatubo eruption in the Philippines in June 1991, which warmed its 'local' stratosphere by three degrees Celsius, released aerosols and spread them to middle and high latitudes within a matter of months.

As its name implies, SAGE-2 was the second such device to reach orbit; its predecessor had provided near-global observations of aerosol extinction, together with ozone and nitrogen dioxide concentrations, from 1979 until 1981. The resultant long-term, stable, data-gathering capability has thus spanned more than two full decades and has been continued by SAGE-3, launched aboard the Russian Meteor-3M satellite in December 2001, and proven invaluable for establishing trends in global ozone levels. In fact, today, SAGE provides key evidence for the United Nations' ongoing assessment of environmental change.

"I would have to believe that SAGE-2 is a fairly big feather in NASA's cap," reflected Science Manager Joe Zawodny of the agency's Langley Research Center in Hampton, Virginia. "While SAGE-2 is probably not the household name that Hubble is, it has had an impact on the average person." That impact has been profound: the instrument's ozone studies ultimately spurred the international community into action with the 1987-signed Montreal Protocol agreement and led to the virtual elimination of CFCs and the development of new, low-emission technologies in air-conditioning, refrigeration and industrial systems.

Added SAGE-2's principal investigator, Patrick McCormick, of the Center for Atmospheric Sciences at Hampton University in Virginia: "The public should appreciate the investment they made in a satellite mission that has exceeded all predictions and hopes of a long life and for its contribution to making Earth a better place now and for subsequent generations." This enormous success was demonstrated by ERBS' other instruments, too. Originally designed to operate for just a couple of years, the scanner did not expire until February 1990 and both the non-scanner and SAGE-2 continue to return valuable data.

Today, the satellite continues to monitor total solar irradiance using its surviving instruments, although budgetary constraints have limited the extent of its observations. At least once every fortnight, the Sun is examined for several, 64-second-long measurement intervals. This ongoing series has proceeded, virtually unbroken, with the exception of brief attitude control, telemetry or battery cell failures in July 1987, September 1992, July–November 1993, March 1998 and December 1999. By July 2002, efforts were afoot to lower ERBS' orbit to permit a controlled, destructive re-entry, although the satellite remains operational to this day.

It was a strange-looking machine: built by Ball Aerospace, it measured 4.6 m wide, 3.8 m high and 1.6 m long. Yet it was, said Dave Leestma, "a beautiful satellite, coated with gold foil insulation and dark, purplish-blue solar arrays". It comprised keel, base and instrument modules, which provided, respectively, structural support, an interface with the Shuttle's payload bay and mounting points for the scanner, non-scanner and SAGE-2. Two large solar arrays generated up to 2,164 watts

Deployment of ERBS. Note the Shuttle's RMS mechanical arm at the top-left of the picture.

of electrical power, supplemented by a pair of nickel–cadmium batteries and a hydrazine propulsion system to perform necessary station-keeping manoeuvres.

For a mission that has yielded such superb results and proven a goldmine of scientific success, the deployment of ERBS – intended to occur eight and a half hours after Challenger's lift-off – did not go entirely to plan. In fact, its solar arrays caused the first hearts to flutter. After a perfect, on-time launch at 11:03 am on October 5th 1984, the deployment effort was under Sally Ride's supervision, with Dave Leestma backing her up. Like LDEF before it, ERBS would rely on Challenger's Canadian-built mechanical arm for removal from the payload bay.

"We trained a lot together and spent a lot of time in the simulators and going to Canada," Leestma recalled years later. "It became a little bit of a contest of who could do it quicker or better. All those competitive games were played in everything we did. Sally was very good at the arm, so I learned an awful lot by just watching how she went through the training. When it came time to deploy the satellite, she had let me actually pull the arm out, do the checkout and then grapple it."

During this time, ERBS' systems were activated, pre-deployment checks executed and, raised high above the payload bay, the procedure to extend the solar arrays and other appendages – including communications hardware – duly got underway. "The solar arrays," said Leestma, "were folded up to the sides of the satellite, so we were getting ready to put them out and the ground checked to make sure they were getting current and everything was powered up and looking good."

All five NASA crew members, by this time, were crowded into Challenger's tiny flight deck: Ride and Leestma at the RMS controls, Sullivan handling the cameras, McBride flying the spacecraft and Crippen, in his own, rather understated, words, "sitting back and managing". The two 'non-career' Payload Specialists, meanwhile, had been confined to the Shuttle's middeck during the deployment effort.

Upstairs, through their headsets, the astronauts heard the voice of fellow astronaut Dave Hilmers, telling them from Mission Control to release the arrays. "We sent the command for the first solar array to deploy and it went up," said Leestma, "but when we hit the command for the second one, nothing happened!" Several more tries, including one initiated from the ground, to unfurl the stubborn array were also fruitless. Next, they attempted to 'jostle' ERBS by rolling the mechanical arm's end effector, without success, and finally McBride oriented Challenger's payload bay towards the Sun to thaw out possibly frozen hinges. "We were talking inside the cabin, of course, about what we could do to free this solar wing," Leestma continued. "This was back before we had all the TDRS coverage, so we went through long periods of time where we didn't have to talk to the ground or they couldn't see data. We were getting ready to come up over Australia and through the Canberra station and talk to the ground and then we would have a 15 to 20 minute period before we'd come up over the States; a big loss of signal time."

As the crew awaited reacquisition of signal, Ride and Leestma considered trying again to shake the array open with the RMS. "We changed the payload identification, which tells the arm what's on the end of it," said Leestma, "and changed the payload in the software to 'zero', which meant there was nothing on the end of the arm. Now we could go to the max rates on the arm and play with it." After receiving authorisation from Crippen – on condition they did not break ERBS – Ride moved the arm as sharply as possible from left to right and back again.

During her second attempt, Leestma suddenly noticed something move. Ride put the satellite back into its deployment position and the balky array slowly juddered, then stopped, juddered again and finally sprang open. "We came up over the States," Leestma exulted, "and the ground said 'Okay, we're with you'. I don't remember the exact quote, but they asked 'What did you guys do?' We said 'We aren't going to tell you, but just check it out and make sure that it's ready to deploy'."

The glitch, which was later attributed to "thermally induced problems", delayed ERBS' deployment from Challenger's sixth orbit to her ninth circuit of the globe. During this time, Goddard Space Flight Center controllers uplinked new telemetry data to the 2,307 kg satellite to activate its attitude control system. For Bill Holmberg, the keeper of STS-41G's crew activity timeline for the mission, their carefully choreographed, minute-by-minute schedule had been swept into disarray. However, he later told Henry Cooper that it was easier to rewrite an already extant plan than to write one from scratch.

After the satellite left the RMS at 10:18:22 pm, Crippen and McBride pulsed the OMS thrusters to separate and, two days later, ERBS' attitude control jets fired in the first of a series of 'burns' to establish it in a 560 km orbit. The flow of the mission, it seemed, might not be adversely impacted at all, for another deployment option had already been added to the timeline. However, the next problem cropped up almost immediately as the crew began operations with their second major payload – the Shuttle Imaging Radar (SIR)-B, part of OSTA-3 – which experienced difficulties transmitting data through Challenger's Ku-band antenna.

This radar, which resembled an enormous, eight-panelled rectangular dining table, measured 11 m long by 2.1 m wide and had already proved something of a headache in pre-mission simulations because of its flimsy nature. Some scientists wanted it to commence radar observations of Earth as Challenger flew at her higher, 350 km ERBS-deployment altitude and continue to do so as she lowered her orbit to SIR-B's 260 km operating altitude. However, since reducing the Shuttle's orbit necessitated two OMS burns, it was feared that the shock could impart structural damage to the radar.

Three years earlier, on its first flight, SIR amply demonstrated its ability to gather data in support of geographical, geological, hydrological, oceanographic, vegetation and ice-monitoring applications, acquiring imagery of more than 40 million km^2 at resolutions of just 40 m. However, on STS-2, a planned five-day mission was halved due to a fuel cell failure and Challenger's flight would be the first time it could be exploited for an 'extended' period of time. It consisted of a side-looking synthetic aperture radar, which illuminated Earth's surface with horizontally polarised microwaves transmitted at the L-band wavelength of 23 cm.

During typical science gathering activities, it radiated pulses of this microwave energy and measured the characteristics of the reflected 'echoes', thus enabling ground-based scientists to determine surface textures and types. Even its limited imaging time on STS-2 was significant, penetrating dry sand dunes in the Sahara Desert of northern Sudan and leading to the identification of long-dried-up river channels. These, in turn, guided archaeologists to the detection of ancient oases and Stone Age settlements. In anticipation of greater gains with SIR-B, its resolution was enhanced to 25 m and engineered to 'tilt' at angles between 15 and 57 degrees.

Consequently, on STS-41G, the device was no longer restricted to just recording the ground-track directly underneath Challenger's orbital path; moreover, by varying its 'look' angles, it became possible to assemble 'mosaics' of adjacent surface features observed over several days. Indeed, when the newly refurbished SIR returned from its second mission, it had greatly outperformed STS-2: yielding data to build three dimensional models of subtle geological features on California's Mount Shasta, permitting contour modelling of parts of eastern and southern Africa and mapping intricate structural features such as faults, folds, fractures, dunes and rock layers.

NASA envisaged that SIR-B would acquire 42 hours' worth of digital data and eight hours of optical measurements during the course of the mission; however, although the optical requirements were met, due to unforeseen problems only seven and a half hours of digital data was acquired. Three main obstacles were responsible for this: problems with Challenger's Ku-band antenna, lost communication links to the Tracking and Data Relay Satellite and depleted power in SIR-B's own transmission system.

Shortly before ending their first day in space, the crew deployed the radar and, for about two minutes, it began transmitting scientific data through the Ku-band antenna to the doddery but invaluable TDRS-1 and from thence to the White Sands ground terminal in New Mexico. Then, abruptly, at 11:54 pm, it stopped. Engineering analysis quickly determined the antenna had lost its 'lock' on the geosynchronous-orbiting TDRS, due to a failed motor in its 'beta' gimbal. One axis by which the Ku-band dish could move was effectively dead and the other – the 'alpha' gimbal – swung backwards and forwards, operating sporadically.

Affixed to the starboard payload bay wall and able to lean out over the sill, the antenna provided space-to-ground communications between the crew and Mission Control, but also acted as a 'rendezvous radar', as it had done during the Solar Max retrieval. However, it could not accomplish both functions simultaneously. During ascent, an S-band link supported voice and data communications, after which the higher rate Ku-band antenna was deployed to support the remainder of the mission. Unfortunately, it was often difficult for TDRS to acquire the antenna's narrow beam, so the orbiter typically employed its larger beam width S-band link to 'lock' the Ku-band into position.

To correct the gimbal problem, early on October 6th, Mission Control directed Ride and Leestma onto Challenger's middeck to unplug a wire that routed power to the antenna's motors. It was hoped that, if the wire was removed at the correct time, just as it swung out at right angles to the spacecraft, the astronauts could reorient Challenger such that the Ku-band was once more focused on TDRS-1. The wire was situated behind a row of lockers, requiring the astronauts to remove them and wait for Crippen to inform them of the right time to unplug it.

Peering at the waving antenna through the aft flight deck windows, as soon as it looked to be in the proper position, Crippen told Ride to pull the wire. The Ku-band antenna stopped its erratic motions and could thenceforth only move slightly in response to external forces, such as OMS firings. At this point, Sullivan – who, as a professional geophysicist, was responsible for SIR-B and Challenger's other Earth resources gear – retracted the delicate radar to enable Crippen and McBride to lower their ship's orbit to some 260 km.

Using aft flight deck controls, Sullivan attempted to fold the two outermost antenna 'leaves' onto the central section and close the assembly into a storage canister. The procedure should have gone smoothly, but, as she watched the clunking radar components through the window, it became clear that SIR-B was improperly stowed. Sullivan tried shutting it with backup controls, with no success. A third option was to fire pyrotechnics, thus slamming it closed, but rendering it impossible to re-open; essentially terminating the option of using it on STS-41G. This was not ideal on only the second day of an eight-day voyage.

Flight rules dictated that the antenna had to be closed before an OMS burn could take place, for fear that it might be damaged during the orbit-lowering manoeuvre. Already, when she first deployed SIR-B, Sullivan had noticed that it wiggled and writhed around in what Henry Cooper later described as "a classic case of dynamic instability". It seemed likely that simply conducting the OMS manoeuvre with the antenna still partially open would inflict damage on it. Bob Crippen, whose responsibility as the Commander was the safety of his crew and payload, wanted to avoid taking this option.

Sally Ride's dexterous handling of the RMS proved the saviour of the day, when she employed its end effector to push the antenna leaves firmly into place. Mission Control was unhappy with this technique, because there was no way of accurately gauging the amount of force imposed on the fragile panels, but neither the arm, nor

Challenger's robot arm is used to push one of the SIR-B antenna's leaves shut, prior to an orbit-lowering manoeuvre.

the radar, appeared dented or scraped. Ride also earned brownie points with Leestma and Sullivan, whose three and a half hour spacewalk – scheduled for October 9th, but later postponed until the 11th – might have been cancelled had SIR-B not been latched back into place.

After the OMS manoeuvre, the remainder of the mission was spent in the lower orbit, whereby the radar and two other Earth-monitoring instruments could acquire their best results. The Measurement of Air Pollution by Satellite (MAPS) measured the abundance of carbon monoxide in the troposphere on a global basis for the first time. Meanwhile, the Feature Identification and Location Experiment (FILE) provided data to automatically classify surface materials into one of four categories: water, vegetation, bare ground or cloud and snow. NASA hoped that this would lead to the development of more advanced sensors on future Earth-watching satellites.

With the exception of one niggling problem, in which MAPS experienced thermal fluctuations in its coolant loop, these two experiments performed admirably. FILE acquired 240 images across a broad range of different environments, successfully classifying their composition. SIR-B's data, too, proved of high quality and the radar took advantage of an unexpected opportunity to monitor Hurricane Josephine, detecting wave patterns associated with its motion and speed. Soil moisture content was measured as part of efforts to identify new water sources, support agricultural monitoring and crop forecasting and even, during a pass over Bangladesh, highlighted hidden breeding grounds of malaria-carrying mosquitoes.

Plant types in Florida and South America were successfully discerned, ocean

Hurricane Josephine as viewed from STS-41G. In Challenger's open payload bay, her RMS mechanical arm (bottom-right) and SIR-B (bottom-left) are partly visible.

waves measuring more than 20 m high were recorded and polar ice flows and evidence of oil spills detected. Despite the importance of these observations for 'real world' applications, however, SIR-B's claim to fame on STS-41G was its involvement in the 'discovery' of a lost city on the edge of the Empty Quarter in southern Oman. Archaeological exploration of the site later concluded it was most probably the legendary settlement of Ubar, a major ancient hub on the frankincense-trading route, first founded around 3000 BC.

"I was surprised to find that we were able to readily detect ancient caravan tracks in the enhanced Shuttle images," admitted geologist Ronald Blom of NASA's Jet Propulsion Laboratory (JPL) in Pasadena, California. "One can easily separate many modern and ancient tracks on the computer enhanced images, because older tracks often go directly under very large sand dunes. We could never have surveyed the vast area where Ubar may have been, nor could we be confident of its location without the advantage of computer enhanced images from space."

Excavation began in the summer of 1990 and found a remote well, together with towers, rooms and artefacts from at least 2000 BC. A Los Angeles filmmaker named Nicholas Clapp brought the possibility of using SIR-B to find Ubar to NASA's attention in 1983 and the radar and archaeological exploration confirmed the city – which, according to myth, was 'swallowed' by the desert – had indeed collapsed into an underground cavity. Corrupted, it is said, by the wealth it acquired from the frankincense trade, Ubar's destruction sometime between the first and fourth centuries AD thus represented God's punishment of its 'wicked' inhabitants.

Its discovery, 'on the ground', was delayed by the first Gulf War, but in November 1991 archaeological teams returned to the Omani desert and employed subsurface radar to aid exploration. As digging progressed, Ubar's remains closely matched the Koran's description: an octagonal fortified city with ten-metre-tall towers and thick walls, containing a variety of buildings, including storage rooms, together with frankincense burners and pottery sherds. "People have written about Ubar for thousands of years," said Ronald Blom, "and they hunted for it in the desert without any luck, but we cracked the case sitting here in Pasadena!"

For the STS-41G crew, operating SIR-B over their week-long mission in October 1984, the exciting discovery of the ancient Omani trading city was still in the future. In fact, apart from their efforts to close the antenna with the Canadian-built mechanical arm, the astronauts had little interaction with the radar, MAPS or FILE. "We turned them on and off, changed the parameters and settings and did some fine-tuning," said Sally Ride. "We changed data tapes for SIR-B. Our direct involvement was not really as scientists, but as operators."

The three instruments, attached to a British Aerospace-built Spacelab pallet in Challenger's payload bay, formed the most visible component of NASA's Office of Space and Terrestrial Applications package. The U-shaped pallet measured three metres long by four metres wide and provided a mounting location for each of the instruments. Behind it was a truss-like Mission Peculiar Equipment Support Structure (MPESS), housing another important OSTA-3 experiment: the Large Format Camera, which, as its name implies, was capable of conducting high-quality orbital photography for cartographic mapping and land use studies.

In terms of accuracy, the camera's 305 mm focal length was capable of acquiring images at a resolution of ten metres from a 200–250-km orbit. As planned, 2,280 photographs were returned to Earth, each monitored by the Shuttle's General Purpose Computers, including high-priority coverage of Mount Everest and the Dead Sea, oblique angle shots of Hurricane Josephine off the United States' eastern seaboard and even contrails left by aircraft travelling between New York and Europe. In fact, its resolution was so good – detecting buildings, houses and streets – that some pictures were immediately classified by the Department of Defense.

This remarkable resolution was achieved by a state of the art lens, high-resolution film and motion compensation system. Moreover, its position in the payload bay eliminated the distorting effects of Shuttle windows. Upon landing, its imagery of the Great Barrier Reef helped update Australian maps and the topography of national forests in Maine were plotted with greater accuracy than previously possible. Fossil fuel deposits were found in the Middle East, water sources identified in southern Egypt and Ethiopia and even geological evidence that blocks of land in China were being forced into the Pacific Ocean along the Kunlan fault line.

However, the Ku-band antenna failure threw a spanner into the works in terms of how much SIR-B data could be downlinked. Since the dish was now rigidly fixed in one place, Challenger had to firstly point the radar Earthward, tape record as much data as possible, then reorient herself to face the Ku-band towards the Tracking and Data Relay Satellite for playback. This repetitive process slowed down how much radar imagery SIR-B could acquire. Seven tapes were aboard Challenger – each capable of storing 20 minutes' worth of data – but playback, unfortunately, took just as long as recording.

Although some steps were taken to maximise scientific return, such as 'dumping' data through TDRS-1 whilst over the oceans so that the radar could be refocused when the Shuttle approached landmasses, it proved a laborious process. Then, on October 8th, the satellite lost its own attitude control capability for almost 16 hours and, worse, lost its lock on the White Sands ground terminal, meaning it could not be commanded. With both these problems in mind, it is quite remarkable that so much valuable data was successfully returned from the OSTA-3 mission.

HAZARDOUS HYDRAZINE

NASA managers, though, had already postponed Leestma and Sullivan's three and a half hour spacewalk from October 9th to the 11th to enable the Earth resources instruments to acquire additional data. Moreover, they would attempt a repair job on the Ku-band antenna to enable it to be properly stowed for re-entry. In spite of the problems, the crew's attitude was "good" and Crippen's absence during early training had no detrimental impact. In fact, the agency declared that, as long as Commanders were experienced, there did not seem to be a problem with them joining crews at a relatively late stage.

One of the objectives of Leestma and Sullivan's excursion was to test hardware for the refuelling of satellites in low-Earth orbit. Already, the Department of Defense had expressed interest in having the Landsat-4 Earth resources platform refuelled and repaired in 1987. NASA, too, hoped to refill its Gamma Ray Observatory (GRO) with station-keeping propellant in 1990. Mounted at the rear end of Challenger's payload bay for STS-41G was the Orbital Refuelling System (ORS), containing highly toxic hydrazine fuel, some of which the spacewalkers would transfer between two spherical tanks.

"Satellites have standard refuelling ports that engineers connect up when they're on the ground," explained Leestma. "One at a time, you very carefully have to handle the hypergolic fuels that go into it, because they're pretty dangerous. Hydrazine is very much like water, but it's got different properties, one of which is that it blows up if it's not handled right! Crip and the safety folks were very concerned that we shouldn't do this with hydrazine; we should just do it with water. The heat transfer properties of water and hydrazine are very similar and that's what we really wanted to know. NASA was worried about 'adiabatic detonation', which is – there's no convection in space – as fluids flow through ducts and into tanks, there's no real mixing of the temperature. As a tank is starting to fill up, if you're refuelling a tank in a constrained volume, there's less volume, so the pressure goes up inside the tank. As the pressure goes up, the temperature also goes up. There's really no big deal on Earth, because there's convection and this heat mixes around. In space, there's no convection. It is possible that all that heat will go into one very minute area – just a few molecules get heated – so that they could very rapidly get very hot and reach the detonation point of blowing the whole thing up. You want to be very careful as you flow the fluid, in that you don't get to this adiabatic point and detonate the fuel. I think Crip thought I was a little too cavalier, because I insisted that we should do it with hydrazine. Crip sent me to White Sands [Test Facility in New Mexico], so I spent about ten days there, watching them do adiabatic detonation tests, watching all kinds of things blow up. I came back with a real appreciation for the capabilities of this deadly stuff. You can't breathe it. If you get it on your skin, you can get poisoned. There were concerns that if we used hydrazine and it sprung a leak or even got on our suits, how are we going to get back in the airlock? We didn't want to bring this stuff back in."

Bob Crippen was not at all happy with using 'real' hydrazine in the ORS tests. He knew that the volatile substance could explode at temperatures above 230 degrees Celsius; temperatures which could easily be reached in the intense sunlight of orbital daytime. On the other hand, in orbital darkness, it could freeze, contract and then flow back, over-pressurising and rupturing its fuel lines. If Leestma and Sullivan got hydrazine on their spacesuits, Crippen and McBride would have had to reorient Challenger's payload bay towards the Sun so they could 'bake' it out.

In such a dire eventuality, they would have had to scrub their suits with towels and detergent, seal them in airtight bags, purge the airlock's atmosphere and pipe in fresh air, then remove their helmets. The effects on the spacewalkers were not Crippen's only concern. In an interview with Henry Cooper, he felt that 80 kg of hydrazine was enough "to take off the back end of the vehicle" if it exploded. Although he wanted to use water for the demonstration, his suggestion was rejected because, said NASA, using the real thing would permit tests of 'real' safety procedures.

Crippen was also assured that, although the ORS tanks and fuel lines had not undergone shaking tests equivalent to the stresses of launch and maximum aerodynamic pressure, they were designed with stiffness and robustness in mind. His four NASA crewmates, though, had already become comfortable with the experiment during their months of training without him and, at length, he was won over. In fact, Leestma and Sullivan had already successfully argued for a manual system to control temperatures and pressures in the tanks, on the grounds that no-one knew the exact parameters of hydrazine under different conditions.

"Crip finally agreed to have us do it with hydrazine," said Leestma, "because he had watched me several times in the WETF, doing the whole procedure and how careful we were. We had triple containment of all the liquids at all times. It's a very tedious task, using small tools and lots of arm and hand manipulation that you had to do to do this task." To achieve 'triple containment', three independent valves were placed in each of the coupling 'halves' and three seals were provided at the interface between the fluid path and the astronauts during the refuelling operation.

Before the flight, they had encountered problems – narrowly averted – with the tools they would use. Leestma insisted on testing them before they were sent to Florida for packing, although the engineers were not anxious for him to do this. When he did conduct his tests, a 'ball valve' – a pipe with a rotating valve that he would have screwed into an ORS fuel pipe and worked through it to open other valves – had an extra component fitted which made it two centimetres too long! Had it flown unchecked, the tool would have not have worked correctly.

Leestma also recalled problems with a new type of grease applied to several tools which, he was assured, was slightly different to that used in previous tests, but which should work in the same way. Leestma took the tools home that night and put them in his freezer; by next morning, the grease had frozen solid, the test was cancelled and the 'original' grease applied instead. "But that was still on my mind," he said, "that if something changes, you'd better make sure that those people know that they've looked at all the different things that can go wrong. I don't think it was so smart on my part. It was just the training that they put into you to kind of question everything. A lot of people don't like the astronauts, because they're always asking those silly, dumb questions, but sometimes those silly, dumb questions are appropriate and that one turned out to be okay."

The excursion began at 3:38 pm on October 11th, when Leestma pushed open the outer hatch and entered Challenger's payload bay. He would later tell Henry Cooper that the difference between being inside the Shuttle and outside on a spacewalk was "like the difference between sitting at a desk in a big room and sitting at a desk in the middle of a prairie – you can see so much more". Both astronauts needed about 30 minutes to fully acclimatise to their surroundings, learning how to move and how different their suits 'felt' in space, compared with the WETF pool.

"I grabbed the handhold and pulled out and when I first saw Earth, my heart rate went real high," Leestma said later, "and the docs later confirmed that, because my electrocardiogram reading went real high, 'This is when you came out of the hatch'. I said 'Yeah, no kidding!' I had a tumbling sensation – came out of the hatch and felt I was going to fall! I think my handprints are still in those payload bay handholds, because I just stopped for a short period of time and had to get my heart rate back down and then continue."

However, the excursion went perfectly and six hydrazine transfers were completed overall, without incident. Already, on October 6th, a series of automatic transfers of small quantities of the toxic propellant had been conducted between the two tanks, using controls on Challenger's aft flight deck. Then, during their spacewalk, Leestma and Sullivan modified the piping with the ball valve, successfully leak tested it and transferred around 50 kg more hydrazine through the fuel lines. In fact, because there was no weight on the thread, Leestma found it much easier to attach the ball valve in space than during pre-flight training.

The overall procedure did take somewhat longer than on Earth – a full, 90-minute circuit of the globe, rather than an hour – because Leestma had to stop work periodically so that Sullivan could photo-document the task. Their work closely

Dave Leestma during his spacewalk.

mirrored what spacewalkers were expected to do on the Landsat-4 mission and, in the summer of 1990, on a flight to refill GRO with fresh reserves of hydrazine. The observatory, it was intended, would be launched in May 1988, loaded with 1,800 kg of the highly toxic fluid and fitted with a specially designed standardised refuelling coupling to support the procedure.

Refuelling GRO, therefore, would mark the first time a fully functional satellite had been refilled with propellant whilst in orbit. Contracts to develop the coupling mechanism for its hydrazine transfer unit were awarded by NASA in December 1984, for completion and delivery just 15 months later – barely six weeks after the Challenger tragedy. It was even optimistically envisaged that high-pressure helium and nitrogen, and even cryogenic fluids, would flow through refuelling lines on subsequent Shuttle missions. In spite of the threefold safety mechanisms, many astronauts breathed a sigh of relief when STS-51L terminated such plans.

With, arguably, the most hazardous portion of the spacewalk over, Leestma and Sullivan's next step was to tend to the Ku-band antenna to ensure it could be retracted and stowed for re-entry. To do this, they had to move it by hand, such that a 'pin', activated from the aft flight deck, locked it securely in place; if they could accomplish this, they would leave the dish open so it could continue relaying data from the Earth-watching instruments. If, on the other hand, they could not get the locking pins in place, they would manually close, deactivate and latch the antenna.

Obviously, for the sake of maximum data return from SIR-B, it was hoped that the second option could be averted. Moreover, if the antenna could not be retracted at all, the crew would be forced to jettison it overboard in order to close the payload bay doors for re-entry. That, said Leestma, was equally unthinkable. "The Ku-band assembly and digital avionics was worth a million dollars," he said, "so it would have been a very big loss to the programme if we had to jettison it."

The repair involved not only the spacewalkers, but also their colleagues inside the cabin. In fact, because of her role in the effort, Sally Ride 'missed' watching most of the three and a half hour excursion. After she and Jon McBride had unplugged the wire to the antenna's electrical motors on October 6th, they also disabled a mechanism that drove the pins to 'lock' the alpha and beta gimbal axes into place. Early on the day of the spacewalk, they rigged a 'jump wire' that would allow them to reconnect power to the pins, though not the motors.

Unfortunately, both plugs in the jump wire were 'female' and they had to quickly rig up a new, 36-pin 'adapter'. As Ride laboured in the middeck, Leestma manually moved the Ku-band in one axis, then the other, while Sullivan radioed her crewmates when the pins were correctly lined up with the holes they were meant to slot into. Crippen, meanwhile, told Ride when to plug in the two ends of the jumper. Working the current in pulses – plugging and unplugging the cable, such that the pins were 'hammered' into position – the attempt succeeded.

Difficulties with SIR-B's data gathering, though, continued.

"Now, that caused us problems, orbiter-wise," admitted Leestma, "because to use the Ku-band, which the SIR-B required, we had to reorient Challenger so the antenna was pointed towards TDRS-1 and make the orbiter rotate. We'd take data and then do data 'dumps' and point the orbiter at the TDRS; then we'd go back and do data 'writes', rather than being able to take data the whole time and point the antenna and dump it. The SIR-B scientists didn't get all the data that they wanted, but the mission was not a loss and they got almost everything."

Prior to returning inside Challenger's airlock at 7:05 pm, Sullivan took a long look at SIR-B, in an attempt to discover why it had proven so difficult to automatically latch into position. It looked, Henry Cooper wrote later, "like an overstuffed sandwich"; its thermal insulation having billowed in space to make it 'thicker' than it should have been. This pure white blanketing had thus frustrated previous efforts to close it. "The insulation is billowing enough," she told her crewmates, "to interfere with a single motor closing and you don't need to miss by much to keep the latch from shutting."

"FIVE PLUS TWO EQUALS SEVEN"

Challenger's sixth mission, in view of the complex tasks already accomplished, nearly detracted from the fact that there were not five people onboard, but seven. Spending

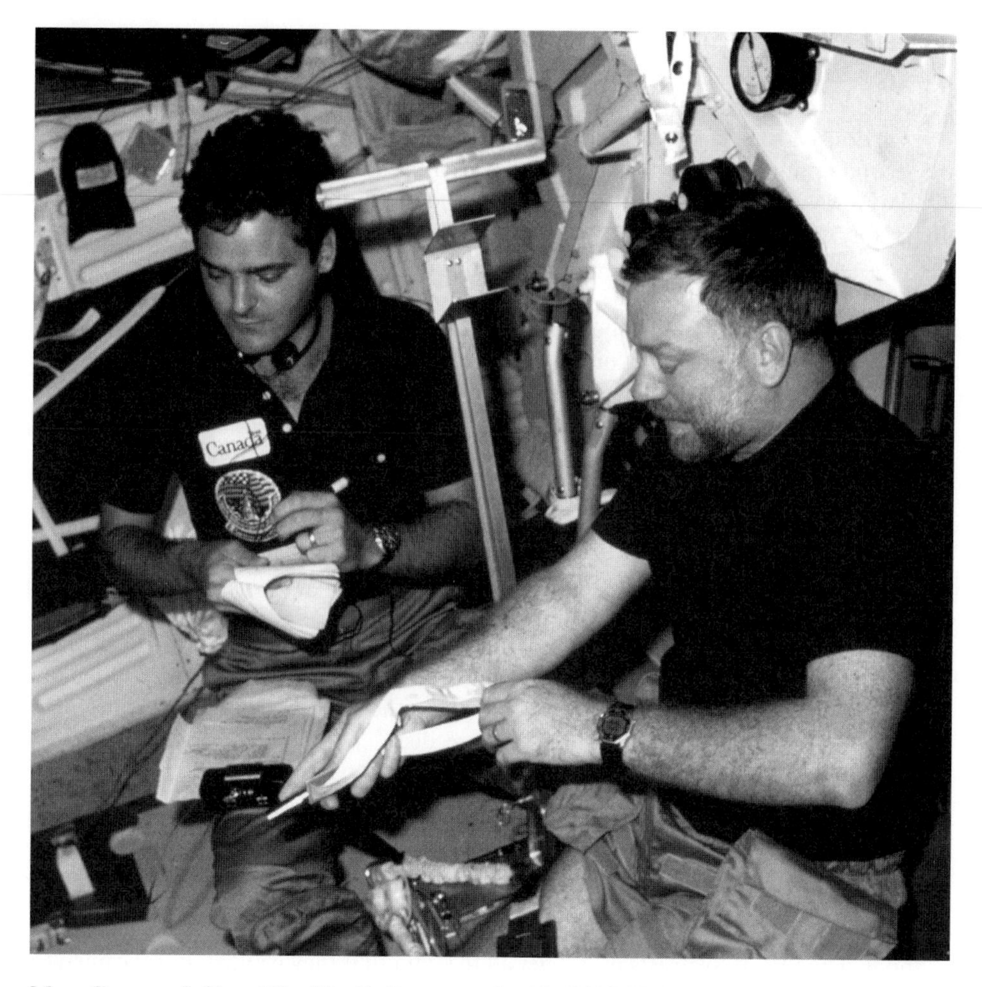

Marc Garneau (left) and Paul Scully-Power work with CANEX-1 experiments on Challenger's middeck.

much of their time conducting their own experiments, Garneau and Scully-Power were fitting in exceptionally well. Obviously, they were not 'career' astronauts, although Crippen treated them as team members enough to invite them to help prepare meals or change carbon dioxide scrubbing lithium hydroxide canisters. "We got along fine with Marc and Paul," remembered Dave Leestma. "They were great guys, but still, five plus two equals seven, and that's a crowd!"

Operating mainly on the middeck, the two Payload Specialists, whom Jon McBride mentored and took under his wing, helped each other out with their work. This proved useful for Garneau, whose ten investigations occupied much of his time. Known as CANEX-1, these Canadian experiments encompassed three major disciplines – technology, space research and life sciences – of which two of the major foci were developing a space vision system and exploring the impact of the harsh microgravity environment on a variety of advanced composites. This would have led to a CANEX-2 mission, run by Canadian astronaut Steve MacLean, in early 1987.

The space vision hardware was among the most important, for its descendant is today used routinely to allow astronauts to 'line up' and install components onto the International Space Station, as well as deploying and retrieving satellites. Garneau used the device to take video recordings of ten targets attached to the Earth Radiation Budget Satellite – four on its solar arrays and six on its sensor base – during its deployment, which were then transmitted to Mission Control. It was hoped the system would accurately calculate the position, orientation and rate of movement of ERBS, relative to Challenger, 30 times per second.

More than 90 per cent of the vision system's objectives were accomplished and, later in the flight, Garneau even made further video recordings with the camera on the RMS wrist to trace the outlines of instruments in the payload bay. Several other experiments under his supervision were attached to the mechanical arm itself. Known as the Advanced Composite Materials Experiment (ACOMEX), it comprised a number of samples, which, on October 7th, Sally Ride positioned in the Shuttle's direction of travel to assess their degradation under 'maximum' atomic oxygen exposure.

Earlier missions had already shown that composites tended to deteriorate in the space environment, with the originally bright red Canadian flag on Columbia's mechanical arm eventually turning brownish. Furthermore, the shiny film on thermal blankets for payload bay television cameras quickly became 'dull' and 'flat' and the insulation itself lost around 35 per cent of its overall mass. For a day and a half, the ACOMEX samples were pointed in the direction of travel and, after every six hours, Garneau examined them with binoculars and reported his observations of their performance.

Elsewhere, the Measurement of Optical Emissions experiment, known as 'OGLOW', provided data in support of a Canadian-designed imaging interferometer, scheduled for launch in 1988 to measure high-altitude wind temperatures. However, it was known since Columbia's STS-3 mission in March 1982 that the Shuttle developed a strange, reddish-orange 'glow' around its extremities in orbit and concerns were raised over its impact on sensitive optical detectors. In addition to this Shuttle glow research, OGLOW gathered data on the Southern Lights (the 'aurora australis'), atmospheric airglow at night and the bioluminescence of the oceans.

Other experiments included a solar photometer to measure atmospheric constituents and, specifically, the extent of sunlight-scattered dust, moisture, pollution and acidic haze with great precision. One of the instrument's main tasks was to examine the density and distribution of the cloud from the El Chichon volcano in Mexico, which had erupted in March 1982, before it fully disappeared. Finally, Garneau tended to a variety of space adaptation investigations. These focused on the effects of head motion, deterioration of sensory functions, awareness of position, space sickness, microgravity induced optical illusions and changes in the taste of different foods.

Bob Crippen, photographed during Challenger's fiery re-entry on October 13th 1984.

The Payload Specialists worked together on these tasks, with Scully-Power acting as a 'recorder' for many of the adaptation experiments and as a 'subject' to some of the vestibular investigations, while Garneau helped him with his observations of the oceans. Unfortunately, the planet did not entirely co-operate: there was around 70 per cent global cloud cover, instead of the 30 per cent anticipated before launch, although the Mediterranean Sea was visible in its entirety and Scully-Power was able to view tightly interconnected spiral eddies moving from one end of it to the other.

One of the few problems seemed to be Garneau's lack of conversation over the airwaves, which, as Canada's historic first man in space, frustrated many journalists who were covering the mission from his country. In fact, one member of the Canadian media even referred to him as 'The Right Stiff', a play on words of Tom Wolfe's description of the Original Seven Mercury astronauts as having 'the right stuff'.

Yet, Garneau was by no means stiff; nor would this be his only chance to fly into space. Seven and a half years later, in March 1992, he was selected by NASA to train as a fully-fledged Mission Specialist for the Canadian Space Agency and ultimately rode two more Shuttle flights. The last of these, STS-97 in the winter of 2000, brought some of his STS-41G work full circle, by using the 'operational' space vision system to dexterously install the first US-built set of electricity generating solar arrays onto the International Space Station. Moreover, his backup for STS-41G – Canadian medical doctor Bob Thirsk – is actually, at the time of writing, immersed in training to fly a long-duration expedition to the space station in late 2007 or early 2008.

Challenger's mission, meanwhile, at just over eight days, was the longest she would achieve in her short life; eclipsing by about six hours her previous personal

Challenger swoops like a bird of prey onto the KSC runway.

record set at the end of STS-41B. After checking her preparedness for re-entry and successfully stowing the troubled Ku-band antenna, the OMS engines were fired at 3:30 pm on October 13th, beginning a hypersonic dive to KSC. Touchdown was perfect, at 4:26:38 pm on Runway 33, with the orbiter rolling 3,000 m to a halt. For Crippen and Ride, it was also a personal achievement as they finally made landfall at the Florida spaceport.

The triumphant touchdown was not entirely without blemishes. Astronaut Dave Hilmers, sitting in Mission Control, jovially radioed congratulations to Crippen on making it to the East Coast. However, he alerted them that, judging from the Commander's track record for making successful Floridian landings, their 'welcome home' case of beer had been delivered to Edwards Air Force Base by mistake ...

5

The Untouchables

UNEQUAL PARTNERSHIP

In spite of his desire to fly as often as possible, astronaut Mike Mullane considered himself fortunate to have been aboard none of Challenger's 1985 missions, for all three carried a fate worse than death: a European-built research facility called Spacelab. "For Mission Specialists, the A-list astronauts were those who flew the Manned Manoeuvring Unit," he recounted in his memoir, 'Riding Rockets'. "At the bottom of the pile were those sorry souls doing actual science in the bowels of a Spacelab. While many scientists enjoyed Spacelab, most of the military Mission Specialists wanted nothing to do with it.

"Piloting an MMU or operating a robot arm had a lot more sex appeal and generated a lot more personal fulfilment than watching a volt meter on some university professor's experiment," he added. "The Untouchables of our strange caste system [in the astronaut office] were those Mission Specialists engaged in Spacelab missions dedicated to life sciences. They collected blood and urine and butchered mice and changed shit filters for primates." Mullane's no-holds-barred summary would prove fitting on Challenger's seventh flight, when squirrel monkeys and rats were carried aloft, leaving STS-51B Commander Bob Overmyer with an unwanted 'gift' under his nose ...

The facility upon which Mullane and others lavished such vitriolic dislike had been outlined by NASA towards the end of 1969, shortly after Neil Armstrong and Buzz Aldrin became the first humans to set foot on the Moon. President John F. Kennedy's challenge to land a man on the lunar surface had been met in spectacular style, but not until the euphoria had died down could serious consideration – and dollars – be devoted to the future. Naturally, Earth-circling space stations, Moon bases and trips to Mars before the end of the century were envisaged, but considered unlikely.

Establishing some kind of semi-permanent, or at least frequent, human presence in space was of vital importance to NASA. During the agency's first eight years of sending people aloft, from Al Shepard's pioneering sub-orbital 'hop' to Apollo 11's triumphant landing on the Sea of Tranquillity, fewer than two dozen astronauts had gazed down on Earth from the heavens. Some of them, admittedly, had flown as many as three times, but two dozen of a United States population of a quarter of a billion was insignificant. If NASA was to maintain public interest, it had to make access to space 'routine'.

A space station called Skylab was already being built from Apollo spares, but the most pressing long-term aim was the development of the Shuttle, for which the United States desperately sought international co-operation. Already, Canada had pledged to construct the reusable spacecraft's robotic arm and, in December 1972, as the final Apollo crew prepared to leave the Moon, the European Space Research Organisation (ESRO) agreed at a ministerial conference in Brussels to build a modular, multi-purpose laboratory for the Shuttle. Initially dubbed, somewhat uninspiringly, the 'sortie can', it was later renamed 'Spacelab'.

The deal between the two agencies called for ESRO – which, in 1974, merged with the European Launcher Development Organisation (ELDO) to form today's European Space Agency (ESA) – to develop the facility for carriage aboard the Shuttle's cavernous payload bay, in exchange for flying its own astronauts on specific missions. Spacelab, it seemed, would permit NASA to neatly sidestep the biggest quandary from its 1971 financial battles with Congress: how to have both the Shuttle and at least a temporary space laboratory in a decade of steadily decreasing budgets. Unfortunately, the reality proved an unhappy prelude to the project's eventual success.

For Europe, involvement in Spacelab initially proved the consolation prize, after two other proposals were rejected by NASA and the Department of Defense. During 1971, ESRO and ELDO invested $20 million in a series of studies to develop one of three components for the Shuttle: its payload bay doors, the sortie can or a reusable 'space tug'. Consensus favoured the latter, although it was turned down by NASA in June 1972, apparently because the Pentagon – envisaged to be the tug's main user – was reluctant to have a 'foreign entity' building a booster to place its national security payloads into orbit.

Constructing the clamshell-like doors was also promptly removed from the options table because, said NASA, the Europeans lacked expertise or organisation to make such a vital contribution to the orbiter's structure. The sortie can, on the other hand, required less sophisticated technology, had a well-defined interface with the Shuttle and schedule slips or budgetary overruns would not hamper the reusable spacecraft's ability to fly. When NASA offered the sortie can option to Europe, it received a lukewarm reception. Many ESRO member states hesitated to participate, questioning what they had to gain from an effort that demanded a $250 million investment and yet would be used principally by the United States.

These lingering doubts were overwhelmed, said political scientist John Logsdon, by a desire to "pursue co-operation on almost any terms, no matter how one-sided". In truth, it provided western Europe with its best chance of advancing its ambitions in space and sending its own astronauts aloft. Indecision gave way to approval of involvement with the sortie can in December 1972; moreover, ministers agreed to form a single European space organisation. In January 1973, ESRO's member states voted in favour of building the sortie can and, from May, began working with NASA on the details of a formal Memorandum of Understanding.

On September 24th, contracts between the two organisations were finally signed in Washington, DC by NASA Administrator James Fletcher and ESRO Director-General Alexander Hocker and the wheels of the largest international colloborative venture in space so far were set in motion. By now renamed 'Spacelab' – the preferred choice of the Europeans – the majority of its funding (54.1 per cent) came from West Germany, with Italy ranking second at 18 per cent. In the terms of the contract, ESRO assumed responsibility for the "definition, design, development, manufacturing, qualification, acceptance testing and delivery" of an engineering model and space-worthy flight unit to NASA by 1978.

For its part, NASA would operate Spacelab, fly European astronauts as Payload Specialists on selected missions and possibly procure additional hardware. Project management in the United States went to the Marshall Space Flight Center (MSFC) in Huntsville, Alabama, and its growing involvement in selecting and training scientists for Payload Specialist positions on Spacelab missions led to an expression of concern from Houston. At the Johnson Space Center (JSC), Director Chris Kraft was reluctant to permit MSFC to choose and train Payload Specialists, arguing that they should be "selected from the present corps residing in Houston".

The decision, unfortunately for Kraft, had already been made by NASA Headquarters, but it highlighted the different way in which Spacelab would operate, compared with other flights. Unlike pilots and Mission Specialists, who were selected formally as 'career' astronauts by the agency, Payload Specialists would be picked by an Investigators Working Group (IWG), whose panel included principal scientists responsible for Spacelab experiments. For the first time, a centre outside Houston was infringing on JSC's territory by providing mission-specific training. In all other areas, however, the two centres remained separate: JSC having responsibility for Shuttle operations and MSFC for Spacelab payloads.

Difficulties with the European partnership also arose when it became clear that NASA would not purchase more than one additional flight unit; in its 1973 plan for Shuttle utilisation, the agency envisaged buying five Spacelabs to support two dozen research missions annually. However, a multitude of technical problems pushed the reusable spacecraft's first flight beyond 1980, only two units had been commissioned and, as ESA's budget declined, it became worryingly clear that the Europeans would have little funding available to even use Spacelab for their own experiments. The $250 million facility, they feared, would inevitably slip under American control.

Perhaps the most important defining phase came in March 1978, when NASA and ESA initiated a nine-month critical design review, which provided a final opportunity to incorporate significant technical changes. Concerns remained, however. The first three Spacelab missions would all exceed the Shuttle's cargo-carrying limit as a result of orbiter-supplied equipment. MSFC opted to upgrade the orbiter's landing capability as a possible solution, although NASA Headquarters felt this would set

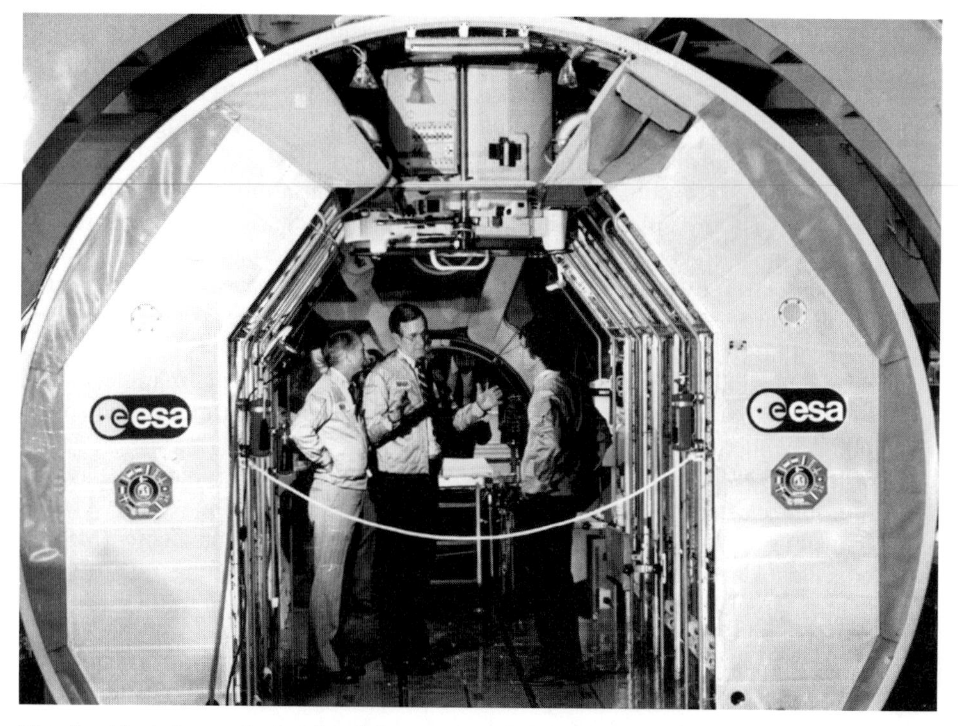

Vice-President George Bush talks with astronauts Owen Garriott (left) and Ulf Merbold (right) in the first Spacelab module, shortly after its dedication at the Kennedy Space Center's Operations and Checkout Building on February 5th 1982. Note the module's cylindrical shell, into which experiment racks are positioned.

a bad precedent. Nor was reducing Spacelab's payload weight desirable. Ultimately, ways were found to absorb the weight excess without seriously impacting each of the missions.

Ongoing budgetary problems caused further woes. The European member states initially pledged up to 120 per cent of their individual financial commitments to accommodate cost overruns, but even those had been consumed by September 1979. After protracted deliberations, ESA was obliged to propose increasing members' funding to 140 per cent. Only Italy refused to accept the new cost ceiling and by the end of the decade both the Europeans and Americans had reason for disappointment: the former resented paying more than expected, while the latter was disappointed that ESA was unprepared to take risks or bear full responsibility for Spacelab's early missions.

Inevitably, these worries impacted the flight schedule. When ESA delivered two Spacelabs to the United States – one under the terms of the original Memorandum of Understanding, the second as a follow-on procurement – NASA's budget was slashed in the first year of Ronald Reagan's presidency. This imposed one-year delays to several Spacelab missions. "Over the past four years," lamented James Harrington, director of the Spacelab project at NASA Headquarters, "the Spacelab-1 launch has slipped three years! Additionally, the manifest of Spacelab flights has been reduced from four to five flights per year to the current two flights per year through 1986."

As the problems of this increasingly unequal partnership were being thrashed out, the miniature space laboratory gradually took shape throughout the mid to late 1970s. Its bus-sized pressurised module comprised two components: a 'core' segment, which housed data-processing equipment, a workbench and a set of air-conditioned research racks lining its walls, and an 'experiment' segment, providing additional room for scientific operations. Although the core could be flown on its own, this configuration was never used and all module flights employed both segments joined together, with a pressurised volume of 75 m^3.

By January 1976, full-duration simulated 'missions' were being undertaken by astronauts and researchers at JSC in full-sized mock-ups of the module. During one such test, physician Story Musgrave – later to fly STS-6 and Challenger's Spacelab-2 mission in 1985 – lived and worked for seven days with nuclear chemist Robert Clarke and cardiopulmonary physiologist Charles Sawin to verify its habitability, perform a number of physical and biomedical experiments and evaluate procedures for 'real' flights. A year later, in May 1977, physicians Bill Thornton, Carter Alexander and Bill Williams undertook a similar demonstration, exclusively for life sciences research.

The 'long module' – the combined core and experiment segments – flew 16 times between November 1983 and May 1998, supporting investigations into life sciences and fluid physics, technology and microgravity research. Of those flights, two were undertaken by Challenger. When one considers the dimensions of the core-only configuration, it is clear why NASA opted to employ the longer version: the latter was 7.1 m long, almost twice that of the former, and virtually doubled the amount of rack space in which to house experiments. However, the core module was expected to support a number of missions equipped with unpressurised, instrument-laden 'pallets', including Earth observation and plasma physics flights, in the mid to late 1980s.

The racks were American-refrigerator-sized facilities that could be loaded with scientific investigations and 'rolled' into the module's cylindrical aluminium shell. Also provided was a central aisle for mounting experiment hardware and two ceiling openings for a window and scientific airlock. Closed at each end by a pair of truncated cones, the module was held in place by three longeron fittings on the payload bay walls and one in its floor. The facility was enshrouded with passive thermal insulation and situated at the midpoint of the payload bay to avoid violating the Shuttle's centre of gravity constraints during ascent or re-entry.

In order to provide access to the module from the airlock built into the middeck of the crew cabin, astronauts traversed a 5.8 m long tunnel; also, because Spacelab's hatch was 1.5 m 'higher' than the airlock hatch, a 'joggle' section was included in the tunnel to compensate for this vertical offset. Built by McDonnell Douglas, this 1,030 kg connecting passageway also provided a miniature airlock to support emergency Extravehicular Activities (EVAs). In such an eventuality, the tunnel could easily be closed off without depressurising the module or the Shuttle's crew cabin, although crew members would not be permitted inside Spacelab during the spacewalk.

Such lengthy tunnels were needed because the long module had to ride towards the 'end' of the payload bay, in order to preserve the Shuttle's centre of mass during descent and landing. If only a core module was flown, it would have been located further 'forward' in the bay, with a shorter tunnel and its pallet 'train' at the rear end.

"The racks were pretty much standard," remembered NASA's former director of life sciences, Gene Rice, of the research accommodations aboard the module. "You either had a drawer in a rack or a whole rack – or you might have a double rack – depending on the magnitude of the experiment. We would help [customers] through the process of designing their experiment, integrating it into a rack, doing the testing that they needed to do and getting it to a NASA centre. They would have to show that they met the safety requirements to put it into the Spacelab and to fly it."

Each rack contained air ducts to cool its experiments, together with power switching panels. On the first 'operational' Spacelab mission, flown aboard Challenger on STS-51B in April 1985, the module contained 12 racks, of which the two closest to the entrance were devoted to command and control subsystems. The ceiling of the core segment provided a 0.3 m wide opening for a high optical quality Scientific Window Adaptor Assembly (SWAA), through which Earth-watching instruments could be directed. This window was not needed on STS-51B, although it had been evaluated successfully during the first Spacelab test flight in November 1983.

In the ceiling of the experiment segment was provision for a Scientific Airlock (SAL), into which samples requiring exposure to the harsh environment of space could be mounted and easily retrieved. On STS-51B, a French-supplied very-wide-field camera was destined to spend the first few hours of the seven-day mission conducting a general survey of ultraviolet radiation across a large part of the celestial sphere. It had previously flown aboard Spacelab-1 and proved fairly successful, yielding 40 per cent of its planned exposures, although on Challenger's mission its fortunes declined, complications arose and it returned no images.

Original plans called for a pair of verification flight tests of Spacelab: on STS-9 in November 1983, a six-man crew, including West German astronaut Ulf Merbold as a Payload Specialist, evaluated the performance of both the long module and an unpressurised pallet in Columbia's payload bay. Next, Spacelab-2 would employ a 'train' of three interconnected pallets and demonstrate a new Instrument Pointing System (IPS) for astronomical and solar physics observations. When both test flights were completed, Spacelab-3 – Challenger's primary payload for STS-51B – would mark the first 'operational' mission, supporting no fewer than 15 investigations in microgravity science and fluid physics.

Unforeseen problems with the development of the IPS, which Spacelab's MSFC project manager Jack Lee described as "new and different and proposed requirements that we hadn't done before", eventually led to Spacelab-2 being postponed until after the Spacelab-3 mission. The latter incorporated a long module and an external platform in the payload bay, known as a Mission Peculiar Equipment Support Structure (MPESS), both of which had already been tested without incident on earlier flights. It was felt that the module and MPESS, at least, had proven themselves sufficiently to be declared 'operational' and Spacelab-3's mission would thus leapfrog that of Spacelab-2.

'ROUTINE' FLIGHTS

STS-51B was to usher in an era of 'routine' Spacelab flights. Planned to run for seven days and support 15 investigations provided by American, European and Indian researchers, Challenger would circle the globe in a 'gravity gradient' attitude – with her tail pointing towards Earth and her right wing in the direction of travel – to ensure a stable microgravity environment and minimal number of thruster firings. It was recognised that particularly vibration-sensitive materials science or fluid physics experiments could be adversely affected by periodic bursts from the Shuttle's Reaction Control System (RCS) jets. The gravity gradient orbit sidestepped this concern.

Preparatory work in support of Spacelab-3 commenced in December 1983, when the module returned to the Kennedy Space Center (KSC) in Florida, following STS-9. Over a period of several weeks, its racks were removed and the few modifications required for its second flight were made. The SWAA would not be needed, was removed and covered with an aluminium panel, whereas the SAL would be kept for Spacelab-3 to house the French very-wide-field camera. In March 1984, a mission sequence test verified the compatibility of the 15 experiments with each other and with simulated Shuttle subsystems.

Each of these tasks was performed inside the Operations and Checkout Building and, unlike many of Challenger's previous cargoes, the Spacelab-3 facility would be installed into her payload bay in a horizontal position, rather than in a vertical configuration out at the launch pad. With each of the experiment racks loaded into the Spacelab module and two additional sensors – the Atmospheric Trace Molecule Spectroscopy (ATMOS) and the Studies of the Ionisation of Solar and Galactic Cosmic Ray Heavy Nuclei ('Ions') – attached to the MPESS platform, the complete payload was moved to the Orbiter Processing Facility (OPF) on March 27th 1985 and installed aboard Challenger.

It had been an eventful six months since the orbiter returned from her previous flight, STS-41G. For astronaut Karol 'Bo' Bobko, it also heaped further disappointment on an already frustrating year for his four-member crew. Originally slated to command STS-41E in August 1984, Bobko's mission had been snatched from him by an anxious NASA in the wake of Space Shuttle Discovery's dramatic main engine shutdown on June 26th a few seconds before it was to have lifted off. Reassigned to lead STS-51E early the following year, he and his crewmates were only days away from launch when their four-day flight was cancelled.

On her seventh mission – during which she would have eclipsed Columbia's record of six trips into orbit – Challenger was to deploy two important payloads: the Anik-C1 communications satellite for Canada and NASA's long-awaited second Tracking and Data Relay Satellite (TDRS-B). After rollout to Pad 39A on February 15th 1985, these payloads were installed and launch was set for March 7th. With just six days to go, however, NASA issued a press release to completely cancel the mission, citing the need to repair a problem with one cell of TDRS-B's 24-cell battery and to attend to a 'timing issue' with the satellite.

The latter issue had already been noted with TDRS-1, although it was thought to be a telemetry problem, rather than an inherent design flaw. By February 26th 1985,

When they finally sat down to eat breakfast on launch morning, April 12th 1985, Karol 'Bo' Bobko's crew would fly under a different mission number. Originally scheduled to fly Challenger on STS-51E in March, their flight was cancelled and (with the exception of Patrick Baudry being substituted for Charlie Walker) they were reassigned as the 'new' STS-51D crew. From left to right at the breakfast table are Rhea Seddon, Don Williams, Charlie Walker, Bobko, Jeff Hoffman, Dave Griggs and Jake Garn.

though, the satellite's contractor, TRW Defense and Space Systems Group of Redondo Beach in California, advised NASA that it was sufficiently serious to warrant postponing STS-51E. With the benefit of hindsight, it is ironic that such action was taken by the space agency with respect to serious concerns about a payload and yet no action was taken on January 27th 1986 with respect to concerns over the safety of a launch ...

Both payloads were promptly removed from Challenger and TDRS-B was transferred to its processing facility for repairs. In the meantime, the Shuttle was rolled back to the Vehicle Assembly Building (VAB) on March 5th, destacked from her External Tank and Solid Rocket Boosters and returned to the OPF to await a new flight opportunity. Anik-C1, meanwhile, was hurriedly added to Discovery's next flight – STS-51D, scheduled for early April – at the expense of dropping one of its own mission tasks: retrieval of the Long Duration Exposure Facility (LDEF) from orbit.

Possibly in recognition of their long wait, the 'new' STS-51D mission was given to Bobko and his crew – Pilot Don Williams and Mission Specialists Jeff Hoffman, Dave Griggs and Rhea Seddon – although their two Payload Specialists were changed. Originally, STS-51E would have carried a French-built echocardiograph, accompanied by Frenchman Patrick Baudry, and Republican senator Jake Garn, a member of the Senate appropriations committee responsible for NASA's annual budget, who was flying as the first congressional 'observer'.

The 'original' STS-51D crew, consisting of Commander Dan Brandenstein, Pilot John Creighton and Mission Specialists John Fabian, Steve Nagel and Shannon Lucid, was reassigned – with Baudry – to a subsequent flight, STS-51G, in June 1985. Consequently, the French Echocardiograph Experiment (FEE) flew on STS-51G, generating two-dimensional imagery of changes in the astronauts' hearts, lungs and blood vessels during exposure to the microgravity environment. The results from both missions provided insight into the major cardiovascular changes that occur during the first 24 hours aloft.

Most notably, the left side of the heart, responsible for propelling blood through the circulatory system, reached its maximum size, as did the blood volume, during the crew's first days in space. The right side of the heart, however, which collects blood returning from the rest of the body, proved to be typically smaller than when imaged on the ground. By Flight Day Two, the entire heart had grown smaller, subsequent changes progressed more slowly and the reduction in the left-side volume remained unchanged for at least a week after landing. Investigators concluded, based upon FEE data and results from an American device carried on STS-51D, that the human cardiovascular system adjusted rapidly to fluid shifts and blood volume loss during orbital flight. However, they identified the need for more extensive testing to determine if the decrease in heart volume was associated with any reductions in its performance.

Of STS-51E's other experiments, the most noteworthy in the eyes of the media was a selection of toys, including a yo-yo and flipping mouse, which the crew would have used to demonstrate microgravity on mechanical behaviour. The results, NASA announced in a February 11th 1985 press release, would be videotaped and become part of a curriculum package for elementary and junior high schools. "Through the proposed filming of simple generic motion toys in the zero-g environment," said Carolyn Sumners of the Houston Museum of Natural Sciences, "students will discover how different mechanical systems work without the constant tug of gravity."

In spite of the disappointment at losing STS-51E, Bobko found humour in the sheer number of crew patches he and his colleagues had to design, as their orbiter changed from Challenger to Discovery and their Payload Specialist complement changed from Baudry and Garn to Garn and McDonnell Douglas engineer Charlie Walker.

The mayhem of flight changes and cancellations during this period is highlighted by the differing patches of Bobko's crew. Originally, STS-41E was supposed to have been the 13th Shuttle mission and its astronauts incorporated a 13-star Betsy Ross flag as their centrepiece. When STS-41E was cancelled and renamed STS-51E (the 16th flight), the flag made little sense anymore, but was retained, nonetheless. Next, the Payload Specialists changed: Baudry's surname was on a small tab at the bottom of the patch and, with Garn's assignment, a new tab had to be added as the crew swelled from six to seven members. Finally, when STS-51E was cancelled, yet another patch resulted, substituting the name of the orbiter 'Challenger' with that of 'Discovery' and again modifying the Payload Specialists' name tab from Baudry and Garn to Garn and Walker ...

The period since October 1984 had been eventful for Challenger herself, as well as her would-be crews. Originally scheduled to fly Ken Mattingly's top-secret STS-51C mission in December of that year, she had been hurriedly substituted for Discovery when it was found that almost 5,000 of her delicate thermal protection tiles had become debonded during the STS-41G re-entry. One tile had separated from her airframe completely, which, although not a catastrophic problem in itself, revealed a far more worrying issue. A vulcaniser material known as 'screed', used to smooth metal surfaces under tile bonding materials, had softened to such an extent that its 'holding' qualities had been severely impaired. Subsequent investigation revealed that repeated injections of a tile waterproofing agent called 'sylazane', coupled with six high-temperature re-entries, had caused degradation in the bonding material. By the time STS-51B flew, the use of sylazane had been scrapped.

A longer than desirable six and a half months on the ground was not providing NASA with the 'routine' and 'regular' access to space that it had promised Congress a decade earlier. In fact, by the time STS-51E was cancelled, the agency had succeeded in launching just one of the 13 missions it had planned for 1985, due primarily to Challenger's tile problems. However, said NASA's Associate Administrator for Spaceflight Jesse Moore, "schedule is a secondary priority. Mission safety and success are top priority". While Moore's words were undoubtedly sincere and reflected the view of the majority within the agency, they would prove bitterly ironic in 1986 when the Rogers Commission investigated instances of 'cannibalism' of Shuttle parts, the widespread acceptance of critical problems and, in astronaut Mike Mullane's words, "the normalisation of deviance". Schedule was indeed the primary priority and, as missions continued to fly, seemingly safely, it would only intensify.

After the cancellation of STS-51E, Challenger's time in flightless purgatory was short. Following the removal of her remaining flight hardware, she was reconfigured to support the Spacelab-3 mission towards the end of April 1985. After the 8,300 kg module and MPESS were inserted into her payload bay, she returned to the VAB on April 10th for stacking and eventually to Pad 39A on the 15th. It proved a remarkable month: Discovery had vacated the same pad with Bobko's STS-51D crew only three days earlier and, with Challenger's own lift-off on April 29th, a new record of just two weeks between launches would be set.

At face value, this increased flight rate was trumpeted by NASA as proof that the Shuttle could indeed support 'routine', fortnightly launches. Not until the following spring, during the lengthy Rogers Commission investigation into the technical and managerial causes of the Challenger tragedy, would the reality be exposed. Nor was STS-51B immune from that reality. Little did Commander Bob Overmyer, Pilot Fred Gregory, Mission Specialists Don Lind, Norm Thagard and Bill Thornton and Payload Specialists Lodewijk van den Berg and Taylor Wang know at the time, but their own lift-off that April day brought them within just milliseconds of disaster.

SEVEN MEN, TWO MONKEYS AND TWO DOZEN RATS

Twenty hours before Challenger's 4:02:18 pm launch, the first members of the STS-51B crew boarded their home for the next seven days. They consisted of two

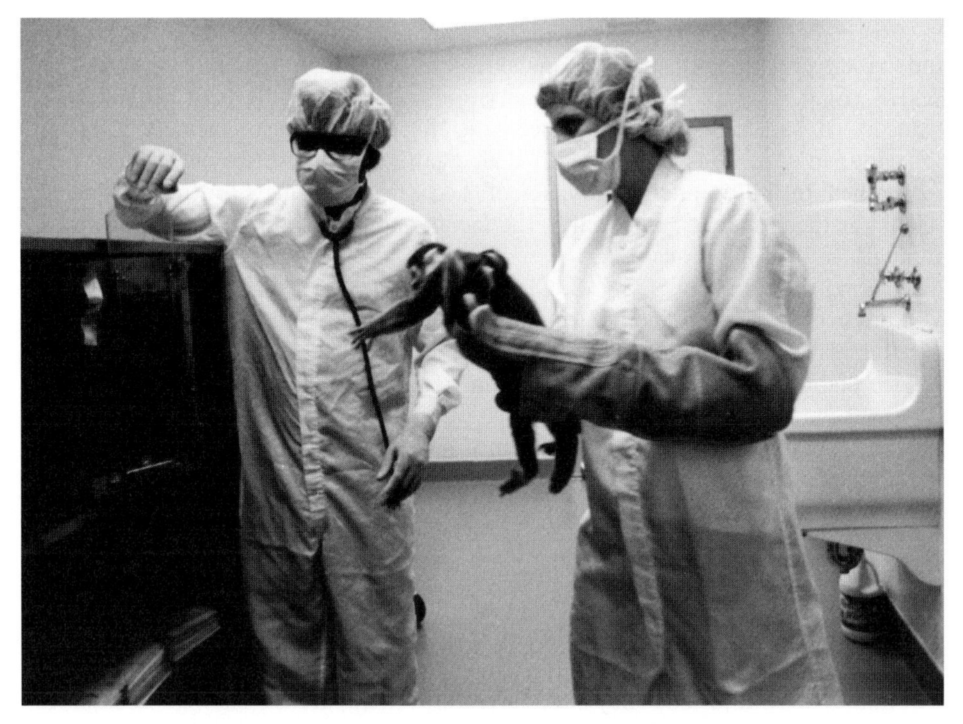

One of the two squirrel monkeys is prepared for STS-51B.

nameless male squirrel monkeys (*Saimiri sciureus*), described by Lind as "cute", and 24 "not so cute" male albino rats (*Rattus norvegicus*). Animal welfare concerns, coupled with the requirement to move the primates and rodents during their 'awake' time to avoid causing undue stress, made it important to wait until the final part of the countdown before loading them into their cages aboard Spacelab-3.

It proved an interesting event, worthy of comment, particularly as the Shuttle was oriented vertically on Pad 39A. Working from Challenger's middeck, two technicians were gently lowered, one at a time, in sling-like seats down the tunnel into the module. One stayed in the joggle section, while the other entered the laboratory to await the cages, which were lowered on separate slings. The delicate, two-hour procedure was problem-free and the cages were installed into dual Research Animal Holding Facilities (RAHFs) on the module's port side wall. The monkeys occupied single Rack Five, while the rats lived in double Rack Seven.

Developed by NASA's Ames Research Center of Mountain View, California, the facility was originally intended to be carried in a middeck locker and transferred to the Spacelab module in orbit. However, as its design progressed, it became clear that moving the bulky unit down the tunnel in space would prove difficult; nor could it be mounted easily in the centre aisle. Ultimately, it was decided to install individual cages in the rack-mounted RAHF whilst the Shuttle sat vertically on the launch pad, which meant their animal occupants would be resting on the cage 'sides' during landing.

Spacelab-3's primary focus was microgravity research, specifically in fluid physics and crystal growth, but an additional life sciences thrust evaluated how well the RAHF could support animals in an environment comparable to a ground-based vivarium. It had long been recognised that effective studies of primate or rodent behaviour in space was impossible if their health and well-being were improperly maintained. In addition to the provision of food – rice-based bars for the rats, banana pellets for the monkeys – and water, the facility supplied lighting, temperature and humidity control functions.

During the course of STS-51B, Challenger's crew worked in two 12-hour shifts, with physicians Thagard and Thornton assigned to separate teams to keep watch on the animals around-the-clock. Depending upon the RAHF's performance on its maiden flight, NASA hoped to use it again to support several rodent-based experiments on the Spacelab-4 life sciences mission in early 1987. Also under test was a Dynamic Environment Measuring System (DEMS) to record acceleration, vibration and noise in the cages during ascent and re-entry and a Biotelemetry System (BTS) to transmit physiological data to the ground from a series of implanted sensors.

"The squirrel monkeys adapted very quickly," said Lind. "They had been on centrifuges and vibration tables, so they knew what the feeling of space was going to be like. Squirrel monkeys have a very long tail and if they get excited, they wrap the tail around themselves and hang onto the tip. If they get really excited, they chew on the end of their own tail! By the time we got into the laboratory, about three hours after lift-off, they were adjusted. They had, during lift-off, apparently chewed off a quarter of an inch of the end of their tails!"

Both monkeys were free of various specified pathogens and, six months before launch, it was mandated that they should also be free of antibodies to the *Herpes saimiri* virus. Although the latter was not known to cause disease in either squirrel monkey or human carriers, it was noted that problems had been documented in other species and a worldwide search found five *Herpes saimiri*-free primates. Due to time limitations, NASA only had the opportunity to prepare two of them for microgravity exposure and properly train them to reach the food pellets and activate the water taps in their cages.

The possibility, however remote, of all seven men becoming infected by herpes was hungrily pounced upon by their peers at JSC, according to Mike Mullane in 'Riding Rockets'. Some US Navy astronauts suggested that as long as the US Marine Corps and Air Force members of the crew – a none-too-subtle jab at the respective military services of Overmyer and Gregory – did not "screw the monkeys", they would be fine.

In orbit, one of the primates exhibited the same space sickness symptoms – lethargy and loss of appetite, but no observed vomiting – as humans for the first half of the STS-51B mission, being hand-fed by Thagard and Thornton at one stage, before recovering completely for the final three days. The second monkey, on the other hand, displayed no ill effects. Both primates proved to be much less active in space than on Earth, although both they and the rodents grew and behaved normally, were free of chronic stress and differed from their ground-based 'controls' only through gravity-dependent variables.

The monkeys, in particular, were spoiled, too.

"I think the environment they had come from was a place where they received a lot of attention," remembered Fred Gregory. "Norm and I would look into the Spacelab and see Bill Thornton attempting to get these monkeys to do things, like touch the little trigger that would release the food pellets. I could tell they expected Bill to do that for them, even though he was outside, looking in. We looked back one time and could see that the roles were kind of reversed and Bill was doing antics on the outside of the cage and the monkeys were watching!"

Thornton and Thagard could view the primates through a window in each of their cages, while a perforated opening gave them limited access to the interior. The rodents' enclosures were similar to those of the squirrel monkeys, with the exception that they housed two occupants per cage, separated by a partition. Half of the 24 rats were rapidly growing, eight-week-old juveniles and the remainder were mature 12-week-old adults. Four of them had been implanted with transmitters three weeks before launch, which enabled continuous monitoring of their heart rates and deep body temperatures to be transmitted through the BTS.

The data returned from the implants, which also measured muscle activity and other parameters, actually proved of such high quality that it was possible to monitor one of the rats for indications of stress. Neither of the monkeys was outfitted with BTS sensors, although their cages included provision for this to be included on future flights. Typically, implant data was transmitted via a dedicated computer to scientists at the Payload Operations Control Center (POCC) in Houston.

Although both the monkeys and the rats were maintained in healthy conditions throughout their seven days in orbit, the latter proved not quite as 'savvy' as the former in terms of their adaptation to microgravity. "They hadn't learned that this was going to last a while," explainded Lind, "and, when we got into the laboratory, they were hanging onto the edge of the cage and looking very apprehensive. After the second day, they finally found out if they'd let go of the screen, they wouldn't fall and they probably enjoyed the rest of the mission."

In spite of their slowness adapting to their new environment, the rats showed no obvious signs of space motion sickness, although post-flight dissection and analysis identified a marked loss of muscle mass and an increased fragility of their long bones. Investigators speculated that this was probably caused by the influence of microgravity, rather than the stress of living in the RAHF cages. Nonetheless, the monkeys and rats were all recovered in good physical condition, healthy and free of microbiological contaminants.

However, the STS-51B crew returned the facility to Earth with a number of concerns, one of which had floated without warning under Bob Overmyer's nose as he sat in Challenger's flight deck – the RAHF leaked crumbs of food, monkey and rodent faeces and unpleasant odours. "The later analysis was that primarily it was food," admitted Gregory, "though there may have been some contaminants in it. It was a passing issue; not something that would have caused any disruption in the current activities."

On the ground, however, it became a big news story. "One anecdote involved this bit of animal dung that escaped from a cage and made its way from the Spacelab

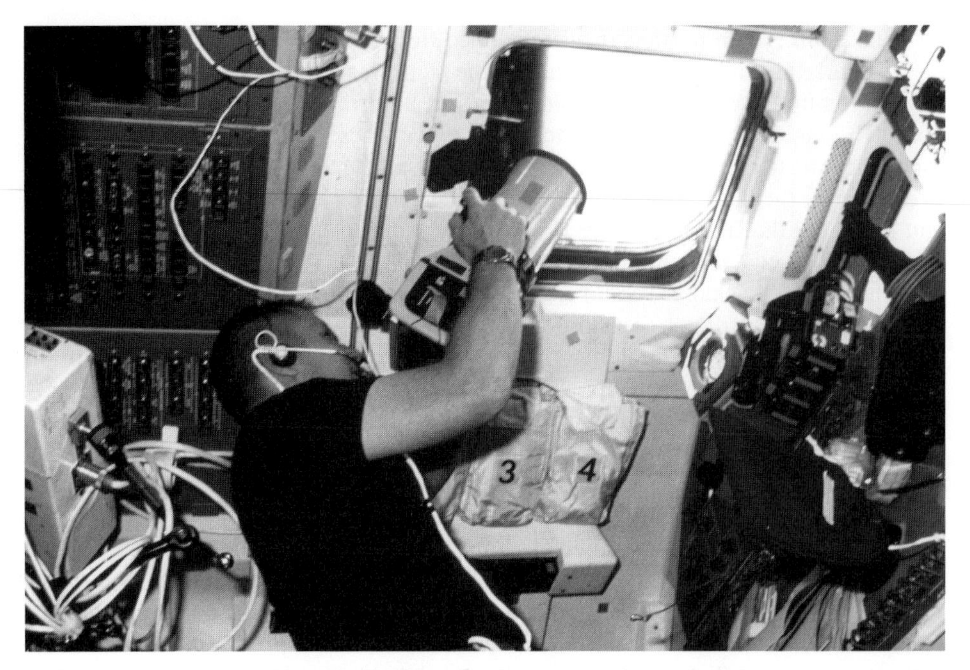

Perhaps trying to distract his attention from floating monkey faeces, Bob Overmyer undertakes Earth resources photography through Challenger's overhead flight deck windows, using a Linhof camera.

module to the flight deck," Thagard told me in an email correspondence. "Bob Overmyer made a comment about it that prompted an editorial page cartoon that appeared in some newspapers. The cartoon depicts a Shuttle astronaut saying to a crewmate words to the effect of, 'I'm not upset, I'm just glad we didn't have elephants on board!' "

Behind the humour of the incident, however, such issues needed to be resolved before the RAHF could be declared operational and flown aboard Spacelab-4. After landing, the rats and monkeys proved to be in good health and good spirits and strikingly calm when handled, although the former were found to have an extensive coating of dried urine and food powder on their coats. It was believed this had been caused by a variable flow rate in their cages, which prevented some of the urine, faeces and food powder from being properly deposited in their waste collection trays.

Overall, Spacelab-3 proved that the new facility provided a suitable animal habitat, aside from concerns about leaking food, faeces and odours. Time was of the essence, however, to resolve these concerns in time for Spacelab-4 which, according to the 'Newsletter of the American Society for Gravitational and Space Biology' later that year, had been scheduled for February 1987. NASA hoped to fly at least one RAHF on that mission, the report read, housing 24 rodents and transferring them, in space, to a new unit called the General Purpose Workstation (GPWS). This made adequate containment of particulate debris even more crucial.

In the wake of the Challenger disaster, the near-three-year downtime enforced on the Shuttle was used by Ames Research Center to modify the animal holding facility and a 12-day 'biocompatibility' test was undertaken in August 1988 to verify a number of adjustments in time for Spacelab-4. Its ability to contain debris – particularly food bar crumbs and faeces – and deal with odours and micro-organisms were identified as key issues. A single-pass auxiliary fan was added to assist the RAHF's environmental control system and follow-on tests in March 1989 confirmed it had indeed overcome the main problems experienced on Spacelab-3.

When the dedicated research mission eventually flew, under the new name of Spacelab Life Sciences (SLS)-1 in June 1991, tests confirmed it could indeed capture crumbs, flecks of rodent hair and faeces (simulated, fortunately, by black-eyed peas) and no noticeable odours or other contaminants were emitted. Moreover, when the SLS-1 crew moved rats from the RAHF to the GPWS at one stage in their mission, it marked the first time that rodents had floated freely in space, as well as allowing scientists to observe their behaviour and performance outside their cages.

"THEY LEFT OUT THE 'WOW'!"

In spite of the RAHF problems, it should be remembered that Spacelab-3 was a test flight of the hardware. The main focus of this first 'operational' mission was fluid physics and crystal growth and STS-51B's two Payload Specialists were chosen as experts in these fields. Taylor Wang, a Shanghai-born physicist employed by NASA's Jet Propulsion Laboratory of Pasadena, California, would operate his own experiment in the Drop Dynamics Module (DDM). Meanwhile, Lodewijk van den Berg, a Dutch materials scientist working for EG&G Energy Management Corporation of Goleta, California, focused on the crystal growth investigations.

They became the second pair of 'career' scientists to fly as Payload Specialists aboard the Shuttle and were chosen by the Spacelab-3 Investigators Working Group in June 1983, along with two backup candidates: metallurgical engineer Mary Johnston and Vietnamese-born fluid physicist Gene Trinh. All four received two basic types of training, known as 'dependent' and 'independent'. The former was directly associated with the specific Spacelab experiments, supported by their principal investigators, while the latter focused on practical skills needed to live and work safely aboard the Shuttle.

"Spacelab was an interesting assignment," Fred Gregory said years later, "because it was a '24/7' assignment. We had two shifts. Bob Overmyer was the Commander of a shift and I was the Commander of the second shift and while one shift worked, the other slept. We had enclosed bunks on the middeck of the orbiter and that's where the 'off' shift would sleep, so we never saw them, really. There was a handover period, but once we began working, they were sleeping and we just wouldn't see them.

"How did we train? There was a common portion of the training, and that was the ascent and re-entry, so Norm Thagard, myself and Bob Overmyer were always involved in the ascent and landing portion of the training. I'd say 75–80 per cent of the training was on ascent and re-entry. The intent was to try to get us three in a kind of mindset like a ballet without music – individual, but co-ordinated activities that resulted in the successful accomplishment of these phases, regardless of the type of failures or series of failures that the training team would impose on you. There were 2,000 switches and gauges and circuit breakers, any number of which we would involve ourselves with during ascent or re-entry. The intent was for us to learn this so well – understand the system so well – that we could brush through a failure scenario and 'safe' the orbiter in the ascent, such that we could get on orbit and then have time to discuss what the real problem was and correct it." Gregory's words would prove prophetic, for on Challenger's very next mission, STS-51F Commander Gordon Fullerton and Pilot Roy Bridges would be obliged to do just that ...

"Re-entry was a phase that, prior to the Columbia accident, would have been considered the easier part of the training," added Gregory. "In any scenario, you would have a series of failures, but all those failures would allow you to safe it, come home and land." Nonetheless, potential disaster was at the heels of every mission. One particularly disturbing incident occurred as Bo Bobko brought Discovery to a halt on KSC's Shuttle runway on April 19th 1985 – just ten days before STS-51B's lift-off – when a locked brake on the inboard right-side wheel resulted in severe damage and a burst tyre.

In his 2006 memoir, Mike Mullane recounted the seriousness of the episode. "The Shuttle is completely dependent on brakes," he wrote, "and it lands 100 miles per hour faster than airplanes of a comparable size. When the Shuttle touches down, it is a hundred tons of rocket, including several tons of extremely dangerous hypergolic fuel, hurtling down the runway at 225 miles per hour. While the Shuttle runways at KSC and Edwards, at three miles in length, are sufficiently long for stopping, they are only 300 feet wide. A perfectly landed Shuttle is only 150 feet from the edge – an eye blink in time at those speeds. It was a minor miracle that Discovery didn't experience directional control problems as a result of the blown tyre and careen off the runway."

Only five days after Bobko's touchdown, NASA issued a press release, announcing its intention to change Challenger's landing site from KSC to Edwards Air Force Base in California's Mojave Desert. "The decision will provide more safety margin for the Challenger's tyres and brake system," read the release, "because of the availability of the unrestricted lakebed and the smoother surface. The Spacelab-3 payload will be a heavy return weight for an orbiter. The decision to land at [Edwards] for the next flight only will enable engineers to determine what corrective actions are appropriate before returning to KSC for normal end-of-mission landings."

As circumstances transpired, no more orbiters would return to the Florida spaceport until Atlantis made landfall at the end of her STS-38 mission five years later. Improvements to the Shuttle's brakes throughout 1985 culminated in a successful nose wheel steering test on Edwards' runway during the STS-61A landing in November and East Coast touchdowns were expected to resume in January of the following year. The loss of Challenger and NASA's insistence on bringing subsequent missions into Edwards for safety reasons contributed to the lengthy hiatus. Even Atlantis' landing in November 1990 only came about because of unacceptable weather in California.

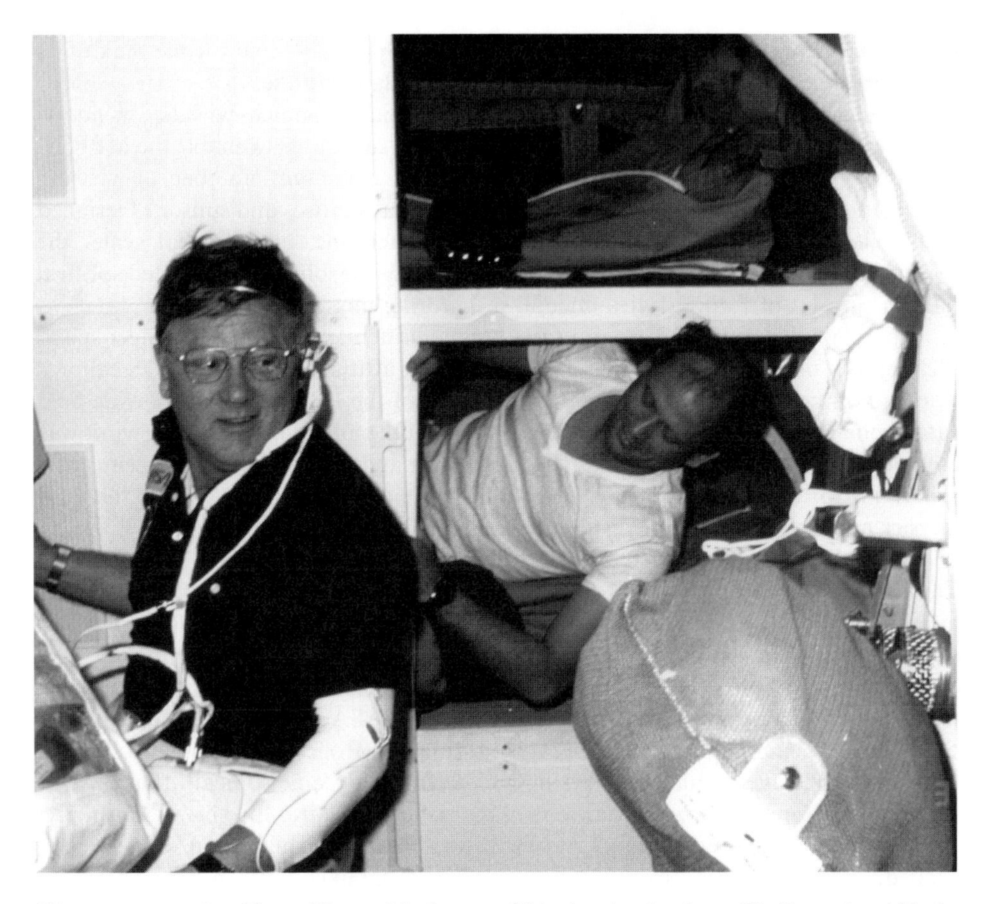

Silver team member Norm Thagard bails out of his sleeping bunk on Challenger's middeck, while gold team counterpart Don Lind works on the Autogenic Feedback Training experiment.

In spite of landing concerns, preparations for Challenger's launch on April 29th continued, with the seven-man crew 'sleep shifting' during their final days on Earth to prepare themselves for dual-shift operations in orbit. On the 'gold' team were Overmyer, Lind, Thornton and Wang, while their 'silver' counterparts were Gregory, Thagard and van den Berg. "I was responsible for all the support systems that keep the orbiter functioning," said Gregory of his role as the silver shift's leader. "Norm and I had respective jobs on board, but we, in essence, were the folks who supported the work of the Payload Specialists."

Although, as the flight engineer, Thagard was technically part of the orbiter crew, his work tended to cross over with that of the scientists working in the Spacelab module and, as already mentioned, one of his main scientific responsibilities was caring for the rodents and primates on his shift. Lind, a high-energy astrophysicist who had waited a record 19 years – longer even than Bruce McCandless – for a flight, was in charge of the activation and deactivation of Spacelab-3 and for the bulk of its experiments, one of which had dictated STS-51B's launch time.

Challenger had scarcely an hour available in which to launch, with her 'window' to the heavens opening at precisely 4:00 pm. This was calculated to enable the MPESS-mounted ATMOS instrument to accomplish the maximum number of viewing opportunities of Earth's atmosphere during 72 orbital sunrises and sunsets. Designed and built at JPL, its two main foci were to determine, on a global scale, the composition of the upper atmosphere and acquire high-resolution, calibrated spectral data in support of future environmental monitoring missions.

Spacelab-3's sensitive microgravity experiments demanded that Challenger spend the majority of her week aloft in a gravity-gradient attitude, so the ATMOS calibration and observation timeline, together with that of the French-built very-wide-field camera, had been 'front-loaded' into the first day of the mission. When their work was completed, about 18 hours after launch, Overmyer and Gregory would reorient their spacecraft for almost six full days in a suitably quiescent environment for the fluid physics and crystal growth investigations.

With the exception of a hydrogen leak during operations to load the External Tank with propellants, the countdown proceeded smoothly until 3:56 pm, when, four minutes before lift-off, a front-end launch processor failed and prevented the liquid oxygen system's replenishment valve and vent hood from closing automatically. The clock was held for just over two minutes as the valves were manually repositioned and Challenger's thunderous ascent was described by NASA as "nominal". It was not quite 'entirely' nominal, though, because during the Rogers inquiry the following year, Bob Overmyer would discover how close his crew came to death that day.

For Fred Gregory, who became one of the last of the Thirty Five New Guys to fly, the fear – for now – evaporated and gave way to sheer exhilaration. "I was very excited," he recalled. "I think I was probably anxious, but certainly not afraid. It was similar to the simulations, but they left out the five per cent, and that was the 'wow'! I remember the feeling inside when the main engines started; how it was almost a non-event. I could hear it and I was aware of it, but I looked out the window and saw the tower move back.

"At least that's what I thought, but then I realised the orbiter was moving forward and then back," Gregory continued, referring to the 'twang' effect of the main engine startup sequence, "and when it came back to vertical, that's when those solids ignited and there was no doubt about it: we were going to go someplace pretty fast! I just watched the tower kind of drop down below me and was probably laughing during this timeframe. Since we had trained constantly for failures, I anticipated failures and was somewhat disappointed that there were no failures. That was Challenger and she went uphill, just as sweet as advertised. The sensation of zero-g was like a moment on a roller coaster, when you go over the top and everything just floats. Once we got there, it was business as usual, just as we had practiced and performed on the ground." For Gregory, Overmyer and Thagard, the first order of business was pulsing their spacecraft's twin Orbital Manoeuvring System (OMS) engines to position themselves in a 360 km circular path, inclined 57 degrees to the equator. One of the main reasons for this high inclination was to provide greater observation coverage for ATMOS.

For Don Lind, who had waited since April 1966 for a mission, the reality was surprisingly close to the training. "The simulations are spectacularly accurate," said Lind, whose first task was to leave his seat on the flight deck and photograph the just-jettisoned External Tank as it tumbled Earthward. "With the motion-based simulators, you even got some of the visceral sensations, because they can move the machine around and give you the sense of onset of zero-g. You can't hold it indefinitely, but we had flown hundreds of parabolas in the KC-135 aircraft, so we were quite accustomed to those things."

Fred Gregory felt that he was well prepared, almost, for weightless, "but it took about half a day to adapt to microgravity. The body very quickly adapted to this new environment and it began to change. You could sense it when you were on orbit. You learned that your physical attitude in relation to things that looked familiar to you – like walls and floors – didn't count anymore as you translated floors and ceilings and walls to your head are always 'up' and your feet are always 'down'. That was a subconscious change in your response: it was an adjustment that occurred up there. You also learned that you didn't go fast, that you could get from one place to the other quickly, but you didn't have to do it in a speedy way. The only referencing system that you have are your eyes, so you can look at something and establish it as a reference that you use."

GRAVITY GRADIENT

There was little time to admire their surroundings, however, for the flight plan took precedence. As Overmyer and Gregory busied themselves with readying Challenger for seven days – and, potentially, up to nine, in the event of weather-related delays to landing – in space, Lind and Thagard set to work opening the hatch to the Spacelab-3 module and activating the first of its research facilities. These included the ATMOS instrument on its MPESS platform in the payload bay, which successfully completed 19 of its planned 72 observations before a power supply leak disabled its internal laser.

Nonetheless, during each three-minute data-gathering period, it acquired 150 independent spectra, each of which contained more than 100,000 measurements of atmospheric constituents between the altitudes of 16 km and 280 km. Detailed infrared solar spectra was gained and five molecules – dinitrogen pentoxide, chlorine nitrate, carbonyl fluoride, methyl chloride and nitric acid – whose existence in the stratosphere had hitherto only been suspected were definitively identified. ATMOS analysis of the lower mesosphere showed it to be considerably more 'active' than previously supposed, with many 'minor' gases typically being split by sunlight to trigger other chemical reactions.

The instrument's spectrometer measured changes in the infrared component of sunlight as it passed through the 'limb' of the atmosphere. Since each of the trace gases under scrutiny by ATMOS investigators was known to absorb sunlight at very specific infrared wavelengths, it was possible to determine their presence or absence, concentration and altitude by identifying which wavelengths had been absorbed from the data. Furthermore, the instrument's sensitivity and ability to detect these trace

Diagram of the gravity gradient attitude adopted by Challenger during the bulk of STS-51B orbital operations. The Spacelab-3 facility, consisting of the long module and MPESS, are clearly visible in her payload bay.

gases in concentrations of less than one part per billion, meant its data could be exploited reliably to test theoretical models of atmospheric physics and chemistry.

Human influence on the continuous changes to the atmosphere was one of the primary reasons for the decision to build and employ ATMOS. In the wake of the Challenger disaster and the resumption of missions, the instrument rode aboard three dedicated Earth-watching flights in the early 1990s. On the first of these voyages, in March 1992, it examined the effects of the previous year's Mount Pinatubo volcanic eruption in the Philippines and recorded large amounts of crustal material and sulphur-based aerosols in the stratosphere.

Additionally, many of the Spacelab-3 science team's predictions of atmospheric change between the first and second ATMOS missions were vindicated when chloroflurocarbon (CFC) quantities were shown to have increased dramatically and their role in atmospheric photochemistry had become more pronounced. When the two sets of results were compared, they highlighted an increase in inorganic chlorine levels from 2.77 to 3.44 parts per billion, together with a fluorine rise from 0.76 to 1.23 parts per billion; the latter confirmed that the source of the increased chlorine level was indeed industrial CFCs.

Other studies of Earth focused on its aurorae, one of the only natural visible manifestations of our planet's highly charged magnetosphere. By examining changes in its form and motion, it was hoped to derive greater insights into changes in our planet's electromagnetic field. In the case of Spacelab-3, observations were conducted from much closer range – in low-Earth orbit – than had been possible with previous, higher circling missions. Five hours of video recordings and more than 270 still photographs were acquired in such a fashion that they could be 'overlapped' and viewed stereoscopically.

The results included features never seen before, including the first views from beyond the 'sensible' atmosphere of thin, horizontal layers of enhanced aurorae. Previously considered to be rare, these layers were recorded on two of Challenger's three orbital passes over the aurora, thus eliminating earlier suspicions that ground-based observations may have been optical illusions caused by atmospheric refraction. Of the 21 scheduled opportunities for studies, 18 were achieved and STS-51B marked the first time since Skylab, which orbited at an inclination of only 50 degrees, that auroral observations had been undertaken in such depth.

This experiment proved particularly satisfying for Lind, who had proposed it and served as its primary operator. "Before our mission, the aurora had only been photographed by some slow scan photometers," he explained, "which gives you a blurred picture, like trying to take a picture of a waterfall. We found out that there is a different component to the mechanism that creates the aurora, involving microwaves, that was not understood before."

The second time-critical experiment for the first day of STS-51B was the French very-wide-field camera, which Lind set up in the SAL in the Spacelab module's ceiling for around 12 hours of ultraviolet observations of very young, massive stars at one end of the celestial scale and their ageing counterparts at the opposite end. Such wide-field observations were considered more rapidly made and more readily interpreted than scanning many points over a large area, as well as permitting constant comparison with the background sky and 'reference' stars.

The camera, provided by the Laboratoire d'Astronomie Spatiale in Marseilles, yielded promising results on Spacelab-1 with 48 exposures of ten astronomical targets, including a superb ultraviolet image of a 'bridge' of hot gas between the Large and Small Magellanic Clouds, two satellite galaxies of our own Milky Way. Had it not been for a bent handle on the SAL, it should have duplicated or exceeded this achievement on STS-51B. Ground controllers examined the airlock and decided that a maintenance procedure by the crew would be inappropriate. This was a pity, because during its initial extension into space, the camera acquired its first target and took a brief exposure.

Eighteen hours into the mission, as planned, Overmyer and Gregory duly manoeuvred the Shuttle into her gravity gradient attitude to support six days of fluid physics and crystal growth research. "I was Laboratory Director," explained Lind, a title that in the post-Challenger era is roughly equivalent to that of 'Payload Commander'. "We had five scientists on the crew: myself, two doctors and two Payload Specialists, who were visiting scientists." Supporting them from the Payload Operations Control Center (POCC) in Houston were the two alternate Payload Specialists, Johnston and Trinh, and a network of principal investigators for each of the 15 major experiments.

These experiments – 12 provided by American scientists, two by European researchers and one by an Indian team of astrophysicists – had been carefully selected for inclusion in the Spacelab-3 payload through a competitive peer review process. After responding to an initial announcement of opportunity and receiving approval, the principal designers of experiments typically formed an Investigators Working Group, chaired by NASA's mission scientist (George Fichtel of MSFC in the case of Spacelab-3). The group then worked with the Shuttle and Spacelab offices to identify requirements for their experiments, propose candidates for Payload Specialist postions and help train the crew members.

A natural candidate for the STS-51B mission was Taylor Wang, whose study of the behaviour of rotating and oscillating liquid droplets utilised a new facility called the Drop Dynamics Module (DDM). The latter, housed in the double Rack Eight on the Spacelab module's starboard side, offered fluid physicists their first opportunity to levitate and manipulate drops in a microgravity environment.

It had already been theoretically demonstrated that space research could lead to advances in new materials technology, including glasses, crystals, ceramics and alloys, whose properties exceeded those of their predecessors in terms of overall quality. However, chemical mixtures of some materials were known to be highly reactive to the walls of their processing chambers and contamination levels as minute as a few parts per billion could seriously degrade the final product. The DDM, explained Wang, had potential applications in the development of future 'containerless' materials processing methods which could significantly reduce such flaws.

Certain fluoride glasses – particularly attractive for their infrared transmission properties – could be manufactured in ground-based laboratories, but imperfections introduced by their containers prevented them from attaining their theoretical performance levels. 'Effective' containerless processing, in which acoustic and electromagnetic forces were applied to suspend and manipulate fluid droplets, could only be practically achieved in space: the influence of terrestrial gravity made it impossible to levitate liquids without introducing forces that masked the very phenomenon that physicists were attempting to examine.

For the DDM's first flight, the fluids carried were water and glycerin, but when Wang attempted to activate it during his shift on April 30th, he was stunned when it promptly shorted out and failed. "Not only that, but I was the first person of Chinese descent to fly on the Shuttle," he wrote later, "and the Chinese community had taken a

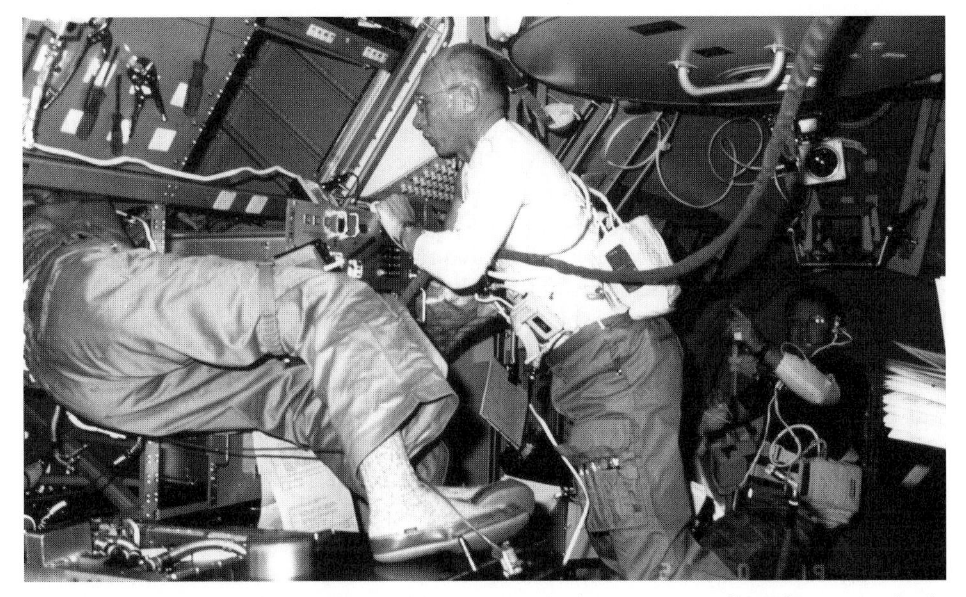

Bill Thornton assists Taylor Wang, whose upper body appears to have been completely swallowed by the Drop Dynamics Module, during the repair of his experiment.

great deal of interest. You don't just represent yourself – you represent your family – and the first thing you learn as a kid is to bring no shame to the family. When I realised my experiment had failed, I could imagine my father telling me, 'What's the matter with you? Can't you even do an experiment right?' I was really in a desperate situation." After asking Mission Control for permission to attempt a repair on the DDM, he quickly got to work, opening the Spacelab rack, isolating the fault and completely rewiring part of it. Several dramatic photographs, snapped by his crewmates, showed Wang's legs sticking out into the module as the DDM rack appeared to completely swallow his upper body.

He had already threatened not to return home if NASA refused to allow him to fix the DDM, so it proved fortuitous that his bluff was not called. "I hadn't really figured out how not to come back," Wang told a Smithsonian interviewer years later. "The Asian tradition of honourable suicide – *seppuku* – would have failed, since everything on the Shuttle is designed for safety. The knife on board can't even cut the bread. You could put your head in the oven, but it's really just a food warmer. If you tried to hang yourself with no gravity, you'd just dangle there like an idiot!"

With the facility successfully repaired, however, there was no time for suicide and he worked virtually non-stop to complete almost all of his scheduled experiments in the last three days of the flight, assisted by his crewmates. The results confirmed several age-old assumptions about the behaviour of liquids in a microgravity environment, although others proved somewhat unexpected: the 'bifurcation point', for example, when a rotating droplet takes the shape of a dog bone in order to hold itself together, occurred earlier than predicted under certain conditions. Another dog bone returned to a spherical shape and stopped spinning much more rapidly than anticipated, apparently from internal differential rotation.

During typical experiment runs, Wang would position freely suspended liquid drops under the influence of their own surface tension and gently manipulate them with acoustic speakers inside the DDM; Challenger's stable gravity gradient attitude kept thruster-induced accelerations to a minimum and avoided unnecessary disturbances. After a drop had been observed as 'stable' and spherical, it was set into rotation or oscillation by acoustic torque or modulated radiation pressure force. In spite of its delayed start, the experiment proved highly successful and, seven years later, an improved version was operated by Gene Trinh aboard another Spacelab mission in the summer of 1992.

Nineteen months after STS-51B landed in California, Taylor Wang's hard work was rightly rewarded with NASA's Exceptional Scientific Achievement Medal, presented in recognition of his "contributions to microgravity science and materials processing in space and for his exceptional contributions as Payload Specialist on Spacelab-3".

Although he would not fly into orbit again, Wang played an important role in the future drop dynamics experiments on the two United States Microgravity Laboratory (USML) missions in June 1992 and October 1995. On both flights, a second-generation Drop Physics Module (DPM) employed speakers to assess the response of water, glycerin and silicone oil to external forces and successfully injected droplets of sodium alginate into calcium chloride drops.

It was ultimately hoped, Wang said after the USML-2 work, that such research could lead to improved techniques of 'encapsulating' living cells to treat hormonal disorders into polymer shells to protect them from immunological attack and provide timed releases. Instances in which such methods would be useful included the treatment of diabetes, perhaps by injecting a pancreatic cell to secrete insulin into the patient's body. Clearly, the potential applications of Wang's original experiment were far more expansive than materials processing alone.

Elsewhere in Spacelab-3, located in Rack 11 on the port side, close to the module's aft cone, was the Geophysical Fluid Flow Cell (GFFC) experiment, provided by John Hart's team from the Department of Astrophysical, Planetary and Atmospheric Sciences of the University of Colorado at Boulder. This investigation, which flew again on the USML-2 mission in late 1995, sought to simulate fluid flows and better understand convective processes in terrestrial oceans, together with the atmospheres of the Sun and giant gaseous planets, particularly Jupiter.

Simulations of atmospheric dynamics were first undertaken in the early 20th century, using oil and water in rotating pan experiments, but since they were cylindrical their effectiveness proved somewhat limited. Supercomputers of the 1960s and 1970s offered greater advances by numerical modelling, although even they had severe imperfections. Even in ground-based spherical models, cold fluids flowed 'downhill' and ended the simulation; the only practical method of largely eliminating this effect of terrestrial gravity was to conduct the experiment in space.

Before the flight, "there was a question of whether you could get convection patterns and wind distributions that resembled those on a gas giant planet," Hart

View of the Drop Dynamics Module during its operations later in the mission.

recalled years later. This question was partially answered through Spacelab-3's research, by creating and observing so-called 'banana cells' – rapidly rotating columns formed as differential heating was increased – which were thought at the time to be a key feature of Jupiter's atmospheric structure. Not all of these phenomena were fully investigated, however, because of time and film limitations, together with an inability to interact on a 'real-time' basis with the experiments.

"The first flight of the GFFC was a little like running an experiment in the lab with the lights off," said Fred Leslie, a co-investigator of the device, who operated it as a Payload Specialist on its second mission. "We had no indication how the fluid was responding to the inputs. On the second flight, not only did we have a real-time video camera to observe the flows, but we also had a computer interface through which the crew could interact with the experiment."

Nevertheless, on STS-51B the facility operated perfectly, completing all of its computer-run scenarios during a period of 84 hours; an additional unscheduled 18 hours' worth of operations were also undertaken, yielding 46,000 images in total for post-flight analysis. Ten years later, during the 16-day USML-2 mission, the GFFC undertook more than 180 hours' worth of experimental runs, revealing that the long-term evolution of convecting flows in slowly rotating spherical shells depends on initial conditions. "Even under the same external conditions, like rotation and heating," said Hart, "small variations in initial conditions can lead to different end states."

The 'heart' of Hart's GFFC was a pair of 'hemispheres' – a baseball-sized one, made from nickel-coated stainless steel, mounted inside a larger, transparent one of sapphire – which were both affixed to a turntable. A thin layer of silicone oil filled the gap between the two hemispheres. During typical operations, the temperatures of both hemispheres, together with the rotation speed of the turntable, were minutely adjusted by the experiment's computer, which also introduced thermally driven motions into the oil. This enabled physicists to model fluid flows within the atmospheres of stars and planets.

One of the primary reasons for the success of both the DDM and GFFC was the high-quality microgravity environment established by Challenger's gravity gradient orbital path, which was described by NASA as "quite stable and conducive to the performance of delicate experiments in materials science and fluid mechanics". Each of the experiments requiring this environment – which also included two crystal growth facilities – were clustered around the Shuttle's centre of mass, roughly from the midpoint to the aft end of the Spacelab-3 module.

The first of these crystal growth facilities shared the same Spacelab rack as GFFC and was provided by French researcher Robert Cadoret of the Laboratoire de Cristallographie et de Physique in Les Cezeaux. His experiment, which also flew aboard Spacelab-1 in November 1983, processed six cartridges of mercury iodide crystal seeds at different pressures for 70 hours at a time, using a two-zone furnace. As with the geophysical flow cell experiment, this facility operated under computer control, with the astronauts monitoring it for problems.

Mercury iodide samples were also grown in the Vapour Crystal Growth System (VCGS) on the opposite side of the Spacelab module, in Rack 12. This experiment was provided by Wayne Schnepple of EG&G Energy Measurements Incorporated of Goleta in California and 'grew' crystals by vaporisation and recondensation at approximately 120 degrees Celsius in a specially designed furnace. In general, the returned mercury iodide crystals – which have considerable practical significance for gamma ray and nuclear radiation detectors – had a lower number of defects than their terrestrial-grown counterparts.

It was also optimistically hoped that space-produced crystals would allow for the construction of such detectors to operate at more ambient temperatures, rather than having to be cooled to near-cryogenic levels. Typically, a crystal the size of a sugar cube was grown from a 'seed' 20 times smaller in the VCGS. The facility carefully controlled the growth process at less than three millimetres per day over a 104-hour period. Its success led to a reflight on the first International Microgravity Laboratory (IML-1) mission in January 1992, confirming the more 'uniform' molecular structure of space-grown crystals over those produced on Earth.

Moreover, electronic measurements verified that the IML-1 crystals were more efficient, thus improving their characteristics as X-ray or gamma ray detectors. Lodewijk van den Berg, the 'materials expert' on the STS-51B crew, concluded from these results that vapour crystal growth could be effectively employed in

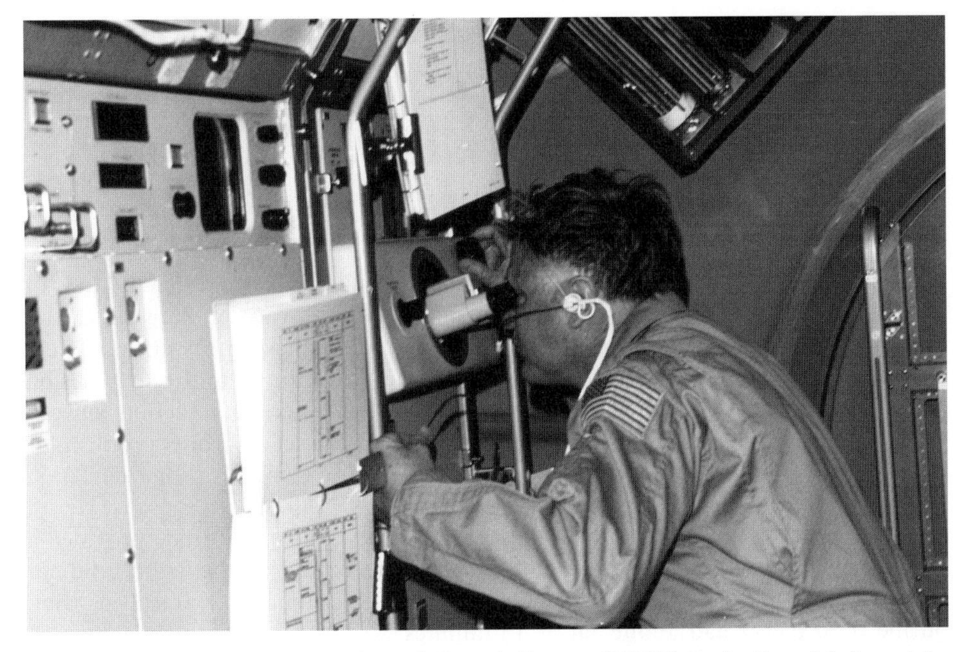

Don Lind observes the Vapour Crystal Growth System (VCGS) in the Spacelab-3 module.

space, where higher quality specimens with better electronic properties could be grown.

Three crystals of triglycine sulphate were also produced in the Fluid Experiment System (FES), located elsewhere in the Spacelab-3 module, yielding the first three-dimensional laser holograms and video recordings of their growth process in space. Visual observations by the science crew provided invaluable real-time descriptions of the crystals, whose potential applications include detectors for astronomical telescopes, Earth observation cameras, military sensors and infrared monitors. Furthermore, they do not require cryogenic conditions under which to operate and could perform well at ambient room temperatures.

On STS-51B, the crystals were grown by slowly extracting heat at a controlled rate through a seed crystal of triglycine sulphate, suspended on an insulated 'sting' in a saturated solution of the same substance. Variations in liquid density, solution concentration and temperature around the steadily growing crystal were carefully monitored. By extracting heat from the crystal in this manner, it was possible to maintain saturation at its 'growth interface', permitting slow but very uniform processing and a higher degree of perfection than possible on the ground.

Images were relayed directly to the ground and the astronauts viewed the developing crystals through an onboard microscope. This allowed them to be tracked through each growth stage and scientists were able to make changes to parameters such as temperature in order to adjust the experiment and reduce defects. An improved version of the FES hardware was flown on IML-1, using polystyrene spheres as markers to aid in characterising residual accelerations in the Spacelab module.

"Lodewijk van den Berg and I ran the crystal-growing experiments, so we would brief each other on what was going on," said Don Lind, who was the Dutch astronaut's counterpart on the gold team. "He'd brief me and then he'd go to sleep and when he woke up, I'd brief him on what I'd done during the last shift. That was pretty well worked out ahead of time."

"I don't think there was competition," said Fred Gregory of the relationship between the silver and gold teams, "because the two shifts did two different kinds of science. Taylor Wang did a lot of drop dynamics. Lodewijk van den Berg did crystal growing. Each shift had its own area of interest and would pick up any unclosed item from the shift preceding them, but would very quickly transition to the activities on orbit. There were really about four hours a day when there was an interaction between the two. During that time, it would just be a kind of status brief on orbiter problems or issues, any review of notes that had come up from Mission Control or some deviation to the anticipated checklist that we had."

For Lind, the first Mormon astronaut, the gravity gradient attitude provided a unique perspective of his home planet. "For the first two days of the flight, I did not take one single minute away from the timeline to just be a tourist," he recalled, "but, on the third day, I had about ten or 15 minutes with no immediate assignment. I floated down to the flight deck. We were flying in an orientation with the tail always pointed toward the Earth and one wing always pointed forward in the velocity vector.

"That oriented the windows on the flight deck from the zenith to the nadir and from horizon to horizon, so it was like a Cinerama presentation. Both my wife and I are amateur oil painters. The sensation in space is that you are always right side up, no matter how you're positioned – 'up' and 'down' are just meaningless in space! Intellectually, you know you're moving very fast, so that orbital velocity will cancel gravity, but the sensation is that you are stationary and the world is rotating majestially below you.

"I thought, 'Could I ever paint that?' The answer is: absolutely not. Grumbacher doesn't make a blue that's deep enough for the great ocean trenches! You look tangentially through the Earth's horizon and you see many different layers of intense blue colours – about 20 different layers of cobalt and cerulean and ultramarine – and then the blackest, blackest space you can imagine.

"When you go over the archipelagoes and the atolls in the Pacific and down in the Bermudas, you see the water coming up from the deep trenches and it appears as hundreds of shades of blue and blue–green up to a little white line, which is the surf and another brown line, which is the beach. Nobody will ever paint that. I looked down and was overwhelmed with the sense of beauty. It was so impressive that it brought tears to my eyes.

"Now, in space, tears don't trickle down your cheeks; that's caused by gravity. In space, they stay in the eye socket and get deeper and deeper and, after a minute or two, I was looking through a half inch of salt water! I had a spiritual feeling, because several scriptures popped into my mind: the 19th Psalm, 'The heavens declare the glory of God'. One of the Mormon scriptures is, 'If you've seen the corner of heaven, you've

Granted considerably more 'space' than previous Challenger crews, the STS-51B astronauts stretch their legs in the Spacelab-3 module. In the light-coloured (or 'gold') shirts are gold team members Bob Overmyer (bottom left), Don Lind (top left), Taylor Wang (top right) and Bill Thornton (bottom right). In the grey (or 'silver') shirts are Fred Gregory (left), Norm Thagard (centre) and Lodewijk van den Berg (right).

seen God moving in his majesty and power'. I thought, 'This must be the way the Lord looks down at the Earth'."

As Lind and the scientists worked aboard Spacelab-3, the 'orbiter' crew kept watch on Challenger's systems. In a gravity gradient attitude, with few thruster firings, this left them with little to do but observe and conduct photography. Fred Gregory found the heavens and Earth fascinating. "You immediately realise you are either a 'dirt person' or a 'space person'," he said. "I ended up being a space person. It was a high-inclination orbit, so we went very low in the southern hemisphere and I saw a lot of star formations that I had only heard about and never seen before.

"I also saw the aurora australis, which is the Southern Lights. If you were a dirt person, you were amazed at how quickly you crossed the ground; how, with great regularity, every 45 minutes, you'd either have daylight or dark – how quickly you crossed the Atlantic Ocean. The sensation that I got initially was that, from space, you can't see discernable borders and you begin to question why people don't like each other, because it looked like just one big neighbourhood down there.

"The longer I was there, the greater my 'a citizen of' changed. The first couple of days, I was a citizen of Washington, DC, but Overmyer was from Cleveland and Don Lind was from Salt Lake City and Norm was from Jacksonville and Lodewijk was the Netherlands and Taylor was Shanghai, so each had their own little location for the first couple of days. After two days, I was from America and after five days the whole world was our home.

"You could see this sense of ownership and awareness. We had noticed with interest the fires in Brazil and South Africa and the pollution that came from eastern Europe, but it was only with interest. Then, after five or six days, it was of concern, because you could see how the particulates from the smokestacks in eastern Europe circled the Earth and how this localised activity had a great effect. When you looked down at South Africa and South America, you became very sensitised to deforestation and how it affected the ecology."

Not only were the astronauts watching countries from space, but India in particular was observing Challenger's seventh mission with interest, for one of the experiments was provided by an astrophysical team from the Tata Institute of Fundamental Research in Bombay (now Mumbai). Led by Sukumar Biswas, it was a study of the ionisation of solar and galactic cosmic ray heavy nuclei and known alternatively as 'Ions' or 'Anuradha'. Like ATMOS, the Indian study was mounted at the rear of the payload bay on the MPESS and examined the composition and intensity of energetic ions from the Sun and galactic sources.

It was a refined version of a similar experiment flown aboard the Skylab space station in 1973 and after Challenger returned to Earth, its data was analysed to identify the cosmic ray ions of carbon, nitrogen, oxygen, neon, calcium and iron and their ionisation states, intensities, energy spectra, arrival times and directions. Despite an initial refusal to respond to commands to rotate its detector stack, a maintenance procedure conducted by the crew enabled it to return to normal operations and it completed two-thirds of its planned observations.

New data on the ionisation states of solar heavy nuclei was of particular interest in developing a clearer understanding of the acceleration and confinement of energetic nuclei in the Sun. The experiment's detector consisted of stacks of thin sheets of special plastics, such as cellulose nitrate and lexan polycarbonate, which were efficient low-noise receptors for heavy nuclei. It was possible to determine the identity and energy of particles from measurements of the geometry of the tracks and the ranges transversed in the stacks.

Aside from his Earth observations, Gregory had little involvement in the Spacelab research. One of his tasks, however, was to monitor the deployment of two small satellites from a pair of Getaway Special (GAS) canisters in the payload bay. Unfortunately, only one of these – the North Utah Satellite (NUSAT) – was actually released into space; the other experienced a battery failure and was rescheduled for another mission in October 1985. Their carriage on STS-51B, however, marked the first occasion that miniature satellites had been deployed from GAS canisters.

NUSAT was an air traffic control calibrator, designed to measure antenna patterns for ground-based L-band radars operated in the United States and member countries of the International Civil Aviation Organisation. It was mounted in its canister by means of a V-band clamp and pedestal. At the instant of deployment, the full-diameter motorised door assembly on top of the GAS canister was sprung open and the satellite was ejected by a compression spring at about 1.1 m/sec.

The concept for NUSAT originated in 1978, in response to a suggestion by the

Federal Aviation Administration in Utah's Salt Lake City to create a means of providing a stimulating educational opportunity for the United States' students and demonstrating a space-based technique for improving the safety of the travelling public. After several years of definition and review, the project finally got underway in 1982. The satellite was built by Morton Thiokol – the Utah-based company also responsible for the Shuttle's Solid Rocket Boosters (SRBs) – and consisted of a 26-sided polyhedron, measuring 48.2 cm in diameter and weighing 520 kg.

NUSAT's communications payload comprised six antennas, a transmitter receiver and telemetry and tracking command equipment. It also housed photodiodes for attitude control, a probe for potential and electron temperatures and strobe lights. During eight months in orbit, a typical 'day' began with a command sent from Weber State College in Ogden, Utah, to 'code' its onboard processor and enable it to discriminate against all illuminating radars except one selected for calibration. A clock was then started to command NUSAT's six L-band receivers to turn on simultaneously as it was about to come over the horizon of the selected radar installation.

The latter transmitted a unique pulse position code during the calibration interval, which would then permit the satellite to distinguish between its signal and others. After passing below the horizon, its receivers were turned off to await its next day of operations. In spite of the successful deployment of NUSAT, the second miniature spacecraft, a Department of Defense payload called the Global Low-Orbiting Message Relay (GLOMR), proved a dismal failure due to battery problems. Nonetheless, it was retained in its GAS canister and was rescheduled to fly aboard another Challenger mission later in 1985.

It has often been remarked by astronauts on dual-shift research flights that the only times the entire crew really got together were shortly after launch and just prior to re-entry. "I think on that particular mission, it may have been anticipated that we would prepare a meal and everyone would eat at the same time," said Gregory. "In reality, that's not what actually happened. I called it 'almost grazing'. You would go down and perhaps get a package of beefsteak and heat it and cut it open and eat it. You may stay on the middeck or you may go back up to the flight deck or you would go back into the laboratory and eat as you were doing your other routine duties. The only time I really had a crew in one place eating would have been on some of my later flights, where I spent two Thanksgivings on orbit and all of us had our Thanksgiving meal together with all the food prepared on the trays."

Many of Spacelab-3's results would require months to fully analyse after STS-51B returned to Earth on May 6th 1985. Their remarkable success would lead to several reflights. However, some scientists have argued that one of the most significant achievements was the mission's contribution to biomedical research, most notably through its studies of the rats' bone and muscle degradation in microgravity.

"It is not surprising that it takes astronauts a few days to recover their pre-flight strength and co-ordination after flight," Kenneth Baldwin of the University of California at Irvine said after a life sciences Spacelab mission in the autumn of 1993, "since their muscles are remodelled by microgravity." Moreover, since muscle protein 'turnover' in rats is much more rapid than in humans, a week or two of microgravity exposure in them was roughly equivalent to two months in us.

Spacelab-3's biomedical research did not solely focus on the rodents and primates, but also on the astronauts – and upon van den Berg and Wang in particular, who served as 'subjects' for the Autonomic Feedback Training (AFT) experiment, closely monitored by Lind and Thornton. This involved employing a number of different techniques to counteract space motion sickness, including the wearing of electronic monitors to record physiological data such as sweat, pulse, heart and respiration rates.

Provided by Patricia Cowings of NASA's Ames Research Center, the experiment provided "encouraging" results. One of the subjects exhibited a low heart rate and little sweating, which was indicative of a lack of stress, although the other did not fare as well, showed less ability to control physiological responses and experienced one episode of space sickness. Nevertheless, the AFT work did offer clear insights into the effects of crew workload and behavioural responses to environmental stress; 'baseline' information which would prove important when planning future long-term space station missions or shorter, high-productivity Shuttle flights.

With the minor exception of a fluctuating water flow sensor, Spacelab-3 was hailed as a tremendous success and – although the pallet-train configuration had yet to undertake its verification flight test – its scientific yield proved more than sufficient to declare the European-built system fully operational. In fact, it has been estimated that some 250 million bits of data were obtained in total from the STS-51B experiments, together with more than three million frames of video footage.

For the pilots, ironically, this proved almost disappointing. "The only flying would be attitude adjustments," remembered Fred Gregory, "and those are generally keypunched in and then executed. In our training, we would simulate failures where you had to do that manoeuvre by hand, and it was quite possible to do it, but not as efficiently as the automatic systems. I don't recall manually flying any of the manoeuvres in orbit and I don't recall Bob Overmyer doing it either. The only time we really put our hand on the stick was in the less-than-the-speed-of-sound descent for landing."

That descent into Edwards Air Force Base in California on May 6th 1985 proved to be among the most dramatic memories of the mission for Gregory, who would later command two Shuttle flights before ultimately serving as NASA Deputy Administrator. "Though it takes eight and a half minutes to get up to orbit," he said, "it takes more than an hour to re-enter and it feels very similar to an airplane ride. You get an excellent view of the Earth. You're going pretty fast, but you are not aware of it, because you're so high. It's an amazing vehicle, because you always know where you are in altitude and distance from your runway. You know you have a certain amount of energy and so you also know what velocity you're supposed to land, and you watch this amazing vehicle calculate and then compensate and adjust as necessary to put you in a good position to land. We normally allow the automatic system to execute all the maneuvres for ascent and for re-entry, but as we slow down for landing, it is customary for the Commander to actually fly it in, using the typical airplane controls."

The deorbit burn, lasting close to four and a half minutes, began at 3:04:48 pm, slowing Challenger sufficiently to drop her out of orbit and set her on course for a touchdown on the west coast of the United States an hour later. After performing a graceful, 193 degree heading alignment circle turn, Overmyer guided the orbiter to a precision landing on Edwards' Runway 17, slowing to a halt in 59 seconds and a rollout of less than 2,700 m. Post-mission inspections of the Shuttle revealed only superficial damage to her thermal protection tiles.

Overmyer and Gregory's apparent ease in setting Challenger down, however, was achieved only following hundreds of practice runs they had undertaken in the Shuttle Training Aircraft (STA) before the mission. "We had participated, in my particular case, in 500 to 700 landing approaches," said Gregory, "and Bob Overmyer, I'm sure, had 400 or 500 more than that! They are flown using the same profile, the same speed, the same sensation of very high sync rates, with a flare about a mile from the end of the runway."

Shortly after draining residual hypergolic propellants from the orbiter, the first time-critical items from Spacelab-3, such as data tapes and film, were removed from the module. About three hours after touchdown, the rats and monkeys were removed – the former to be euthanised – and the remaining samples extracted by mid-afternoon on May 7th. Following transportation back to Florida, Challenger was ensconced in the Operations and Checkout Building to be readied for her next mission in mid-July.

Barely four weeks had elapsed between two Shuttle missions and another was scheduled to be undertaken by Discovery in mid-June. After countless development problems with tiles and main engines, the reusable spacecraft, it seemed, was finally living up to the vision that NASA had promised to Congress in the 1970s: a commercial, reliable, frequently launched 'space truck'. On her next mission, however, Challenger would demonstrate the long-feared fallibility of the main engines and then, as euphoric 1985 wore into tragic 1986, Bob Overmyer's crew would finally come face to face with the disaster that STS-51B very nearly became.

"GOING TO SPAIN"

The flight engineer on a Shuttle crew has arguably one of the most important jobs a Mission Specialist can possibly hold.

Seated behind and between the Commander and Pilot during ascent and re-entry, he or she is responsible for helping to monitor the orbiter's instruments and offering a vital third set of eyeballs in the event of 'off-nominal' events. Typically, during dual-shift Spacelab missions, the flight engineer led one of the two 12-hour teams. He or she "makes sure that all the checklist steps for normal and emergency procedures are performed flawlessly," said astronaut Carl Walz, who served as Columbia's flight engineer on a July 1994 Spacelab mission, "and keeps a log, recording all systems that had problems during launch and landing. Afterwards, he or she figures out how those problems affect future Shuttle procedures. The training is very similar to the training for the Commander and Pilot. In orbit, he or she is the 'traffic cop' who makes up the plans and makes sure they are executed properly."

For Story Musgrave, encumbent of the flight engineer's mantle on Challenger's eighth voyage, STS-51F, his ascent to orbit on July 29th 1985 was arguably the most dramatic of his six-mission career, when the Shuttle suffered a hair-raising main engine shutdown 108 km above Earth. "The ground made the call 'Limits to Inhibit', which is, for us, an extremely serious omen," he recalled years later. "Going to 'Limits to Inhibit' means the ground is seeing problems that are going to shut you down. I'm looking through the procedures book and thinking we're going to land at our transoceanic abort site in Spain. I'm rehearsing all the steps and my hands are moving through the book and I'm thinking 'We're going to Spain. Things are bad!' "

The emergency, which occurred five mintues and 45 seconds after launch, was apparently triggered when temperature readings for the Number One (uppermost) engine's high-pressure turbopump indicated 'above' its maximum redline, resulting in an automatic shutdown by Challenger's General Purpose Computers (GPCs). At this stage of the ascent – three and a half minutes after the Solid Rocket Boosters had been jettisoned and just three minutes ahead of main engine cutoff – the Shuttle was already too high and travelling too fast to return to an emergency landing back at her launch site in Florida.

A shutdown at this point meant one of two things: a Transoceanic Abort Landing (TAL) in Europe or a manoeuvre known as an Abort To Orbit (ATO), whereby Challenger would burn her twin OMS engines to limp into a low, but stable, orbit. Musgrave's initial focus was upon the page of his checklist that dealt with requirements for a touchdown at Zaragoza Air Base – a joint-use military and civilian installation with a nearby, NATO-instrumented bombing range – in the autonomous region of Aragon in north-eastern Spain.

Zaragoza had been designated as a TAL site in 1983 and was pressed into service on STS-51F in view of the mission's flight profile, which would carry the Shuttle to a higher than normal orbital inclination of 49.5 degrees. This placed Zaragoza near Challenger's nominal ascent ground track, thus allowing the most efficient use of available main engine propellant and cross-range steering capability. As Musgrave flipped through the procedures that he would recite to Commander Gordon Fullerton and Pilot Roy Bridges for a Zaragoza landing, fellow Mission Specialist Karl Henize, seated to his right on the flight deck, was becoming nervous.

Henize had good reason for his nervousness. The TAL mode was the second available abort after the Return to Launch Site (RTLS) contingency option and encompassed the six-minute period from shortly after SRB separation until, theoretically, main engine cutoff. It utilised one of three sites – one in France (Istres Air Base) and two in Spain (Zaragoza and Moron Air Force Base) – which, under international agreement, provided emergency support for just-launched Shuttles following orbital insertion inclinations between 28.5 and 57 degrees.

Flight rules dictated that the TAL option would only be selected in the event of a premature main engine shutdown or other major malfunction – for example a significant cabin pressure leak or cooling system failure – which would prevent the vehicle from continuing into space and completing at least one orbit. Had the command been given from Mission Control that day, Fullerton would have placed the abort rotary switch on his instrument panel in the TAL/AOA position and

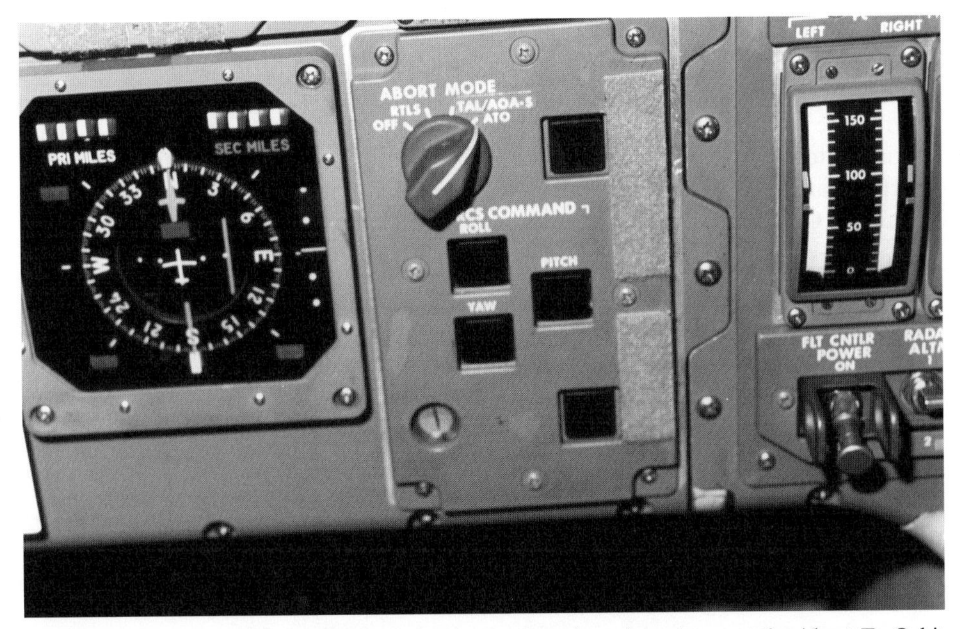

The abort mode switch on Challenger's instrument panel, ominously set at the Abort To Orbit (ATO) option. Commander Gordon Fullerton was obliged to select this abort during the STS-51F ascent, following the shutdown of one of the Shuttle's three main engines.

depressed the abort push button next to the selector switch; Challenger's computers would then have automatically steered the spacecraft toward the plane of the European landing site.

"He didn't know what was going on," Musgrave said of Henize's reaction during those frantic seconds. "He looks over and sees the top of my page: SPAIN. He's looking, poor Karl, and I'm going down the checklist that I'll be reading to the guys in the front seats when we abort. Karl's looking over at me and I sense a really severe stare. Then he dares ask: 'Where we going, Story?' I just looked at the top of the page and said 'Spain, Karl'. Then I quickly retracted it. 'We're close, but not yet'."

Eventually, the call came from Mission Control: "Abort ATO. Abort ATO". Challenger had achieved sufficient velocity and altitude to undertake the next available option: the Abort to Orbit. In fact, she had missed the closure of the TAL 'window' by just 33 seconds! At 9:06:06 pm, some six minutes and six seconds into the climb and hurtling towards space at 15,000 km/h, Gordon Fullerton fired the OMS engines for 106 seconds, consuming 1,875 kg of the orbiter's much-needed propellant, but permitting the Shuttle to continue into a lower than planned orbit. Two minutes later, at 9:08:13 pm, the Number Three (lower right) main engine data indicated excessively high temperatures, bringing it within seconds of being automatically shut down.

Fortunately, this was inhibited to ensure that the STS-51F crew reached an acceptable orbit. The 'inhibit' command effectively instructed the GPCs to ignore the over-temperature signals and prevented them from shutting down the Number Three engine. The two remaining engines, meanwhile, fired for an additional 49 seconds, shutting down nine minutes and 20 seconds after launch. "We never did get the call for the transoceanic emergency landing," said Musgrave, "and we ended up making it to orbit and finishing the mission." According to Flight Director Cleon Lacefield after the launch, Challenger could have achieved orbit without the additional OMS burns, but at the expense of having to jettison her External Tank over north-eastern Africa. "We don't do business that way," he said.

Their eventual orbital path, with an apogee of 230 km and a perigee of 174 km and requiring three OMS firings, was considerably lower than the 390 km originally planned and would inevitably impact several of the astronomical and solar physics instruments that comprised STS-51F's primary payload, the Spacelab-2 observatory. This incident is, to date, the only in-flight shutdown ever experienced in the Shuttle programme and proved to be a bolt out of the blue, for all main engine parameters appeared to be normal during the countdown, ignition sequence and the first few minutes of the flight.

Then, at approximately two minutes into Challenger's ascent, roughly at the same time as the SRBs were jettisoned, data from Channel A – one of two measurements of the Number One engine's high-pressure fuel turbopump discharge temperature – displayed characteristics indicative of the beginning of failure. Its measurement began to drift and, at three minutes and 41 seconds after launch, the Channel B sensor failed. However, its sibling continued to drift, approaching and then exceeding its own redline limit some five minutes and 43 seconds into the flight, which triggered the shutdown.

The high-pressure fuel turbopump discharge temperature data from Channel B of the Number Three engine, meanwhile, began to climb and passed its own redline just over eight minutes after lift-off. Measurements from its Channel A, however, remained within prescribed limits and, read NASA's post-mission report, all other operating parameters relating to the Number Two and Three engines were deemed normal. Post-mission analysis suggested that the problem was not with the Number One engine itself, but with faulty sensors that had incorrectly indicated an overheating situation and issued the shutdown commands.

These faulty sensors, said Bill Taylor, then head of the main engine project at Marshall Space Flight Center, were extremely thin wires, whose electrical resistance changed as they heated up. He added that they had already suffered failures on three earlier missions and that upgraded versions were scheduled to fly aboard Space Shuttle Discovery on STS-51I in August 1985.

Otherwise, the performance of the SRBs in propelling Challenger into orbit for the eighth time was described as "nominal". However, gearbox nitrogen pressures in one of the Shuttle's Auxiliary Power Units (APUs) exceeded their maximum allowable levels and, during a post-launch sweep of the Cape Canaveral beaches, a fragment of spray-on foam insulation, apparently from the External Tank, was discovered. A survey of the orbiter's heat-resistant tiles was performed in space using the Remote Manipulator System (RMS) and found a large number of debris impacts. Further inspection after Challenger's touchdown identified a total of 553 'hits'.

Challenger experiences a dramatic on-the-pad shutdown of her three main engines on July 12th 1985. Seventeen days later, STS-51F would finally fly.

As they settled into orbit and divided themselves into their two 12-hour teams – a 'red' shift led by Bridges, together with Henize and Payload Specialist Loren Acton, a 'blue' shift led by Musgrave, with Mission Specialist Tony England and Payload Specialist John-David Bartoe, and Commander Fullerton working across both – the seven-man crew barely had chance to reflect on what had been not just an eventful day, but a crisis-filled month. Originally scheduled to head for space at 8:30 pm on July 12th, they had been thwarted by the second on-the-pad main engine shutdown, only seconds before lift-off.

"At T−7 seconds," recalled Bridges, who made his first and only spaceflight on STS-51F, "the main engines start with a rumble from far below. As the person in charge of all engines, I watch the chamber pressure indicators come to life and surge towards 100 per cent. I think 'Wait, what's this?' The left engine indicator seems to be lagging behind. Before I can say a word, it falls to zero, followed by the other engines. With less than three seconds before our planned lift-off, we have an abort. The groans from the rest of the crew are now audible. I take a quick look around to see if there's anything else to be done and notice Gordon Fullerton turning to look at me. The thought crosses my mind: 'Gordo probably is thinking I've done something to screw it up'. I show him both hands, palms up, and say 'Gordo, I didn't touch a thing. It was an automatic shutdown'."

The July 12th shutdown, executed because the Number Two main engine's chamber coolant valve was slow in closing from 100 per cent open to the 70 per cent required for startup, necessitated a 17-day wait for a second launch attempt. When one of two command channels failed to execute the closure, fortunately, the backup took over without incident. However, flight rules dictated that both channels had to be fully functional for the countdown and lift-off to proceed.

These timings, however, posed a problem. STS-51F's lift-off time on July 12th lasted barely two hours and was calculated to satisfy lighting conditions needed for particular plasma physics and astronomical instruments aboard Spacelab-2. For five or six days after the pad shutdown, the launch window opened at roughly the same time – 8:30 pm – before becoming unfavourable due to a requirement for orbital dark skies.

Even the July 29th lift-off, originally set for 7:23 pm, was postponed by an hour and a half by an erroneous command to Challenger's backup flight computer.

"WE DON'T HAVE TIME TO TALK ABOUT THIS"

The wait would be worth it and, although the ATO proved a momentary scare, the sensation of launch was later compared by Loren Acton – in terms of its raw power – to the destructive Loma Prieta earthquake of October 1989, which hit the Greater San Francisco Bay area of California and measured 7.1 on the Richter scale.

Acton, a 49-year-old solar physicist from the Space Sciences Laboratory at Lockheed's Palo Alto Research Laboratory in California, was flying with 40-year-old astrophysicist John-David Bartoe of the Naval Research Laboratory in Washington, DC, on a mission whose objectives were primarily devoted to observing the Sun and astronomical targets. In the eyes of the world, though, STS-51F would infamously become known as "the Coke and Pepsi flight". Even today, said Acton, when he visits schools to speak about his mission, children are far more interested in the intricacies of carbonated drinks in space – which turned out to be 'too' carbonated and were not received particularly well by the astronauts – than in the wonders of solar physics ...

"Coca-Cola had gotten permission to do an experiment in space to see if they could dispense carbonated beverages in weightlessness," Acton recalled. "They got approval to build this special can, put significant money into it and were all set to fly it on one of the early Shuttle missions. This was during the 'Cola Wars' when Ronald Reagan was in the White House. Somebody at a high level at Pepsi found out, went to their contacts in the White House and said 'This cannot be allowed to happen – that Coca-Cola would be the first cola in space'. So the Coke can was taken off the mission it was supposed to go on and Pepsi was given time to develop their own can so they could both fly on the same flight. It turned out that our mission ended up getting the privilege of carrying the first soda pop in space."

Although subsequent pictures from the STS-51F mission showed members of Bridges' red shift drinking Pepsi and Musgrave's blue team sipping Coke from the dispenser bottles, the distraction for Acton from what was actually an important

Although both Coke and Pepsi dispensers were ultimately carried on STS-51F, following lengthy 'Cola Wars', the original 'first' sponsor, Coca-Cola, was actually the first to be sampled in space.

mission for astrophysical and solar research proved a extreme irritation. "The morning before the launch, there is always a briefing, during which all the last-minute things that need to be talked about get talked about," Acton said later. "We were about halfway through a briefing on the latest data concerning the Sun, when who should walk in, but the chief counsel of NASA, who began to brief us once again on the Coke and Pepsi protocols. I just blew my stack and said 'We've been getting ready for this mission for seven years. It contains a great deal of science. We have a very short time to talk about the final operational things that we need to know. We don't have time to talk about this stupid carbonated beverage dispenser test. Please leave'. He turned and walked out."

It was true that STS-51F was one of the most important – and difficult – scientific ventures yet flown by the Shuttle. Indeed, years later, it was remarked by the solar physics and astronomical communities that records set during the eight-day mission would probably "stand until the era of the space station, because no payload now under consideration matches the complexity of Spacelab-2, which tested the limits of hardware, software and people everywhere in the system". That success is all the more remarkable in view of the narrowly averted abort and problems experienced with the observatory itself when finally activated in orbit.

As its name implies, Spacelab-2 should have been the second flight of the European-developed laboratory, but was postponed until after Spacelab-3 in view of problems preparing one of its major components – a telescope-aiming device known as the Instrument Pointing System (IPS). It was the second of two so-called Verification Flight Tests of the Spacelab unit and would put the pallet-train-and-igloo hardware through its paces for the first time.

Pallets are U-shaped metal frames, measuring three metres long by four metres wide and covered with aluminium panels onto which large instruments, telescopes or antennas requiring unobstructed fields of view can be attached. On Columbia's STS-2 mission in November 1981, for example, an engineering version of the pallet had been employed to hold a large synthetic aperture radar and several other scientific sensors. Up to five pallets – three of them bolted together in a rigid 'train' – could fit into the Shuttle's payload bay and the versatile platforms continue in service in today's International Space Station era.

The pallet train, which would be used on Spacelab-2 to support the IPS and a number of solar physics and astronomical instruments, was held in place by five attachment fittings – four along the walls of the payload bay and one in the floor – and included aluminium ducts and trays on its port and starboard sides to route cables to and from experiments and subsystems. Thermal control was provided by multi-layered insulation and Spacelab's freon-21 coolant loop, which collected excess heat from the pallet-mounted hardware through a series of 'cold plates' and rejected it into space via the Shuttle's heat exchanger.

Although the pallets had been tested on both STS-2 and STS-3, before being used 'operationally' to support NASA's Office of Space and Terrestrial Applications (OSTA)-3 payload on STS-41G in October 1984, their carriage aboard Spacelab-2 marked the first time that the 'train' and another device, known as the 'igloo', had been utilised for a 'full' scientific mission. The igloo was a 2.1 m tall aluminium alloy

The Spacelab-2 hardware undergoes checkout in the Orbiter Processing Facility in March 1985. Note in particular the egg-shaped Cosmic Ray Nuclei Experiment (CRNE) on the extreme left-hand side of the picture and, just right of centre, the IPS base and the Spacelab igloo canister.

cylinder and was mounted vertically on a crossbeam at the forward end of the pallet train, providing a temperature controlled container to hold subsystems and equipment for the instruments.

Pressurised to 14.7 psi, the 660 kg igloo offered electrical power, cooling and command and data acquisition services for the pallet-mounted experiments; in effect, it supplied many of the services a 'core' Spacelab module would have offered. This is highlighted by plans for an Earth Observation Mission in late 1986, featuring both pallets and a short module. In the wake of STS-51L, the flight was redesigned to achieve many of its original tasks, but – under the new name of 'ATLAS-1' – in a pallet-only mode, with the command and control services of the short module provided instead by an igloo.

On Spacelab-2, the verification hardware comprised a multitude of sensors installed throughout the pallet-train-and-igloo system and Challenger herself, providing data on their combined performance during launch, ascent, orbital flight, re-entry and landing. This equipment verified that the observatory's thermal control system kept temperatures within required limits and prevented condensation or heat leaks. The thermal, acoustic and structural responses of the entire payload during the most critical and dynamic portions of the mission were closely monitored and the astronauts checked their satisfactory operation and ability to communicate and transmit scientific results through the sole Tracking and Data Relay Satellite.

Construction of the Spacelab-2 observatory commenced in 1982, when its three pallets arrived at KSC's Operations and Checkout Building to begin pre-flight testing. During the following year, shortly after the assignment, in February 1983, of Henize and England to the STS-51F mission, additional equipment was attached to accommodate the instruments. In addition to the pallets, a support structure was mounted at the back of the train to hold the University of Chicago's duckegg-shaped Cosmic Ray Nuclei Experiment (CRNE). By the end of 1984, the IPS had arrived in Florida and preparations began to install its solar physics and atmospheric instruments.

Mission sequence testing in the spring of the following year confirmed that each of the components of this complex payload could operate as a single unit and, in May, a 'closed loop' dummy run commanded the entire observatory remotely from the Payload Operations Control Center at JSC in Houston. On June 8th, with Challenger now deconfigured from her STS-51B mission, the Spacelab-2 unit was moved to the Orbiter Processing Facility and installed into her payload bay. Tests of their compatibility, including another POCC run, proved successful, even utilising the Tracking and Data Relay Satellite and high-rate data modes.

Despite its main focus upon solar physics, the mission also encompassed studies of atmospheric and plasma physics, high-energy astrophysics, infrared astronomy, technological research and life sciences. Akin to the verification flight of the Spacelab module, conducted aboard STS-9 in November 1983, the mission was a 'free-for-all', covering virtually all possible areas of scientific inquiry for which the system had been designed. Eleven of Spacelab-2's 13 investigations were developed by United States scientists and two came from the United Kingdom, including an X-ray telescope supplied by this author's alma mater, the University of Birmingham.

"It was about as multi-disciplinary as you could imagine," Loren Acton said later. "One of the things we learned was that we tried to accommodate and carry out a great variety of experiments."

As with previous missions, Spacelab-2 began with an announcement of opportunity issued by NASA and experiments were selected by peer review process on the basis of their scientific merits. When chosen, their principal researchers formed an Investigators Working Group, chaired by Mission Scientist Eugene Urban of the Marshall Space Flight Center. Like Spacelab-3, one of the IWG's tasks was to select and help train the Payload Specialists. Four candidates were picked: Acton and Bartoe would fly, while Dianne Prinz, a Naval Research Laboratory physicist, and solar scientist George Simon of the US Air Force's Geophysics Laboratory would back them up.

SOLAR OBSERVATORY

One of the key verification tests was evaluating the IPS, which had been booked to fly aboard Columbia's STS-61E mission in March 1986 to undertake ultraviolet observations of Halley's Comet. Since the celestial wanderer only frequents the inner Solar System every 75 years, it was essential that the telescope-pointing device could be demonstrated and declared successful. During Challenger's first day in orbit, a set of tests were to be performed on the IPS, with the crew unstowing it from its horizontal position on the pallet train and aiming it at several solar targets to verify pointing capabilities and overall accuracy.

On STS-51F, sadly, its success took some time to achieve.

The IPS, developed by the Marshall Space Flight Center as a means of providing precise alignment of astrophysical instruments, was admittedly the most complicated element of the Spacelab system and would not fully prove its capabilities until two dedicated astrophysics missions in December 1990 and March 1995. Even before its maiden voyage, NASA's Spacelab management was describing it, "in terms of technical complexity, organisational responsibilities, schedule difficulties and cost escalation", as MSFC's most challenging project.

It required drive motor systems to move its instrument payload in three axes, together with mechanisms to secure the gimbals for 'loading' and 'unloading' experiments, an optical sensing system for alignment in relation to the Sun and stars, a device for directional control and stabilisation, support structures, a holding clamp for ascent and re-entry and a means of adequate thermal control. Fortunately, MSFC had built the Apollo Telescope Mount for the Skylab space station in the early 1970s and during the Spacelab definition stage, the Europeans turned to NASA for guidance on developing the intricate pointing system.

ESRO's proposal for the IPS called for a system known as the 'inside-out gimbal', which differed from conventional ring gimbals and had provided gyro-stabilised platforms on several recent rockets, including the Saturn V. The Marshall Space Flight Center, on the other hand, was keen to employ a different design that would satisfy the broader demands of experiment customers. By early 1975, NASA's Spacelab managers expressed concern that "no one IPS design will satisfy all the users' pointing requirements" and "it was very difficult to get designers to agree on a statement of specifications".

Ultimately, when ESA's budget proved restrictive, it was obliged to suggest more conservative specifications, finally abandoning its inside-out gimbal altogether in favour of a cheaper alternative. As cost, schedule and technological problems continued to rise, tensions also grew between NASA and its European partner until, in 1977, ESA suggested removing the IPS from its Spacelab effort in order to find another means of development. When the European members refused to approve additional funding for the pointing system in 1978, prime contractor Dornier Satellitensysteme GmbH (now EADS Astrium GmbH) was forced to make modifications and delay the first flight.

A proposal for a redesigned IPS was submitted to the West German contractor in April 1981 and the first flight unit was delivered to the Kennedy Space Center in Florida in November 1984. Its advertised capabilities included providing precise guidance of instruments weighing up to 7,000 kg and pointing them to within two arc seconds, holding them on target to just 1.2 arc seconds. To achieve this, it comprised a three-axis gimbal onto which instruments were affixed by means of an 'integration ring': one end of the pointing system was mounted onto the Spacelab pallet, the other to the ring.

When operational, the 1,180 kg IPS was capable of manoeuvring telescopes and instruments backwards and forwards, from side to side and could even 'roll' them in a 22-degree arc around its 'straight up' position. Its movements were commanded from the Spacelab subsystem computer and a pair of Data Display Units (DDUs) on the aft flight deck and it could be operated in manual or automatic modes, capable of spending long periods focused on single objects or conducting slow-scan mapping operations. Moreover, its reaction times were much better than those of the Shuttle's attitude control system.

In comparison with the orbiter's pointing precision of one-tenth of a degree at best, the IPS' ability to achieve accuracies of one-thirty-six-hundredth of a degree has been likened to keeping an instrument on the steps of the Capitol Building in Washington, DC, aimed at a coin on the Lincoln Memorial, some 3.5 km away! Even the effects of crew motions, equipment operations or Shuttle thruster firings could be compensated by accelerometers mounted on the IPS and the Sun-watching instruments kept on target.

Solar physics research from the Shuttle was highly desirable for two main reasons: firstly, the spacecraft's orbital path, above the turbulence of the 'sensible' atmosphere, meant it could acquire much better images. Secondly, since orbital 'night-time' lasts only 45 minutes, it proved easier to trace the evolution of phenomena on the Sun's surface and in its atmosphere without long interruptions.

The solar instruments aboard STS-51F focused primarily on three specific portions of the Sun – its chromosphere, 'transitional' region and corona – as part of efforts to better understand their function and the mechanisms responsible for transferring heat from layer to layer. The temperature of the chromosphere, an irregular region above the Sun's visible disk (its photosphere), rises from 6,000 to 20,000 degrees Celsius; in fact, at its high temperatures, hydrogen emits light that produces a reddish colour, evidenced through prominences that project above the Sun's limb during total eclipses.

Sandwiched between the chromosphere and the much hotter corona is the transitional region, in which heat from the latter flows down into the former and produces a dramatic and rapid temperature change from over a million degrees to 'just' 20,000 degrees Celsius. Hydrogen is ionised at such extreme temperatures, making it difficult to see and the light emitted by the transitional region is dominated instead by ions of carbon IV, oxygen IV and silicon IV. Finally, the Sun's outermost 'atmosphere', the corona, produces a glow around the darkened lunar disk during total solar eclipses.

Early coronal observations revealed bright emission lines at wavelengths that failed to correspond with any known materials, prompting some astrophysicists to propose the existence of 'coronium' as the principal gas feeding the outer atmosphere. The mystery endured until it was determined that coronal gases are super-heated to temperatures higher than a million degrees Celsius. At such extremes, both hydrogen and helium are stripped of electrons and even elements like carbon, nitrogen and oxygen are reduced to 'bare' nuclei. In fact, only heavier trace elements like iron and calcium are able to retain some of their electrons in this intense heat.

Spacelab-2's Instrument Pointing System, equipped with four powerful solar physics instruments, comes alive on STS-51F.

On Spacelab-2, four instruments – the Solar Optical Universal Polarimeter (SOUP), the Coronal Helium Abundance Spacelab Experiment (CHASE), the High-Resolution Telescope and Spectrograph (HRTS) and the Solar Ultraviolet Spectral Irradiance Monitor (SUSIM) – were mounted on the IPS. All four, and the pointing system itself, were clamped in a horizontal position in Challenger's payload bay during ascent and re-entry and unlatched in orbit. For safety reasons, there was also provision for the emergency jettisoning of the entire IPS if it was unable to retract properly before the closure of the payload bay doors at mission's end.

Unfortunately, SOUP – provided by Alan Title of the Lockheed Solar Observatory in Palo Alto, California – turned out to be somewhat irksome. This was particularly frustrating for Loren Acton, who, as the instrument's co-investigator, had been specifically picked to fly on Spacelab-2. "About eight hours after its activation," he said later, "it shut itself off and would not accept the turn-on command. The crew did everything it could, but ended up having to forget it."

SOUP was intended to observe solar magnetic field activity in different wavelengths and polarisations of visible light and, when it finally came fully to life later in the mission, it returned dramatic movies of the active Sun that proved far more consistent in quality from frame to frame than previously obtained by high-altitude rocket flights. In particular, it recorded several hours of observations of sunspots and active regions, including 6,400 photographic frames that solar scientists hailed as 'unique' in terms of their extreme stability.

Not until the seventh day of the flight – August 4th – did SOUP awaken. Sadly, Acton had suffered a severe bout of space sickness and the happy news was relayed to him by his blue shift crewmate, John-David Bartoe. "I got sick as a dog," Acton said years later. "Thirty seconds after the main engines shut off, I felt like my stomach and my innards were all moving up against my lungs. I was sick for four days and learned very quickly that you cannot unfold your barf bag as fast as you barf! When Bartoe came to tell me SOUP was alive, I was feeling so bad I didn't even get up to go look."

Post-mission analysis and film processing of the SOUP results determined that bubble-like convective cells, known as 'granules', on the Sun's surface were in almost continuous motion at speeds of up to 4,000 km/h. These granules, which typically measure about 1,000 km in diameter, cover the entire solar disk, except those portions where sunspots are prevalent. SOUP indicated that the granules 'float' like corks atop much larger convective cells – varying between 10,000 and 40,000 km across – in which hot fluid rises from the interior in the bright areas, spreads across the surface, cools and then sinks inward along the dark 'lanes'.

Adjoining SOUP was the British-built CHASE instrument, whose objective was to improve measurements of the solar helium abundance, which, at the time, were uncertain by a factor of three, relative to that of hydrogen. Such measurements were considered important for helping to verify models of the birth of the Universe. Developed jointly by Alan Gabriel of the Rutherford Appleton Laboratory in Chilton and Leonard Culhane at University College London, CHASE recorded ultraviolet emissions from hydrogen and ionised helium, both on the solar disk and in the corona above the Sun's limb.

Accurately accounting for helium in the Universe has proved pivotal to understanding a number of important astrophysical processes. Since all of the helium in the Sun's surface layers is thought to be primordial in origin, data collected by CHASE was of significance to cosmologists, as well as solar scientists. In addition to its primary task, the instrument also examined the properties of the Sun's outer atmosphere, revealing that hot, active-region material typically formed 'bridges' between the corona and the somewhat cooler chromosphere beneath it. Unfortunately, the instrument's capabilities were compounded by Challenger's orbit, since there was sufficient 'free' helium at the lower altitude to partly interfere with its measurements.

Eight hours of video and more than 500 still photographs of the Sun in hydrogen-alpha ultraviolet light were acquired by the HRTS, which had been developed by Guenter Brueckner of the Naval Research Laboratory in Washington, DC. This instrument, which had ridden several high-altitude rocket flights since 1975 before being commissioned for use on Spacelab-2, observed the fine-scale structure of the chromosphere, corona and the 'transition' zone between them, recording over 19,000 exposures of sunspots, spicules (high-speed gas jets shooting up into the corona) and explosive events.

Its main focus was upon how energy could be transported and dissipated to form our parent star's hotter, outermost regions by 'seeing' spectral lines emitted in the chromosphere, transition zone and corona at the highest resolution possible at that time. Eighteen instrument-observing sequences were executed – some many times – and over 23 orbits were devoted to solar studies. Essentially all of the available film was used, exposing roughly 600 full-frame and 18,000 short-frame spectrograph exposures, in addition to the eight hours of video footage.

It "has the ability to zoom in on very small features on the surface of the Sun," said Bartoe in an April 1986 interview. "The primary goal is to try to understand how the Sun makes the solar wind [a stream of charged particles hurtling into space at up to 240,000 km/h]. Some interesting things happen right on the surface of the Sun – for instance, the temperature goes up very dramatically as you go just above the surface. We're trying to look at that region right there, where that sudden transition of temperature takes place. Most of the light emitted there is in the ultraviolet."

Unfortunately, reflected sunlight led to higher than expected temperatures in Challenger's payload bay and pointing problems with the IPS caused complications for HRTS and the other three solar physics instruments. Moreover, it had to be powered down on several occasions to keep the temperatures of its computer and photographic film within limits. By the second orbit of HRTS operations, its spectrograph started to lose sensitivity, which affected its scientific yield to an extent, although excellent results were ultimately achieved. It was one of two instruments provided by the Naval Research Laboratory on Spacelab-2 – the other being the SUSIM ultraviolet spectrometer – and, entirely appropriately, solar scientist Bartoe from that institution flew on the eight-day mission. Watching from the ground was fellow NRL physicist Dianne Prinz, who, at the time of the Challenger disaster, was expected to serve with George Simon and a British Payload Specialist on a reflight of the solar telescopes, known as 'Sunlab', on Columbia's STS-71O voyage in September 1987. Sadly, even after the resumption of flight operations, Prinz' mission never transpired. She died of lymphoma in October 2002.

"The position of Alternate Payload Specialist during the mission is at the top of the pyramid over all the experimenters who are trying to get information up to the crew," Bartoe said in April 1986. "During this mission, Dianne had to listen to five or six or seven telephone conversations at a time, listen to the crew and then try to sort it all out. In my opinion, this is the toughest job, much more difficult than flying. We only had seven people and two telephone lines up there!"

Added Prinz in the same interview: "I suppose the crew was just as tired as we were on the ground, but it was really fatiguing trying to keep track of everything at once. Each experimenter was very interested in getting dedicated information about his experiment, but I'd try to prioritise what we could ask of the crew without overtaxing them. I was the interface for all the experiments – not just the solar ones – and I'd be called into back rooms to iron out problems that occurred. There was almost no time to think. All of the NASA crew was of a like mind when it came to the purpose of the mission, which was to get the science done. Everybody knew what everybody else was doing and there were certain fatiguing tasks they'd do in order to get better results. For instance, the crew controlled 'free drift' to maintain pointing stability on the Sun. It was that kind of understanding of what the intent of the mission was that made the crew such an outstanding group. We had a fantastic relationship with each other; very, very close."

The fourth instrument affixed to Spacelab-2's pointing system (SUSIM) was designed to monitor long-term variations in solar ultraviolet radiation which, although only a small percentage of the Sun's total output, is the main energy source for Earth's upper atmosphere. Its advertised capabilities were to measure solar irradiance from 120–400 nanometres with an accuracy of between six and ten per cent. The difficulty was that solar ultraviolet radiation was also responsible for causing the instruments measuring it to become degraded and lose their accuracy over time. Consequently, long-term solar changes could not be effectively distinguished from instrument changes. This could easily lead to misinterpretations of long-term changes in instrument readings as 'real' trends in the solar–terrestrial relationship.

SUSIM, however, could overcome such problems because it was only destined to fly aboard a week-long Spacelab mission, before being returned to Earth for analysis and calibrations to determine its degree of degradation. Determining its 'absolute sensitivity' was done at the National Bureau of Standards' Synchrotron Ultraviolet Radiation Facility (SURF) in Gaithersburg, Maryland, both prior to delivery of the instrument for integration and testing and again following its return from Spacelab-2. These calibrations, which began in January 1984 and ended in January 1986, were performed by measuring the spectrometer's response to an absolute source of radiation produced by SURF.

The results from SUSIM indicated an approximately 38 per cent loss in sensitivity for the solar spectrometer and a 20 per cent degradation in its calibration spectrometer; both also had a declining loss between seven and ten per cent at longer wavelengths of 400 nanometres. Although this magnitude is quite low, considering the 24-month interval between calibrations, it proved significant when compared with the anticipated small-scale changes in solar ultraviolet output.

More alarming have been the results of SUSIM and more recent Shuttle Solar

Backscatter Ultraviolet (SSBUV) observations, which hint at ongoing depletion of stratospheric ozone levels since the mid-1980s, due partly to solar effects and atmospheric dynamics, but chiefly the release of man-made carbon fluorides. It has been commented that this could prove catastrophic for the continuation of life on Earth, because small atmospheric effects observed over a four-year period, extrapolated over half a century, could lead to a 60 per cent ozone depletion.

"The grant proposal" for SUSIM, said Prinz in an August 1998 interview, "cites the need for longer term measurement possible only from a satellite platform – a purely scientific need – but also mentions the ozone problem: a phenomenon with social repercussions. Global temperature variation is not mentioned in the grant proposal, but is discussed in the SUSIM brochure, published circa 1990. Of course, back in 1978, ozone depletion was a current concern, while global warming became so only later."

All four IPS-mounted instruments proved highly successful, but the prospects seemed initially bad on the evening of July 29th 1985, when STS-51F Flight Director Lee Briscoe told journalists that the pointing system could not track its solar targets smoothly with a built-in Optical Sensor Package (OSP). Instead, it tended to 'wander', or move erratically. "It's like a drunk that can hold it between the ditches," remarked one observer, "but can't stay between the white lines." Karl Henize, a member of Challenger's red shift, was able to fine-tune the tracking system – and, for a time, CHASE and HRTS were used as pointing sensors – but he complained that it did not stay 'locked' on target during manoeuvres.

Indeed, the 'Houston Chronicle' newspaper reported on July 30th that, rather than tracking the equivalent of "a dime at a range of two miles", the IPS "would do good to hit the broad side of a barn". Still, Spacelab-2 Mission Manager Roy Lester told journalists of his confidence that "it will work very well before the mission is over".

In spite of these problems, and the Shuttle's lower-than-planned orbit which affected all of the instruments, it is quite astonishing that Spacelab-2 returned to Earth as a tremendous scientific success. During troubleshooting of the IPS, a number of software changes were uplinked to adjust its optical sensor package and most of the originally scheduled observations were completed.

Unlike the Spacelab-3 mission, the seven men did not have the luxury of a pressurised module and had to work from the relatively small area at the rear of Challenger's flight deck. "I held a joystick control in my hand, like on a video game," recalled Bartoe, "that permitted us to move the solar pointing telescopes around to point at particular features on the Sun. We would have a conference call – a solar conference – just before sunrise on each orbit. This gave us a chance to talk to the investigators so we'd know what we were trying to do on that orbit. We also received about 20 feet of typed messages over the teleprinter, every orbit. This was the first mission where we had to replace the teleprinter paper, because we had to totally replan the mission. One of the advantages of the fact that this mission took years from formation to launch was that there was a long time to fine-tune each instrument's observing plan. That was good, because a lot of things went wrong, but we really understood how the various instruments' observations fit together and it was easy to make quick changes."

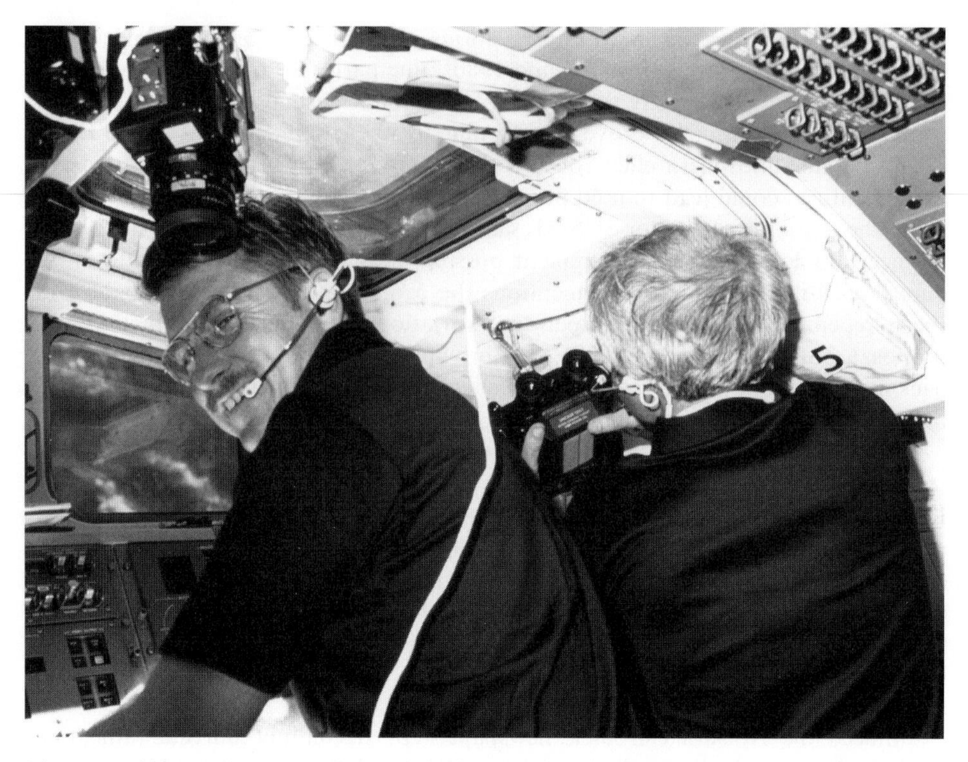

Blue team shift members Tony England (left) and John-David Bartoe at work in Challenger's cramped aft flight deck area.

Despite their lack of a Spacelab module, the dual-shift system made Challenger's flight deck considerably more roomy, with no more than four people on duty at any one time. "During your 12 hours 'on', you ran all these instruments," remembered Gordon Fullerton. "During the 12 hours 'off', you had dinner, slept, had breakfast and then went to work. Two weeks before launch, we set that up. I anchored my schedule to overlap transitions, so if something came up on one shift, I could learn about it and carry it over to the next shift. I also had to stagger things so I got on the right shift for re-entry, so I was in some kind of reasonable shape at the end of the misson. We had the red team sleeping up till launch time, so that once we got on orbit, they were the first one up and they'd go for it for 12 hours. The last week [before launch], we didn't see the other team or I only saw part of one and part of the other myself."

Working alongside Bartoe and, sometimes, Fullerton, on the blue shift was an astronaut who had been at NASA since 1967 and yet was making his first space voyage. Tony England, a 43-year-old geophysicist, had actually left the agency shortly after supporting the Apollo 16 lunar landing mission in early 1972 to join the US Geological Survey in Denver, Colorado. He returned to the astronaut corps in June 1979, saying that he was "looking forward to getting back into training and making a Shuttle flight".

Karl Henize, too, was a 1967 selectee and his wait, like England's, closely rivalled that of Bruce McCandless and Don Lind. At 58 years old, Henize also scored a personal triumph by becoming the then-oldest man in space, beating previous record-holder Bill Thornton. A little over eight years after returning to Earth, in October 1993, he died of respiratory failure whilst climbing Mount Everest; in accordance with his wishes, his remains are buried there. A former astronomer at the Smithsonian Astrophysical Observatory in Cambridge, Massachusetts, one of Henize's tasks on STS-51F, appropriately, was to operate an SAO-built instrument.

Known as the Small Helium Cooled Infrared Telescope (IRT), it was developed under the direction of Principal Investigator Giovanni Fazio and examined infrared radiation from a number of celestial sources. In doing so, it worked in conjunction with the Infrared Astronomy Satellite (IRAS), launched in 1983, which had successfully mapped most of the Milky Way galaxy with a cryogenic detector.

During Spacelab-2, the telescope produced mixed results, meeting more of its technical objectives than its scientific ones. In particular, the performance of its superfluid helium/porous plug cooling system exceeded expectations and demonstrated convincingly that extremely low operating temperatures – down to 3.1 Kelvin – could be established and maintained. Unfortunately, saturation of its mid-wavelength detectors by an intense infrared background compromised some astronomical results. Nonetheless, about half of the galactic plane was satisfactorily mapped at shorter wavelengths than were possible using IRAS.

Additionally, data from the 15.2 cm diameter telescope yielded useful information about the infrared 'background' of Challenger herself and helped to determine the extent to which STS-51F's experimental plasma physics studies and the 'Shuttle glow' phenomenon affected its sensitivity. "We see the IRT back there doing its hickory dickory dock," Loren Acton told journalists of the telescope's sweeping motion in the payload bay. "It makes you think of one of those birds that you dip in a glass of water and it dips, dips, dips."

PARTICLE 'SNIFFER'

The IRT was attached to the third, rearmost pallet in the payload bay, together with two other investigations: a 158 kg Plasma Diagnostics Package (PDP), built by the University of Iowa, and the Superfluid Helium Experiment (SFHE). The former previously flew aboard Columbia's STS-3 mission in March 1982 and consisted of a small cylindrical canister of electromagnetic and particle sensors to 'sniff out' the environment surrounding the orbiter. Its data on the atomic cleanliness of the payload bay proved invaluable in allowing NASA to commit sensitive instruments to Spacelab-2.

During its first venture into space, PDP was hoisted aloft by the RMS mechanical arm to analyse electromagnetic and particle conditions within about 14 m of Columbia. Its data provided, for the first time, detailed insights into the strange ionospheric plasma 'wake' generated as the Shuttle passed, boat-like, through the

The drum-like Plasma Diagnostics Package (PDP) in the grip of Challenger's mechanical arm during a series of experiments on August 1st 1985.

electromagnetic environment of low-Earth orbit. This wake might, it was theorised, complicate the measurements of future detectors and the STS-3 results proved pivotal in planning and developing the Spacelab-2 payload.

The 'ionosphere', whose gases are partly ionised, extends from 60 to 1,000 km above Earth's surface and has shown itself to be an excellent location from which to study ionised gases, otherwise known as 'plasmas'. It had been recognised since the early days of human spaceflight that low-orbiting vehicles are immersed in ionospheric plasma, enabling Shuttle-era scientists to deploy and subsequently retrieve small satellites, directly expose sensors and disturb it with beams of energetic particles to trace, modify or stimulate the environment.

For this reason, the PDP was also used on both flights in conjunction with another experiment called the Vehicle Charging and Potential (VCAP). This had been provided by Utah State University to examine the orbiter's electrical characteristics and its effects on surrounding ionospheric plasmas. It included a fast-pulse 'gun', which fired 100-volt bursts of electrons for durations ranging from 500 nanoseconds to several minutes. It investigated the extent to which electrical charges accumulated on the Shuttle's insulated surfaces and how 'return currents' could be established through a limited area of surface-conducting materials to neutralise active electron emissions.

The VCAP data from STS-3 provided practical experience of using particle accelerators on the Spacelab-1 mission in November 1983. Plans were also afoot, in collaboration with the Italian Space Agency, to build a revolutionary 'tethered satellite', which would be trawled through upper atmospheric plasma on a 20 km conducting cable. The first tethered satellite mission took place in July 1992, several years later than planned, and it flew again aboard Columbia in February 1996. Such tethers, researchers argued, could offer a steady supply of electrical power for future spacecraft or space stations.

For four days of STS-51F, beginning on July 31st 1985, the PDP was employed on the end of the Shuttle's robotic arm, before being released into space to acquire wideband spectrograms of plasma waves at frequencies up to 30 kHz and distances up to 400 m. Due to the reduced level of OMS and RCS propellants, thanks to the Abort To Orbit, original plans to conduct a flyaround survey of the package could not be realised in full. Nonetheless, valuable data was gathered. Two types of interference patterns were subsequently identified in its wideband data: one associated with the ejection of electron beams from VCAP, the other with lower hybrid waves generated by interactions between the neutral gas cloud around Challenger and ambient ionospheric plasmas.

As Fullerton, Bridges and Musgrave executed 'flyaround' manoeuvres of the PDP over a six-hour period on August 1st, a momentum wheel spun the satellite to fix it in a stable enough position for accurate measurements. Among the notable findings from its plasma wave instrument was a region of intense broadband turbulence around the Shuttle at frequencies from a few hertz to ten kilohertz. The highest intensities occurred in the region 'downstream' of Challenger and along magnetic field lines passing close to the orbiter, tending to increase during periods of high thruster activity.

In general, the joint PDP–VCAP observations showed that thermal ion distributions around the Shuttle were considerably more complex than predicted and, frequently, an unexpectedly intense background level of ion current due to incoming hot ions was measured. Surprisingly, these ions often tended to 'change energies'; indicative of high temperatures and turbulent plasma activity and demonstrative of the huge impact of a large, gas-emitting spacecraft on the ionosphere. Indeed, water vapour was detected in the orbiter's immediate vicinity, extending out to a couple of hundred metres, and proved particularly dominant in its wake.

Whilst PDP was extended ten metres into space on the end of the robot arm, Challenger performed a roll manoeuvre to sweep the satellite through this wake. Measurements showed that ions from the 'ambient' ionosphere were accelerated into the wake from 'above' and 'below' the spacecraft; triggered, perhaps, by a strong electric field created by density differences between the two. Ground-based and PDP observations were made of Shuttle thruster firings, yielding faint red airglow emissions which produced a cloud 300 km in diameter. Further tests indicated that even minute thruster firings affected the ambient plasmas in some way.

One particularly interesting experiment, conducted on four occasions as the crew flew above the University of Tasmania's low-frequency radio observatories in Hobart, sought to test the concept of carrying out astronomical measurements through artificial 'windows' temporarily created in the ionosphere by bursts from Challenger's thrusters. Unfortunately, radio waves in the band lower than three megahertz were blocked by the ionosphere, although some cosmic signals were received. In fact, said Eugene Urban after one 30-second firing over Milstone Hill in Massachusetts, the 'hole' produced was "very large and bright" for radio observations. Significantly for future astronomical and plasma physics research, it was determined by the PDP that contaminants released by thruster firings can interfere with measurements of 'natural' plasmas made from instruments in the payload bay.

Alongside PDP on the third pallet of Spacelab-2 was the superfluid helium experiment, designed to explore the properties of this unusual substance and demonstrate the performance of a reusable, space-compatible cryostat. Superfluid helium, in which helium is cooled almost to absolute zero, was tested for its efficiency as a cryogenic coolant in future astronomical or solar physics instruments. Of particular interest to the experiment's investigators were examinations of the behaviour of capillary waves and studies of its sloshing motions and temperature variations.

Superfluid helium moves freely through pores so small that they block normal liquids and conduct heat a thousand times more efficiently than copper. Prior to Spacelab-2, many subtleties of the substance were unknown because gravitational effects disturbed the superfluid state in terrestrial experiments. Early results from the mission provided promising indications that it could indeed be managed efficiently in space using a porous plug cryostat.

Sandwiched between the IPS and its battery of solar and atmospheric physics instruments on Pallet One and the PDP, IRT and superfluid helium experiments on Pallet Three was an X-ray telescope provided by the University of Birmingham in England. Developed by the university's Peter Willmore and mounted on Pallet Two, it

The STS-51F crew poses for their official portrait in front of the Instrument Pointing System. Seated are Gordon Fullerton (left) and Roy Bridges. Standing from left to right are Tony England, Karl Henize, Story Musgrave, Loren Acton and John-David Bartoe.

also proved one of few instruments aboard STS-51F to have been unaffected by the low orbit enforced by the ATO abort.

It employed a coded mask technique to make X-ray images at energies between 2.5 and 25 keV and comprised two co-aligned telescopes. The two masks contained different sized holes, producing different angular resolutions and allowing the higher resolution telescope to make detailed studies of brighter celestial sources, while its lower resolution counterpart examined fainter regions of diffuse emission. By the end of the mission, more than 75 hours' worth of data were obtained, including observations of eight galactic clusters and the Vela supernova remnant.

"It was a great mission," said Gordon Fullerton, who had accompanied the PDP–VCAP investigation on both STS-3 and Spacelab-2. "Some of the missions were just going up and punching out a satellite and then they had three days with nothing to do and came back. We had a payload bay absolutely stuffed with telescopes and instruments." It proved quite remarkable, journalists reported after the flight, that STS-51F had turned into such a grand scientific success after suffering an on-the-pad main engine shutdown, a hairy abort during ascent and the multitude of IPS problems. "We even made up for the fuel we'd had to dump on the way up because of the engine failure and eked out an extra day on it," added Fullerton. "We were scheduled for seven and made it eight!"

In addition to the solar and astronomical telescopes and plasma physics detectors, possibly the most unusual instrument in Challenger's payload bay was the Cosmic

Ray Nuclei Experiment, a giant, duckegg-shaped contraption to count and analyse cosmic rays as much as a hundred times more energetic than any previously studied. Nicknamed "the cosmic egg", it was developed by Peter Meyer and Dietrich Muller of the University of Chicago and mounted on a special support structure at the rear end of the bay, behind Pallet Three. Early results showed that the experiment recorded about 24 million particle events, of which perhaps 30,000 had energies in the formerly unexplored range from hundreds of billions to trillions of electron volts.

MULTI-DISCIPLINARY MISSION

Although Spacelab-2 was a multi-disciplinary flight to continue verifications of the new laboratory configuration, the presence of two life science experiments in Challenger's middeck seemed at complete odds with the astronomical, cosmic ray, plasma physics and solar research conducted elsewhere on this mission.

One investigation, provided by Heinrich Schnoes of the University of Wisconsin at Madison, measured vitamin D metabolite levels of the crew members to gather additional data on the causes of bone demineralisation – loss of density – and mineral imbalances during prolonged periods of microgravity exposure. Astronauts had typically returned from previous missions with evidence of loss of lower body mass, especially in the calves, together with decreases in muscle strength and negative calcium balances. This process has been compared with the initial phases of some bone diseases or the wasting away of muscle observed in bedrest patients.

To undertake the research, three vitamin D metabolites were measured in blood samples taken from four STS-51F crew members before, during and after the mission. Although the levels of two metabolites remained essentially unchanged, a third underwent an interesting pattern: showing a rise in the level of blood samples collected early in the flight, dropping around August 3rd and returning to normal after landing.

The second life sciences experiment, labelled 'Gravity-Induced Lignification in Higher Plants', studied the effects of microgravity on the growth and lignification in oats, pine seedlings and Chinese mung beans in a pair of Plant Growth Units (PGUs) on the middeck. This research had begun on STS-3 and sought to determine whether 'lignification' was a response to gravity or a genetically determined process with little environmental influence. Lignin is a structured polymer, which gives plants the structural strength to maintain a vertical posture despite the effects of gravity, and thus is highly important for the plant's ability to grow properly.

Earlier experiments aboard Skylab and the Russian Salyut space stations throughout the 1970s had revealed that the strange conditions in low-Earth orbit did indeed cause root and shoot growth to become disorientated, as well as increasing their mortality rates. However, little was known about the physical changes within them. Understanding how plants behave and grow in the absence of gravity was – and, with President George W. Bush's 2004 vision for trips to the Moon and Mars, still is – essential for long duration missions, in which astronauts will need to grow their own foodstuffs.

Chinese mung bean, oat and slash pine seedlings were chosen for both STS-3 and

Spacelab-2 because all three could grow in closed chambers and under relatively low-lighting conditions. Additionally, pine is a 'gymnosperm', which means it is capable of synthesising large amounts of lignin, and it was believed that its growth was directly affected by gravity. Unlike the mung bean and oat seedlings, which were germinated only hours before Challenger's launch, the pine samples were germinated four to ten days earlier.

Preliminary results indicated that the mung beans and oat seeds behaved normally and the pine seedlings grew well in space. Some reduced growth, in the order of 15–20 per cent, in the mung beans was observed and both they and the oat roots grew 'above' the supporting medium of the PGUs, indicative of disorientation. Lignin content was significantly reduced in all three species, compared with ground-based controls, providing direct evidence of the important role of gravity in lignification.

In terms of its solar and plasma physics research, Spacelab-2 proved an extraordinary success, snatching triumph from the jaws of what could have proven an abortive mission. Furthermore, the decision to extend the flight from seven to eight days increased its scientific yield substantially. Finally, at 6:43 pm on August 6th, Fullerton and Bridges fired Challenger's OMS engines for 172 seconds to begin the hour-long glide to Edwards Air Force Base in California. The re-entry profile featured a number of test inputs on the control stick, including a manual manoeuvre of the aft body flap at Mach 18.

Touchdown on Runway 23 at 7:45:26 pm was picture-perfect, with the pilots guiding Challenger to a halt in 55 seconds and 2,500 m. Although he had flown with many of the payloads – the PDP, the VCAP, the plant lignification experiments – on STS-3, Fullerton had a very different role on this flight. "It was different because we had a flight engineer, which we didn't have on STS-3," he said later, "so it was really a three-man launch and re-entry crew, which made a lot of difference in how we could do a better job responding to emergencies and trained that way.

"The pressure is higher when you're the Commander: the pressure of making sure that not only you, but somebody else, doesn't throw the wrong switch! During the re-entry, it's your fault if this doesn't come out right. When you're in the right [Pilot's] seat, it's not all your fault; the Commander bears culpability even if you make a mistake. I'm dwelling on this pressure thing because that really is a strong part of the challenge.

"You're really tired after spaceflight," Fullerton continued, "mostly because you elevate yourself to this high level of mental awareness that you're maintaining. Even when you're trying to sleep, you're worried about this and that. It's not like you're just lollygagging around and having a good time. You're always thinking about what's next and mostly clock-watching. Flying in orbit is watching a clock. Everything's keyed to time and so you're worried about missing something [and] being late. We had 270 manoeuvres or something like that.

"Every sunrise and sunset, we had to go to a different attitude to put the right telescopes at the right stars or Sun. Those are all 'typing' exercises – typing long strings of numbers into the computer and the time to start to manoeuvre so it goes to the right attitude. You mess one number and you're going to go to wrong attitude, then you're going to miss that data. Every 40 minutes, you've got a new one."

COMING OF AGE

Spectators at the Operations and Checkout Building beheld an unusual sight on October 30th 1985, as a crowd of blue-clad pilots, engineers and physicists from three nations headed for Pad 39A. Led by snowy-haired skipper Hank Hartsfield, astronauts Steve Nagel, Bonnie Dunbar, Jim Buchli, Ernst Messerschmid, Reinhard Furrer, Wubbo Ockels and STS-8 veteran Guy Bluford were set to make history as the world's first eight-member Shuttle crew. In view of the restricted volume aboard Challenger, it was fortuitous that her ninth and last successful mission carried a Spacelab module and required a dual-shift system of around-the-clock operations.

"The red team of Jim, Ernst and I had to do a circadian rhythm shift," said Bluford, whose four-flight astronaut career involved three missions before which he had to 'sleep-shift', "so, for us, the launch was coming near the end of our work day. While in [pre-launch] quarantine, one team was up while the other was in bed. A new lighting system had been installed in the crew quarters to facilitate the shift in circadian rhythm. Once we got on orbit, the blue team activated Spacelab, while the red team went to bed. We had four soundproof bunks to sleep in while the blue team was at work. The two shift operations worked very well on orbit, with both teams up at the same time during breakfast and dinner, when we transferred Spacelab

The record-sized crew of Challenger's last successful mission pose for an impromptu portrait in front of the Shuttle simulator. In the front row, from left to right, are Guy Bluford, Wubbo Ockels, Hank Hartsfield and Bonnie Dunbar. Behind them are Ernst Messerschmid, Steve Nagel, Jim Buchli, Ulf Merbold (the Alternate Payload Specialist) and Reinhard Furrer.

operations. The simultaneous transfer of responsibility – both on orbit as well as on the ground – went smoothly as we exchanged information and updated our flight data files. Each of the crew shared a sleep bunk with a member from the opposite team. Only Hank had a bunk to himself, which gave him flexibility to work on either shift."

By this time, the European-built pressurised research facility – which, housed in the orbiter's payload bay, provided a miniature space station for a week or more – had amply demonstrated its capabilities on two occasions. Its next flight, known as 'Spacelab-D1' for 'Deutschland', was almost entirely focused on life and microgravity investigations funded by West Germany. This was, in a way, unsurprising, since the latter had a 54.1 per cent financial stake in the reusable laboratory. Even STS-61A's launch time of 5:00 pm was carefully timed, said aerospace engineer Bluford, "so as to give maximum TV coverage to Germany".

By the time Challenger lifted off, he and biomedical engineer Dunbar had been training for the ambitious mission since February 1984. As Spacelab completed its adolescence and began 'operational' flights, it was becoming common practice to name 'science crews' in advance of the three-member 'orbiter teams', in order to give them additional time to iron out payload-oriented operational and training issues with experiment sponsors and principal investigators. Preparations for Spacelab-D1 were considerably more complex than most previous missions because it involved a great deal of travel between the United States, Holland and West Germany.

"Our primary training was conducted at Porz Wahnheide in Germany, a small, very picturesque town, south of Cologne," Bluford recalled. "This European astronaut office housed the ground training units for several Spacelab experiments. These included the Werkstofflabor, the Prozesskamer, the Biowissenschaften and the Biorack. Our Vestibular Sled training was conducted at Massachusetts Institute of Technology in Cambridge. Bonnie and I trained on Spacelab systems at Marshall Space Flight Center and Shuttle procedures at JSC."

Costing $180 million, and including a $62 million Shuttle launch fee, it took five years to prepare Spacelab-D1 from conception to launch and the mission was managed by the Federal German Aerospace Research Establishment (DFVLR) on behalf of the German Federal Ministry of Research and Technology (BMFT). Joining Bluford and Dunbar on the science crew were a record-breaking three Payload Specialists, all of them physicists – West Germans Messerschmid and Furrer, together with Dutchman Ockels (representing ESA, which had contributed about 40 per cent of Spacelab-D1's experiments) – and the quintet were responsible for running 76 investigations in several major facilities aboard the Spacelab module.

With names such as Werkstofflabor, Prozesskamer, Biowissenschaften and Biorack, the West German-supplied research complement sounded like a fearsome medieval torture chamber, yet encompassed a range of studies of the behaviour and processing of materials and fluids and the functioning of biological organisms in the strange microgravity environment. Werkstofflabor, firstly, was a multi-purpose unit which housed three furnaces, a fluid physics module and a crystal growth device to investigate areas of materials processing, semiconductor growth for electronics applications, fluid boundary surfaces and heat-transfer phenomena.

The Spacelab-D1 payload is installed aboard Challenger in KSC's Operations and Checkout Building.

The Prozesskamer (or 'process chamber') was designed to show and measure flows, heat and mass transport, together with temperature distributions during the melting and solidification processes of various materials. Another important facility was the Materials Science Experiment Double Rack for Experiment Modules and Apparatus (MEDEA), which comprised three separate furnaces: one that conducted long-duration crystallisation studies, another that processed metallic crystals at extremely high temperatures using a 'directional solidification' technique and a high-precision thermostat that examined the behaviour of metals under carefully controlled thermal conditions.

By the second day of the mission, although in general experiment operations were running smoothly, the medical and biological investigations were progressing more satisfactorily than their materials science counterparts. In particular, a problem with MEDEA's pressure sensor had to be corrected by an in-flight maintenance procedure and a lamp on the furnace was also replaced. Unfortunately, the hiccup in data-gathering activities meant that many hours' worth of 'run-time' were lost. Discussions to extend STS-61A from seven to eight days were ultimately turned down because Spacelab's power usage levels could not be reduced enough to provide the required extra day aloft.

Elsewhere, the Biowissenchaften and Biorack facilities focused upon life and biological science applications. Results from the latter, built by the European Space Agency, in particular, offered striking evidence of the influence of gravity on bacteria, unicellular organisms and white blood cells. A total of 14 cellular and developmental

Working busily around their 'half' of the 24-hour clock, red team members Guy Bluford, Reinhard Furrer (back to the camera) and Ernst Messerschmid tend to various research facilities in the Spacelab-D1 module.

biology investigations were carried on Spacelab-D1, marking the first occasion on which specimens were 'fixed' and thus preserved in orbit for post-mission analysis.

Two of these experiments confirmed observations made on several previous Shuttle flights: bacteria tend to reproduce more rapidly in space than on Earth, suggesting that astronauts could be exposed to higher risk of infection. Of particular note was an investigation featuring the common pathogenic organism *E. coli*, which has demonstrated an increased resistance to antibiotics in orbit. On the other hand, some of Spacelab-D1's bacteriological research indicated that some cells actually exchanged genetic material through physical 'bridges', perhaps leading to novel techniques for introducing human genes into bacteria to synthesise useful products.

Vestibular experiments involving both humans and tadpoles were also conducted and, in the latter case, revealed pronounced alteration in swimming behaviour upon return to Earth. The tadpoles swam in small circles around fixed centres until their behaviour returned to normal a couple of days after Challenger's landing. Later examinations of the morphology of their vestibular gravity receptors revealed no structural deformities, indicating that they developed normally in space, and corresponded well with earlier studies of amphibians and rodents. Running on rails along the centre aisle of the Spacelab-D1 module was ESA's Vestibular Sled, designed to explore the functional organisation of the astronauts' vestibular and orientation systems and adaptation processes. The accelerations provided by the sled, which could accelerate its subjects at up to 0.2 g along the length of the module, was combined with thermal stimulation of their inner ears and optokinetic stimulation of their eyes.

Not only was STS-61A's large crew unusual, but so too was its distinction of being the first Shuttle flight to be run from outside the United States. Although Mission Control at Houston was in overall command, the German Space Operations Centre (GSOC) at Oberpfaffenhofen, just outside Munich, managed daily research activities. This proved to work exceptionally well, although Oberpfaffenhofen's limited data-transmission capabilities meant that several functions had to be monitored from JSC. When Spacelab-D2 lifted off in April 1993, that situation had been remedied and satellite-transmitted data was received by German ground stations and forwarded directly to the control centre. Moreover, due to the presence of only one Tracking and Data Relay Satellite, Spacelab-D1 received only limited communications coverage for approximately 30 per cent of each orbit. By the time the second mission flew, almost eight years later, four TDRS platforms were fully operational.

According to Pilot Steve Nagel, training in Munich did not differ significantly from the United States. "You could say it's more complex and there are more issues to be resolved when you're working an international programme," he said. "Not having a US mission manager made it more complex, but I see that mission was an early lead-in to the space station. It was hard for Hank to pull together and complicated when you're dealing overseas. We got along fine with the Germans, but we butted heads about things and the long distance part made it more complex."

Despite the focus of the mission, the common language currency was always English, although on a few occasions German was spoken over the space-to-ground communications link, including one opportunity for Messerschmid and Bluford to speak to the head of Bavaria. "The conversation was conducted in German with Ernst doing all the talking," Bluford remembered years later. "Although the mission's dialogue was conducted primarily in English, infrequently, the Payload Specialists would revert to German during on-orbit discussions."

For Nagel, who would later command Spacelab-D2, the year before the Challenger disaster was a pivotal one, for he flew the Shuttle not once, but twice! Originally assigned in November 1983 as a member of Dan Brandenstein's STS-51A crew, he should have flown the following October on the Microgravity Science Laboratory (MSL) mission. That crew, which included astronauts John Creighton, Shannon Lucid and STS-7 veteran John Fabian, would have also inserted a Canadian communications satellite into geosynchronous transfer orbit.

Unfortunately, Nagel's first space voyage was pushed back from 1984 due to a rejuggling of flights in the wake of the STS-41D main engine shutdown. Brandenstein's team found themselves with a new launch target of March 1985 and a new mission number of 'STS-51D'. Then, when Bo Bobko's STS-51E flight was cancelled with barely a week's notice, his crew was given Brandenstein's slot. That left the hapless STS-51D astronauts, now redesignated as 'STS-51G', to fly in mid-June.

Matters had been complicated by the fact that Nagel had also been assigned, along with Bluford and Dunbar, to the Spacelab-D1 flight in February 1984! Peculiarly, as Pilot, he would form part of the 'orbiter' crew, but was actually named in advance of Hartsfield and Buchli, an anomaly that has never been satisfactorily explained. When the rest of the Spacelab-D1 team was named in August 1984, with a launch scheduled for October of the following year, it should have given Nagel a comfortable period of just under a year between his first and second missions.

It did not quite happen that way.

"One kept slipping," Nagel said, "and the other one didn't. When we lost [STS-51D], my two flights were four months apart! Before I thought about that, Dan said 'You're in trouble here'. He went over and talked to George Abbey and negotiated for me to stay on both flights; that I could train for both for a while, then stop training for the second one and finish the first one. I don't think they'd ever do that today, so I owe Dan for the fact that I was able to hang onto both of those."

SUSPENDED IN A GONDOLA

As a result, by the time Challenger ascended from Pad 39A on October 30th 1985, a mere 136 days separated Nagel's two launches. This record would not be broken by another Shuttle crew until, in April 1997, a Columbia team were obliged to cut their Spacelab mission short after only a few days and their mission was reflown in July. Upon achieving a 320 km, 57-degree-inclination orbit, however, Nagel had little time to reflect upon his good fortune: as leader of STS-61A's blue team, he was in charge of configuring the spacecraft for seven days of operations, while Dunbar and Furrer busied themselves with activating the Spacelab module.

Although not strictly attached to either shift, Hartsfield and Ockels tended to align their work schedules with that of Nagel's blue team. "Wubbo decided to freelance," remembered Hartsfield. "He didn't have a fixed shift. His shift would overlap the other two shifts. It was kind of a weird arrangement. He chose to sleep in the airlock. He had a sleeping bag – a design of his own – and the only trouble was people going back and forth would bump him as they went through there."

In honour of the traditions of his Dutch homeland, Ockels also took a large bag of gouda cheese as part of his personal allowance. "The coolest part of the vehicle," said Hartsfield, "was the tunnel that went from the middeck to the lab. He taped that bag of gouda up in the tunnel. It was so convenient that anybody that went there – on the way back and forth – reached in. About the second or third day, he was upset because two-thirds of his cheese was gone!"

Wubbo Ockels, whose shifts overlapped both the red and blue teams, climbs into his own sleeping bag in Challenger's airlock.

In a manner not dissimilar to Spacelab-3, Challenger was oriented in a gravity gradient, 'free drift' attitude to provide a quiescent microgravity environment for the onboard materials processing and fluid physics investigations. "There's a little bit of atmospheric drag, even at those altitudes, and there's a gravity effect from one end of the Shuttle to the other," explained Nagel, "which will cause it to change attitudes, so you get it in a stable attitude before you turn the jets off. This is interesting, because usually you want the long axis pointed at the Earth, either tail to the Earth or nose to the Earth, and the wing oriented in some way that it'll be fairly stable. And we would get it in this attitude, which was nose at the Earth, and the right wing pretty well forward. You 'slide' along like that and get it all stable and turn off the jets, and it would just stay there. It would slowly wander around a little bit and roll over a long period, like half-hour or so, kind of oscillate. It made for very interesting Earth viewing, because you'd go up in the cockpit and look out the front windows – and the Earth is coming by! It's almost like you're suspended in a gondola. We flew that attitude for eight or ten hours a day, and the other time we were called minus-ZLV, which is 'top to the Earth', with tail forward. I don't think you could tell the difference that the microgravity was significantly better when we had the jets turned off, because on the next mission we didn't do that."

Activation of Spacelab-D1 was complete by a little over five hours into the mission and, towards the end of the first work day of Jim Buchli's red team, 73 of the 76 experiments were running. Another key milestone was deployment, at 5:36 am on October 31st, of the Global Low Orbiting Message Relay (GLOMR) satellite from its Getaway Special (GAS) canister on Challenger's payload bay wall. This small, 62-sided polyhedron weighed just 68 kg and was ejected by means of a standard autonomous controller on the aft flight deck.

Upon receiving the proper command, a full-diameter motorised door assembly on the GAS canister opened and a spring-loaded device pushed the tiny satellite into space at a steady rate of 1.2 m/sec. For GLOMR's manufacturer, Defense Systems Incorporated of McLean in Virginia, the deployment came as a moment of triumph; for a previous attempt during STS-51B in April 1985 had been stalled by a battery problem. It was also Challenger's last successful satellite release.

Perhaps displaying evidence of the monotony of Spacelab flights for pilot astronauts, Nagel's job consisted of periodic purging of fuel cells, dumping waste water, taking photographs and preparing meals for the rest of the crew on his shift. "But the good thing about the mission," he said, "was the high inclination. We flew 57 degrees, which means you cover most of the inhabited part of the world. It was just a bonanza of Earth observations. We shot all of our film."

For Hartsfield, the comparatively relaxed pace for the orbiter crew allowed him to indulge in some light-hearted banter, particularly as Halloween coincided with Challenger's second day in space. "I took the back off one of the ascent checklists," he said, "drew a face on it, cut out eye holes, got some string and made a mask! I took one of the stowage bags and went trick-or-treating in the lab. They don't do Halloween in Germany, so they didn't know what I was up to! I decided not to pull any tricks on them, but I didn't get much in my bag. One of the guys took a picture of me with that mask on, and somehow it got released back in the US. About a month

During their shift on the flight deck, Commander Hank Hartsfield (left) and Pilot Steve Nagel (arm partly visible at right) kept watch on Challenger's systems. She behaved flawlessly.

after the flight, I got a letter from a congressman who had a complaint from one of his constituents about her tax money being spent to buy toys for astronauts! I had to explain that nothing was done and it was made in flight from material we didn't need anymore. It was just fun. I never heard any more, so I think maybe that satisfied her."

In addition to the research facilities in the Spacelab module, two devices were attached to an MPESS carrier at the rear end of Challenger's payload bay. The Navigation Experiments (NAVEX), firstly, comprised a pair of canisters and an antenna to develop and test a precise clock synchronisation and evaluating a new method for precise, one-way distance measurements and position determination. The second payload was the boxy Materials Experiment Assembly (MEA), which had previously flown on STS-7, and investigated atomic diffusion and transport processes in various liquid metals.

The final full flight of Challenger passed remarkably quietly and smoothly. One of the few problems experienced was a persistent cabin leak, which triggered alarms on several occasions. "We discovered, later on, the leak was due to one of the experiments inadvertantly venting into space," said Bluford. "We also had a false fire alarm go off on us during flight." Nonetheless, despite the hectic, around-the-clock pace, some time was granted to all eight spacefarers to gaze down on their home planet; particularly the trio of Payload Specialists, for whom the opportunity to fly in space would come only once.

"We were flying into darkness, passing over Tasmania," Jim Buchli told a Smithsonian interviewer years later, "and heading down toward Antarctica. The southern aurora was just unbelievable! It looked like an octopus sitting over the South Pole, with tentacles of light coming out. The orbiter was flying upside down, with the nose pointing toward the pole, and the tentacles shimmered a fluorescent blue–pink. It was like the whole nose was bathed in aurora. Even though we were much higher, you could still see the glow off the front of the nose. I knew what was coming, because I had seen the same geometry when we passed over the pole the day before. I went down to the middeck and literally grabbed Reinhard Furrer, who was on the other shift ... and stuffed him up there in the nose of the vehicle. We're lying upside down, with all the switches and circuit breakers next to our chests, and we're peeking out the front windows, straining to look to the side of the orbiter. For probably ten minutes, we watched these shimmering bands coming off the South Pole. Finally, Reinhard said 'Jim, that was fantastic! That was the most beautiful thing I've ever seen'. Then he went back downstairs to work."

Less than ten years later, in September 1995, Furrer was killed whilst flying back seat in a vintage, Second World War-era Messerschmit 108 at Berlin's Johannisthal airfield. He was just 54.

QUIET TIME

Watching the aurora and the periods of reflection provided some final quiet time for not only the crew, but also, in the wake of the calamity that would befall Challenger's next mission, for the venerable ship herself. She had completed nine flights into space – three more than her siblings Columbia and Discovery and eight more than her recently added sister Atlantis – and spent 62 days aloft. Forty-six astronauts had gazed Earthward from her windows, one of them (Bob Crippen) on three occasions and five spacefarers (Sally Ride, Norm Thagard, Bill Thornton, Story Musgrave and Guy Bluford) had ridden her twice.

She had ferried 12 satellites into orbit – two of them deployed from GAS canisters, five atop Payload Assist Module (PAM)-D boosters, one attached to an Inertial

Upper Stage (IUS) and the remainder released by her Canadian-built mechanical arm – and carried out the triumphant repair of Solar Max and evaluated the Manned Manoeuvring Unit. Several products of her labours remain in space to this day, including the Earth Radiation Budget Satellite, launched by Bob Crippen's STS-41G crew. She validated the Spacelab pallet-train-and-igloo combination and lofted representatives of six nations – Americans, Australians, Canadians, Dutch, Chinese and West Germans – into the heavens.

The confidence in which the Shuttle was now held, as it prepared to conclude its 22nd flight, is exemplified through Guy Bluford's businesslike description of his return to Earth on November 6th 1985. "We closed up Spacelab and readied the vehicle for re-entry as the blue team was getting up," he said. "I rode upstairs in the cockpit, next to MS2 [Buchli], as we came home. Hank Hartsfield and Steve Nagel flew us home to a safe landing at Edwards Air Force Base in California."

After firing the OMS engines for 171 seconds whilst on the opposite side of the planet, Challenger began her hour-long glide to Earth, touching down at 5:44:51 pm on Runway 17 and slowing to a stop in 3,000 m and 49 seconds. As part of preparatory work to resume routine Shuttle landings in Florida, Hartsfield conducted a computerised steering test of the nose wheel during rollout. Until STS-61A, the left and right wheel brakes were applied to steer the orbiter on the runway, although this had regularly caused excessive brake and tyre wear.

On this mission, Hartsfield had the ability to depress either the left or right rudder pedal, signalling Challenger's computers to direct a hydraulic actuator to turn the nose wheel and steer the spacecraft onto the runway's centreline. As she slowed to around 170 km/h, he deliberately steered the Shuttle off the centreline by just under seven metres, before returning to normal as he braked to a halt. "It went very well," he said later. "I didn't get very far off the centreline." The test was hailed a success and STS-61C in mid-December was provisionally set to resume landings at KSC.

In spite of the steering test, however, Hartsfield was not so keen to resume Floridian landings so soon. "As a test pilot, I would like to see a concrete landing at Edwards [for STS-61C]," he added. "One landing does not prove a system." Summing up, the veteran astronaut was happy with his third mission and remarked that he had the most fun of his entire spacefaring career, spending most of his time taking photographs through the flight deck windows and nursing Challenger through a virtually trouble-free week-long trip.

For Steve Nagel, who went on to fly two more missions, including Spacelab-D2 in April 1993, landing the reusable orbiter was almost identical to the Shuttle Training Aircraft (STA), which he now flies as an instructor pilot. "The Shuttle has a flight control system that is one of the early fly-by-wire systems, which means there is no physical linkage between the stick and the controls," he said. "It's all electric commands through a computer that then tell the flight control surface or the reaction control jets to do whatever it needs to do to control the airplane. It's not that you can't learn to fly it; it just takes a while to get used to it, and the fact it has no engines and it's a poor glider, so it's coming down at a real steep angle. The handling qualities and responses you get out of the Shuttle Training Aircraft are very close to what the real orbiter is. If anything, the real Shuttle is a little nicer; a little bit more responsive."

Challenger returns from space for the ninth and final time.

As Challenger settled onto the runway from her ninth mission, she and her crew had good reason for pride in their achievement. The West German sponsors of her many experiments would later describe the mission as "extremely challenging", but would label its results "outstanding". Spacelab, it seemed, had been cleared for ever more ambitious journeys of scientific discovery. In March 1986, thanks to the verification of the pallet-train-and-igloo combination, Columbia was set to take three ultraviolet telescopes into orbit to conduct long-awaited observations of Halley's Comet.

Challenger, too, would play an important role in the sixth year of Shuttle operations, with her heaviest number of mission bookings to date: five. She would finally deploy the second Tracking and Data Relay Satellite in January, followed by a third in July, and also release the joint US/European Ulysses probe on a journey to observe the Sun's polar regions in May. Later that year, in September, she would complete unfinished business by retrieving the Long Duration Exposure Facility from orbit, deploying an Indian communications satellite on the same flight and, in December, would stage her first top-secret Department of Defense mission.

From a public relations standpoint, too, 1986 would be a banner year for Challenger. High school teacher Christa McAuliffe – the first 'private' citizen passenger to fly aboard the Shuttle – was already listed among the seven-member crew for STS-51L and a journalist was tipped to ride STS-61I in September. McDonnell Douglas engineer Bob Wood, Indian astronaut Nagapathi Bhat and US Air Force Manned Spaceflight Engineer (MSE) Chuck Jones would fill Payload Specialist slots on three other flights. In total, around 28 more spacefarers would reach orbit courtesy of Challenger in 1986, at least one of them (astronaut Mike Smith) flying more than once.

With STS-61A, as on STS-41G more than a year earlier, she had also shown that, despite the increased flight rate, individual astronauts could train and fly multiple missions in rapid succession. Just as Bob Crippen had flown three commands in just 16 months, from June 1983 to October 1984, so Steve Nagel had moved from a Mission Specialist's seat to the Pilot's position in an interval of merely 19 weeks. The sole difficulty, he recalled later, was the activities being conducted on each mission and the hardest transition was the training needed to prepare for Spacelab.

"The MS2 is the flight engineer, so you learn the same Shuttle systems that you'd learn as the Pilot or Commander, so the switch from MS2 to the Pilot's seat wasn't that hard," Nagel said. "It was just that I had all the 'head knowledge' and now had to put it into practice. What I had to learn for the Shuttle systems was almost a one-to-one carryover, except for Spacelab ... [which] was something totally different. I'd had some classes on it and learned about it, then I had to stop and hit it really hard before the second flight."

By December 1985, with the completion of STS-61B by Atlantis, the four-strong Shuttle fleet had flown 23 times and was expected to complete 15 missions during 1986. However, beneath its 'routine' veneer, it was faltering: voraciously consuming man-hours in terms of preparing orbiters for launch and the requirement for more spare parts than were actually available led to an increasingly dangerous practice of 'cannibalism' from one vehicle to equip another. "The joke within the astronaut corps," wrote Mike Mullane in 2006, "was a Shuttle could not be launched until the stacked paper detailing the turnaround work equalled the height of the Shuttle stack: 200 feet!"

After several delays and two brushes with disaster, Columbia lifted off on January 12th 1986 to begin STS-61C. Her landing in Florida was cancelled due to bad weather and she came home two days late into Edwards Air Force Base. This had already pushed Challenger's tenth flight, STS-51L, to the end of January. As pressures on the system mounted, and the launch schedule grew ever more fierce, something, it seemed, was bound to break. On January 27th, fortunately, the only thing that snapped was a drill bit; but on the following day, for Challenger, the Golden Age would be over.

6

“Major malfunction”

ULTIMATE FIELD TRIP

“Now don’t break our airplane,” Judy Resnik joked.

Hank Hartsfield promised not to. It was October 1985 when the two astronauts shared their moment of cameraderie. Three months later, the STS-61A Commander would recall Resnik’s light-hearted advice with sorrow; for it was on her flight, rather than his, that the ‘airplane’ – Space Shuttle Challenger – finally broke. Hartsfield had just returned from the week-long Spacelab-D1 mission, sponsored by West Germany, aboard Challenger and had immediately flown to Europe on his crew’s public relations tour. He was aboard a commercial airliner, returning to the United States, when Resnik and the rest of Challenger’s tenth crew blasted off.

The pair had good reason for their closeness: not only did they share the same career, but they had flown together on the maiden voyage of the Space Shuttle Discovery barely a year earlier. “You sort of become family,” Hartsfield recalled two decades later. “We worked together for 13 months, partied together and you do get close.” Whilst in Europe, he had followed with interest his former crewmate’s seemingly fruitless efforts to return to space in January 1986.

Right from the start, launching Challenger’s tenth mission had proved to be an exercise in frustration.

It was, furthermore, a frustration that NASA could ill afford. The planned six-day flight, designated ‘STS-51L’, would feature the first private citizen to fly aboard the Shuttle – a social studies high school teacher from Concord in New Hampshire, named Christa McAuliffe. Picked from over 11,000 applicants for the Teacher-In-Space initiative, she would teach two lessons from space, providing a much-needed public relations boon for the agency as it sought to demonstrate that its reusable fleet of orbiters were truly the spacegoing equivalents of commercial airliners. Indeed, years later, McAuliffe’s mother, Grace Corrigan, would insist that the general atmosphere in the weeks leading up to Challenger’s fateful launch was that the Shuttle was actually far safer than an airliner, simply due to the higher number of precautions taken by NASA. Even McAuliffe herself had expressed jocular confidence that her only 'fear' was a failure of the orbiter's multi-million-dollar toilet.

When the STS-51L crew arrived at the Kennedy Space Center (KSC) in Florida in the third week of January, their launch was routinely postponed by delays in bringing Challenger's sister ship, Columbia, home from her own flight and weather concerns at a Transoceanic Abort Landing (TAL) site in Senegal. More trouble was afoot. Unacceptable weather in Florida put paid to a second attempt and, when Commander Dick Scobee and his six crewmates settled into their seats aboard Challenger on January 27th, they were again thwarted by high winds and a frozen handle on the hatch. That night, temperatures at the launch site plummeted to an unseasonal (for Florida) minus 13 degrees Celsius, forcing technicians to switch on safety showers and fire hoses at Pad 39B to prevent water pipes from freezing. This proved particularly worrisome for the ice inspection team, who began their final 'sweep-down' of the pad area in the early hours of January 28th, and they were obliged to knock huge, 30 cm long icicles away with broom handles as the countdown clock continued ticking towards launch.

As the Sun rose, temperatures climbed slightly to a few degrees above zero Celsius, producing the coldest conditions under which a Shuttle launch had ever been attempted, a fact that would be investigated in depth during the subsequent presidential inquiry into the cause of the tragic events later that day. The copious amounts of ice on Pad 39B then forced an additional two-hour delay to permit the Sun to thaw it. Nonetheless, many of the astronauts' families, including Scobee's wife, June, doubted that NASA would conceivably fly under such conditions. She was partially appeased by her husband's insistence, over the phone that morning, that he felt it was safe to do so. Hank Hartsfield, good-naturedly, had called the crew on a regular basis to jokingly ask what the hell they were up to. Would they ever launch, he wondered?

Tragically, as we now know, Resnik and her colleagues – Scobee, McAuliffe, Pilot Mike Smith, Mission Specialists Ellison Onizuka and Ron McNair and Payload Specialist Greg Jarvis – would indeed launch that frigid Tuesday, with catastrophic consequences. Two decades later, the world is familiar with the technical and human causes of Challenger's loss, but the disaster also put paid to plans for two important satellite deployments, a range of scientific and engineering experiments and a comprehensive survey of Halley's Comet.

This fabled celestial wanderer, which frequents the inner Solar System only once every 75 years, was to be the focus of not only STS-51L, but also two other missions in the spring of 1986. Two weeks before Challenger lifted off, Columbia's STS-61C crew had been prevented from making significant observations, due to problems with their Comet Halley Active Monitoring Program (CHAMP) cameras. In March, another Columbia team on STS-61E was to have employed a battery of ultraviolet telescopes and a wide field camera to analyse the comet. Meanwhile, STS-51L would have utilised CHAMP and deployed a free-flying satellite to explore Halley's tail and the gaseous 'coma' around its peanut-shaped head.

Onizuka, a Hawaiian-born astronaut of Japanese and American parentage, was

Icicles on Pad 39B's launch tower on the morning of January 28th 1986. Note the 'patchwork' of black thermal protection tiles on Challenger's belly and the connecting propellant lines from the External Tank to the orbiter's aft compartment.

The STS-51L crew in the 'white room' on Pad 39B, during their Terminal Countdown Demonstration Test on January 8th 1986. From left to right are Christa McAuliffe, Greg Jarvis, Judy Resnik, Dick Scobee, Ron McNair, Mike Smith and Ellison Onizuka.

responsible for the CHAMP hardware and was to have buried himself on January 29th under a black shroud on the Shuttle's aft flight deck to ensure maximum darkness for his observations. "I will have about two minutes on four different orbits to photograph Halley's Comet in both the visible and ultraviolet spectrum," he told an interviewer. "The objective is to try to get this data as the comet approaches perihelion, which is just as it goes around behind the Sun and starts to head back out. It's a regime where we do not have any data at the present time, so I've been told we'll probably be the only human beings to see it at that time."

As Onizuka worked, his crewmates would have been involved in preparing the Shuttle Pointed Autonomous Research Tool for Astronomy – a mouthful that NASA's finest acronym-makers had somehow carved into the name 'Spartan' – for deployment on January 30th to commence its own series of Halley observations. Built by the agency's Goddard Space Flight Center (GSFC) of Greenbelt, Maryland, the small, boxy satellite had previously flown on Dan Brandenstein's STS-51G mission in June 1985 and was designed to be serviceable and capable of returning to orbit every six to nine months.

On STS-51L, it would have used a pair of ultraviolet spectrometers and two modified Nikon F-3 cameras to study the composition of the comet's dirty-snowball-like nucleus and million-kilometre-long shimmering tail. McNair, with Resnik, was responsible for deploying and later retrieving Spartan with the Canadian-built Remote Manipulator System (RMS). Interestingly, the spectrometers, produced

The official mission emblem for the Spartan–Halley project, showing the small, boxy satellite and highlighting its quest to better understand the celestial wanderer.

jointly by GSFC and the University of Colorado's Laboratory for Atmospheric and Space Physics, were derived from backups for an instrument aboard the Mariner-9 spacecraft, which had begun investigating the Mars atmosphere in 1971.

"Comets happen to be one of the remnants of the creation of the [Solar System]," McNair said before the mission, "and they're just a big mass of ice – of frozen gases – and the last time [Halley] came around, we weren't sophisticated enough to do the type of things that we're doing now. Scientists will be able to analyse the gases [and] emissions by looking at the Sun's reflection and the absorption of sunlight and give some credibility to some of the theories – or possibly tear them down – about the origin of the Universe. Who knows what we're going to find out of this? But these types of observations can change the way you think."

Several other missions, besides Spartan, were also watching Halley at the time: Europe's Giotto, the Soviet Union's twin Vega spacecraft and Japan's Suisei and Sakigake probes were heading for the comet, but NASA believed its Shuttle-borne studies had the 'edge' by conducting observations as it neared 'perihelion', at its closest point to the Sun.

It was during a narrow, five-week 'window', from January 20th until February 22nd 1986, that the agency hoped Halley – then some 225 million km from Earth and only 97 million km from the Sun – would be chemically at its most active and yield the most desirable scientific data. The Spartan mission to explore the comet, codenamed '203', would have got underway on the second day of STS-51L, when Scobee and Smith were scheduled to fire Challenger's Orbital Manoeuvring System (OMS) engines to nudge it to a slightly higher altitude, about 245 km above Earth. The assignment of Spartan-203 had already resulted in some changes to the Shuttle's own launch period; originally targeted for a morning lift-off, it was moved to the afternoon, in order to provide the best lighting conditions for the satellite's observations. However, an afternoon start would delete the option of touching down in Casablanca in the event a TAL abort to Morocco became necessary. Ultimately, as Challenger's launch was pushed into the final week of January, conditions for optimum viewing of Halley, based on an afternoon window, could no longer be met and the lift-off time was shifted back to the morning hours. In a sense, therefore, the delay was actually beneficial.

By January 30th, after Spartan-203's software had been uploaded from NASA's Johnson Space Center (JSC) in Houston, Texas, and voltage and current checks carried out, it would have been 'hung' over the payload bay wall and released into free flight by the mechanical arm. The satellite would then have executed a slow, minute-and-a-half-long pirouette to prove that it was working properly, after which Scobee would have pulsed the Reaction Control System (RCS) thrusters to achieve a maximum separation distance of around 145 km. This would have ensured that sunlight reflected by Challenger's pristine white surfaces did not 'confuse' Spartan-203's sensors.

Following two orbits of further tests, the aperture doors covering the satellite's two ultraviolet spectrometers would have automatically retracted to initiate an aggressive, 40-hour-long phase of free flight, of which more than half would have been dedicated to studies of the photodissociation of water in Halley and analysis of its various nitrogen-, carbon- and sulphur-containing molecules. Meanwhile, its cameras would have offered an ongoing record of the 'large-scale' activity of the comet itself, including outbursts in its nucleus and asymmetries in its coma. Retrieval would have followed on February 1st and Spartan-203 would have been repositioned on its Mission Peculiar Equipment Support Structure (MPESS) carrier in the forward section of the payload bay.

Had Onizuka and his six crewmates survived their violent climb to orbit on January 28th, however, the delicate and tricky Spartan-203 deployment, two days of station-keeping and retrieval would have actually been the secondary task of their mission. By far the largest, most expensive and most powerful payload aboard STS-51L was NASA's second $100 million Tracking and Data Relay Satellite – known as

'TDRS-B' – which, it was expected, would enable future Shuttle astronauts to communicate directly with Mission Control for most of each 90-minute circuit of the globe.

"That's going to be a big improvement," Smith told an interviewer in the weeks leading up to the launch, "not only for the Shuttle, but also for the space station when it gets up later on." Until the early 1980s, US missions had relied on a network of ground stations to relay communications between orbiting crews and Houston-based controllers. The TDRS network of at least two large satellites, positioned in orbits 35,600 km high, would gradually bring this era to a close.

Onizuka, though, was simply thrilled at having the chance to help deploy "one of the largest communications satellites ever!" His words were, to say the very least, an understatement. The 2,540 kg TDRS-B would, when fully functional in its operational orbital 'slot' and numerically renamed 'TDRS-2', resemble a colossal windmill with four 'paddles' extending from beryllium booms affixed to a hexagonal 'bus'. Two of these paddles held electricity generating solar panels, while the others carried umbrella-like S-band and Ku-band antennas. Between the tips of its solar panels, TDRS-2 would have spanned an impressive 12 m when full unfurled in orbit, making it virtually identical to the satellite launched by Paul Weitz' STS-6 crew almost three years before.

In order to achieve its high orbit, it was attached to a Boeing-built Inertial Upper Stage (IUS), whose two solid-fuelled sections would have delivered the satellite, over a period of about seven hours, into its operational location. Deployment of the 14-m-long combo would have consumed most of Challenger's first day in space and, although all five 'career' members of the crew would have been involved, the lengthy procedure would have been conducted under the direction of Onizuka and McNair.

Shortly after reaching space and opening the payload bay doors – thus exposing the folded-up satellite and its booster to the harsh environment of low-Earth orbit for the first time – the two men, located at instrument panels on the aft flight deck, would have run through a series of checks and eventually hoisted the 'stack' to a pre-deployment angle of 29 degrees using the ring-doughnut-shaped 'tilt table'. As Scobee and Smith manoeuvred Challenger into the correct attitude, Onizuka and McNair would have switched TDRS-B over from the Shuttle's electricity supply to the IUS' internal batteries.

Next, they would have commanded the tilt table to raise the combination to an angle of 59 degrees and, precisely ten hours after leaving Earth, spring-ejected it, such that it swept smoothly over Challenger's cabin roof. Nineteen minutes later, Scobee would have fired the OMS engines to create a safe separation distance in anticipation of the IUS' first stage ignition. After computing the stack's correct attitude by taking star sightings, the IUS would have fired its engine an hour after deployment and run for two and a half minutes. An additional burn by the second stage, lasting just under two minutes, would then have inserted TDRS-B into near-geosynchronous orbit.

Whilst still attached to the now-exhausted second stage, the satellite's solar arrays would have opened – "like an insect coming out of a cocoon," astronaut Mike Lounge, who deployed TDRS-C in September 1988, would later remark – and, eventually, so too would its communications payload. Over a period of several months, during a series of extensive tests, it would have gradually drifted westwards to its final orbital position over the Pacific Ocean, directly above the equator south of Hawaii, at 171 degrees West longitude.

It was a complicated task and one for which the IUS itself had made a rather inauspicious start. When TDRS-1 was launched in April 1983, the second stage of its booster had malfunctioned during the circularisation manoeuvre and delivered it into a lower-than-planned orbit; this forced controllers to use three-quarters of the satellite's precious hydrazine fuel to limp into the correct slot, reducing its operational lifespan. The investigation into the embarrassing failure led to the postponement of several other IUS-dependent Shuttle missions and the TDRS-B launch, originally targeted for August 1983, was repeatedly delayed. Additional problems with a timing circuit aboard TDRS-1 pushed it back yet further from March 1985 until the spring of the following year.

However, following his first mission aboard Space Shuttle Discovery in January 1985, which featured the successful deployment of a top-secret Department of Defense satellite affixed to an IUS, Onizuka expressed confidence in the weeks leading up to STS-51L in the training and procedures involved with releasing both the enormous TDRS and its problem-prone booster. "The basic training was the same," he said of the similarities between his first and second flights. "Once we enter the area of payload and mission operations, there were some differences, [but] I'm very familiar with the IUS; very comfortable with it."

Capable of handling up to 300 million bits of information per second – roughly equivalent to processing a couple of hundred 14-volume encyclopaedias every minute – TDRS-B would technically bring the system up to fully operational status. Nevertheless, a third satellite was scheduled to be ferried into orbit by Challenger's STS-61M crew in July 1986 to replace the degraded TDRS-1. Until the arrival of this third member of the network, TDRS-B would operate from an initial 'spare' orbital slot of 136 degrees West longitude, providing much-needed backup services for its prematurely ageing sibling. After the launch of the third satellite, however, TDRS-B was scheduled to be moved to its final position at 171 degrees West longitude.

Despite its important contributions to astronomy and communications, the STS-51L mission naturally attracted media attention, as NASA had intended, thanks to the presence of teacher observer McAuliffe. Explorers, journalists and entertainers were considered in the early 1980s as the agency weighed up options for which profession would yield 'the best' private citizen to send aloft on the pioneering mission. Ultimately, in August 1984, President Ronald Reagan announced that a teacher would fly first. Dick Scobee agreed that it was the right decision.

"Teachers teach the lives of every kid in this country through the school system and if you can enthuse the teachers about doing this, then you enthuse the students and impress on them that's something to expect in their lifetime," he explained in the weeks leading up to Challenger's launch. "Man needs to explore and that's part of the thing we have to do to ensure our future. So as far as I'm concerned, it's a good insurance policy for the human race."

McAuliffe's selection as the primary candidate for the mission, with Idaho teacher Barbara Morgan backing her up, was revealed by Vice-President George Bush in July

1985 and a few weeks later both women arrived in Houston to begin training in earnest. Her tasks included performing two, 15-minute-long lessons: the first, entitled 'The Ultimate Field Trip', was a guided tour of the Shuttle to familiarise students with onboard living and working conditions, while the second, called 'Where We've Been, Where We're Going', focused on NASA's fledgling plans for a permanent space station. Both were to have been aired by the Public Broadcasting System sometime on February 2nd and McAuliffe would have explained the roles of her six crewmates, identified and summarised the experiments aboard Challenger and enthused 'her' students with a vision of the future.

"I think it's going to be very exciting for kids to be able to turn on the TV and see the teacher teaching from space," she said. "I'm hoping that this is going to elevate the teaching profession in the eyes of the public and of those potential teachers out there. Hopefully, one of the secondary objectives of this is students are going to be looking at me and perhaps thinking of going into teaching as professions."

McAuliffe and Jarvis were both 'Payload Specialists' – candidates chosen by their respective companies, agencies or organisations to operate specific experiments, but not 'career' astronauts like their five crewmates – and both joined the STS-51L line-up relatively late in the training flow. Yet both were quickly accepted and grew to become highly respected members of the team.

"It's refreshing to have somebody on board that's really dedicated and enjoys doing what they're doing," Scobee remarked, "but also she goes into the training with a positive attitude and stays out of the way when she needs to stay out of the way, she gets involved when she needs to get involved and does basically all the right things, and so does Greg Jarvis. Both of them, from our standpoint, are good Payload Specialists. They came onboard with a good, open mind, they're accommodating to our system, we try to be accommodating to theirs and it's a nice trade-off."

The level of respect was, of course, mutual and Jarvis, a Hughes aircraft engineer, recalled one particular session as an example of the astronauts' ability to operate seamlessly together. "When you watch them work through the malfunctions they work through, you get very comfortable that they know what they're doing," he said. "One time when we were in the Motion Base Simulator, the lights went out for the visual for the landing. The Commander called down and said 'Aren't the lights out?' And they [Mission Control] said, 'I think so, we'll get back to you on that'. The conversation went on for about two or three minutes and it turns out they had mistakenly turned the lights out on the visuals. The thing you didn't realise was that he made a perfect landing without any lights!"

The arrival of Jarvis in October 1985 had come particularly late in the crew's training period. During the mission, he was assigned to conduct a battery of investigations using spinning, fluid-filled plastic models on Challenger's middeck to evaluate 'optimum' shapes for future satellite fuel tanks. The reason for his late assignment was primarily linked to the fact that payloads for Shuttle missions were in constant flux prior to STS-51L; indeed, the cargo for Dick Scobee's flight had changed several times, as had the identities of 'his' Payload Specialists.

Ironically, one of the main reasons for flying Greg Jarvis was to allow a representative of the Hughes company – which had built several Shuttle-ferried communications satellites – to observe and analyse the physics of an actual deployment in depth. Originally assigned to fly aboard Columbia on STS-61C, he was transferred to STS-51L, ostensibly because the Hughes-built Westar-6S satellite scheduled to ride aboard the former mission had experienced technical problems and been delayed. This sounded perfectly reasonable, but for one thing: STS-51L, also, had no Hughes satellite aboard! The more likely reason for reassigning Jarvis, wrote Mike Mullane, was that Congressman Bill Nelson had requested a Shuttle flight and the space agency had hurriedly complied. "NASA bumped the oft-abused Jarvis one mission to the right," Mullane recalled in his 2006 memoir. "The next time he would pose for a crew photo would be for STS-51L, the mission that would kill him. He would die on a mission that had no Hughes satellite to deploy, the singular event that had been the original justification for his assignment to a Shuttle flight."

Even when the five NASA crew members were assigned in January 1985, with a projected lift-off in November, the mission's payload was changing every few months: first they would deploy an Australian communications satellite and operate a pharmaceutical processing factory, then for a short time 'their' orbiter was switched from Challenger to Atlantis and, finally, back to Challenger again. In fact, one of the issues raised by the Rogers Commission – headed by former Secretary of State William Rogers – was this practice of constantly juggling payloads between missions. In STS-51L's case, this had led to no fewer than six postponements of the critical Cargo Integration Review, an essential meeting at which payload requirements are assessed and the development of final flight products can begin. Although the commission admitted that most payload adjustments were complete by the time the review finally took place in June 1985, it was particularly critical of the late assignment of Jarvis and his experiments, just three months before launch.

"The launch minus five months Flight Planning and Stowage Review was conducted on August 20th 1985," continued the Rogers report, "to address any unresolved issues and any changes to the plan that had been developed to that point. Ideally, the mission events are firmly determined before the review takes place. For 51L, however, Mr Jarvis was not added until October 25th 1985 and his activities could not be incorporated into mission planning until that time. There were changes to middeck payloads, resulting from the addition of Mr Jarvis, that occurred less than three months before launch. The most negative result of the changes was a delay in publishing the crew activity plan. [This] specifies the in-flight schedule for all crew members, which in turn affects other aspects of flight preparation."

Furthermore, Rogers investigators expressed concerns that changes were being made to flights at very short notice – not only Payload Specialist adjustments, but also satellite swaps and experiments being added, delayed or dropped entirely – which, of course, would directly impact the training time available for crews. "Had we not had the accident," said Hank Hartsfield in his testimony to the commission, "we were going to be up against a wall; STS-61H [a Columbia mission, scheduled for June 1986] . . . would have had to average 31 hours in the simulator to accomplish their required training and STS-61K [an Atlantis flight in October] would have to average 33 hours. That is ridiculous. For the first time, somebody was going to have to stand up and say [that] we have got to slip the launch because we are not going to have the crew trained." Training was also affected by the presence of only two Shuttle simulators at JSC, capable of supporting crews for no more than 12–15 missions per annum. "The flight rate at the time of the accident," read the Rogers report, "was about to saturate the system's capability to provide trained astronauts for those flights."

At length, with everything (TDRS-B and Spartan-203) and everybody in place, the STS-51L mission-specific training commenced in the late autumn of 1985, with the astronauts averaging 49-hour work weeks to ensure proficiency in robot arm operations, Spartan deployment and retrieval activities, IUS systems, ascent and re-entry procedures and each of the experiments crammed into Challenger's middeck. The mission itself was deemed "moderately complex" in view of the Spartan commitment, although both it and a TDRS deployment had already been 'baselined' on previous flights.

Still, despite a hectic six days in space, all seven astronauts intended to spend some moments appreciating the uniqueness of where they were. "We have a fairly busy timeline and it's nice to have time to go look out the windows," Scobee, who had flown once before on Challenger in April 1984, said during one of his last interviews. "I guess one of the things that pleasures me most is to have a quiet time where you can go look out the windows, turn out the lights and look at the stars and Earth and thunderstorms. Just the sheer joy of doing it is probably the most fun part because it's hard to single out one thing, but even the hard work of it is generally fun. I enjoy the flying. I enjoy the excitement and thrill of the ascent, because it is really dramatic. Entry is fiery – just an amazing light show – and the fires of hell are burning outside your window and you're sitting there nice and comfortable watching all this go on and it's just a neat feeling."

Nonetheless, Scobee had already announced before setting off that STS-51L would be his last space mission; doubtless, he intended this one to count even more so than his previous flight. His last comments of encouragement to his crewmates over Challenger's intercom in the final seconds of the countdown were words that conveyed enthusiasm, dedication, professionalism, childlike wonder – and an uncanny, though unwitting, preview of what would happen.

"Everybody strap in tight," he told them cheerily. "We're about to go for the ride of our lives."

THE GOLDEN AGE ENDS

That ride began at precisely 4:38 pm on January 28th 1986. Six and a half seconds before lift-off, Challenger's three main engines thundered to life and, as the countdown clock touched zero, the assembled spectators at KSC were greeted by the ear-splitting staccato crackle of her twin Solid Rocket Boosters (SRBs). It proved to be the failure of both primary and secondary O-ring seals at the base of the right-hand booster, Rogers investigators would later conclude from photographic, physical and other evidence, that was directly and solely responsible for the destruction of STS-51L and the loss of her crew.

The response from the astronaut corps was one of astonishment: many knew nothing of prime contractor Morton Thiokol's concerns over the integrity of the O-ring seals, a problem which had first drawn engineers' attention during Columbia's STS-2 ascent in November 1981. In fact, the astronauts most feared a failure of the Shuttle's main engines, as evidenced by the STS-41D on-the-pad shutdown and the hairy Abort To Orbit endured by STS-51F. The boosters, on the other hand, were deemed uncontrollable whilst firing, yet they 'worked' and were regarded as 'big' and 'dumb' ... and totally reliable. Clear evidence of their fallibility, made public for the first time by the Rogers report, occurred serendipitously when, 0.678 seconds after lift-off, a video camera mounted close to Pad 39B captured "a strong puff of grey smoke ... spurting from the vicinity of the aft field joint of the right Solid Rocket Booster".

The camera had identified the tell-tale result of both the primary and secondary O-rings – meant to stop searing hot gases from escaping between the joints of the booster segments – failing, disintegrating and streaming away in the moments after ignition. More significantly, the point of failure directly faced the External Tank (ET) and its two million litres of highly volatile propellants. Any flame from the compromised booster could now play on the ET like a blowtorch, igniting its contents in a fireball and destroying Challenger, together with the entire launch complex. Years later, Thiokol structural engineer Roger Boisjoly would express astonishment that the vehicle did not explode on the pad; by an incredible sequence of events that can scarcely be attributed to 'good' or 'bad' luck, a chunk of solid fuel temporarily plugged the O-ring hole and the first minute of Challenger's ascent proceeded normally.

The temporary plug, however, was just that: temporary. It would not hold.

Several more puffs of increasingly denser, blacker smoke – further indicative that the products under combustion were indeed the grease, insulation and rubberised O-ring material from the joint seals – were recorded by other ground-level cameras between 0.836 and 2.5 seconds after lift-off, as the boosters' hold-down posts were severed and the Shuttle commenced its climb for the heavens. As each puff was left behind by Challenger's upward trajectory, the next fresh puff could be seen close to the level of the joint. The frequency of these emissions, disaster investigators would later determine, was directly related to flexure within the structure of the SRB as the gap in its joint opened and closed. The last incidence of smoke above the joint was timed at T + 2.733 seconds. In the moments that followed, a combination of atmospheric factors and exhaust from the boosters made it difficult to determine if any more smoke was emerging from the failure point.

A little under eight seconds into the STS-51L mission, as planned, the vehicle cleared the tower and began her 'roll program' manoeuvre, moving onto the correct flight azimuth for an intended 28.45 degree inclination orbit, then pitching onto her back under the control of her five General Purpose Computers (GPCs). Shortly thereafter, at T + 19 seconds, to prepare herself for passage through a period of maximum aerodynamic turbulence, Challenger's main engines were throttled down from 104 to 94 per cent, and later 65 per cent, of their rated thrust. Thirty-seven seconds into the ascent, she encountered the first of several high-altitude wind shears,

First evidence of a puff of smoke from the right-hand Solid Rocket Booster, milliseconds after STS-51L's lift-off.

lasting until just past a minute after launch. In its inquiry, the Rogers report noted that the Shuttle's guidance, navigation and control system immediately detected and compensated for these conditions and – although STS-51L's aerodynamic loads were higher than previous missions in both the yaw and pitch planes – the SRBs, too, responded effectively to all commands.

It is possible that the mission may still have proceeded normally, had the plug of solid fuel remained jammed into the O-ring breach. However, by an incredible stroke of cruel luck, Challenger happened to pass through the most severe wind shear ever encountered by an ascending Shuttle; a shear which dislodged the plug somewhere around a minute into the mission. After passing through maximum aerodynamic turbulence, at 51 seconds into the climb, her main engines were throttled back up to full power; shortly afterwards, 58.788 seconds after lift-off, a frame of video recorded the first evidence of a flickering flame from the right hand SRB's aft joint. The temporary plug of solid fuel had gone and, although they were oblivious to anything amiss, the crew's fate was now sealed.

The flame rapidly established itself, growing into a well-defined plume within barely half a second. Exactly a minute into the mission, downlinked telemetry pointed to an unusual chamber pressure differential between the left and right boosters – the pressure of the latter was some 11.8 psi lower than the former, indicative of a steadily growing leak in its aft joint. As the flame increased in size, Challenger's aerodynamic 'slipstream' deflected it backward and circumferentially by the protruding structure of the upper ring which linked the SRB to the External Tank. As a result, the flame was focused directly onto the surface of the ET.

Sixty-two seconds into the ascent, the left booster's thrust vector control moved to automatically compensate for the yaw motion caused by the reduced thrust from its right-side counterpart. A couple of seconds later, however, came the first visual manifestation that the flame from the damaged booster had breached the lower segment of the External Tank: an abrupt change in its shape and colour, indicating that it was now mixing with leaking liquid hydrogen. Moreover, pressurisation data telemetred from the ET at around this point reinforced the fact that its liquid hydrogen tank was indeed ruptured.

In the seconds that followed, an incredibly rapid sequence of events occurred and concluded with the destruction of the External Tank, the separation of both boosters and the structural disintegration of Challenger into several large pieces. Seventy-two seconds after lift-off, the flame from the right SRB finally burned through the lower of two struts holding it onto the External Tank; pivoting around its upper strut, the top of the booster impacted the instrument-laden inter-tank and the base of the liquid oxygen tank, breaching them both. Nearly simulaneously, around T + 73.1 seconds, clouds of white vapour were spotted at the top of the ET and around the area of its bottom dome: the former was clearly indicative of the ruptured liquid oxygen tank, the latter conclusive evidence that it had suffered structural failure.

Almost immediately, at T + 73.6 seconds, came a massive – "almost explosive", read the Rogers report – burning of both the hydrogen leaking from the lowermost tank and the oxygen from its uppermost section. At this point, STS-51L was at an altitude of 14 km over the Atlantic Ocean, travelling at almost twice the speed of

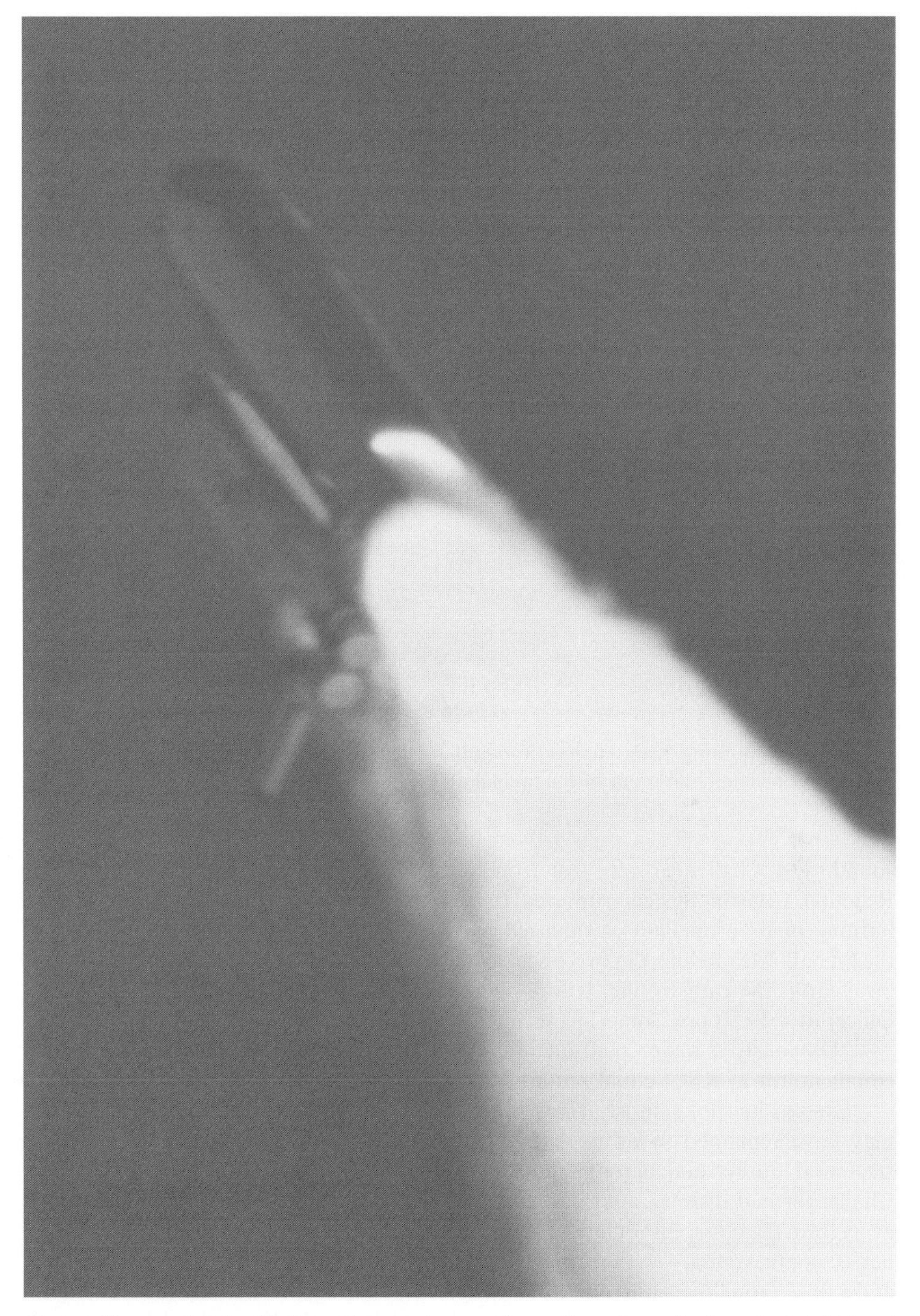

Almost 59 seconds into Challenger's ascent, an ominous flame from the breached right-hand Solid Rocket Booster begins to impinge on the External Tank.

Challenger's still-firing main engines, forward fuselage – containing the crew cabin – and residual propellants hurtle out of the fireball at T + 78 seconds.

sound, and Challenger was lost from view in the resultant explosive burn. Her Reaction Control System ruptured during this period, setting off the hypergolic burning of its propellants, evidenced by a reddish-brown hue around the edge of the fireball. Meanwhile, the two boosters, now released of their loads, rapidly climbed away from the catastrophe, before being remotely destroyed by the Range Safety Officer at 4:39:50 pm, some 110 seconds after launch.

"Obviously a major malfunction," was all Steve Nesbitt, the stunned launch commentator at KSC, could remark.

Even as he spoke, most, if not all, of the STS-51L crew were still conscious and may have remained so as they tumbled to Earth. Now placed under significant structural duress and aerodynamic loads for which she had not been designed, Challenger had disintegrated into several large fragments. Clearly visible, tumbling away from the blossoming cloud of debris that had swallowed the External Tank, were her aft compartment – with main engines, briefly, still firing – together with one wing and a fuzzy, roughly triangular blob: the forward fuselage, containing the crew cabin, trailing a jumble of umbilical lines ripped from beneath the payload bay floor. It continued, for a time, on an upward trajectory, reaching a maximum altitude of around 18 km, before terrestrial gravity began to pull it inexorably on a curving, ballistic arc towards an Atlantic grave. Tumbling at close to 400 km/h at the point of impact, ocean water would have been as unyielding to Challenger as solid ground.

A PREVENTABLE TRAGEDY

Six weeks later, on March 7th, the crew cabin was found by divers in less than 30 m of water, some 27 km north-east of KSC, and recovered by a team from the USS Preserver. It "was disintegrated, with the heaviest fragmentation and crash damage on the left side," read the Rogers Commission's final report. "The fractures examined were typical of overload breaks and appeared to be the result of high forces generated by impact with the surface of the water."

Tellingly, US Navy spokeswoman Deborah Burnette told a 'Washington Post' journalist shortly after the discovery that "we're talking debris, and not a crew compartment, and we're talking remains, not bodies". Mike Coats, one of several astronauts directed by NASA to examine the wreckage, describing it as looking "like aluminium foil that had been crushed into a ball". It contained the remains of the crew, but their horrific condition could be guessed from pathologists' difficulty in identifying them: a few strands of Judy Resnik's hair and a necklace were all that was left of Mission Specialist Two. Indeed, in the months after the disaster, all astronauts were required to submit a clip of hair and a footprint to NASA for identification. In the case of the STS-51L remains, apparently, even dental records were insufficient for positive identification ...

In his memoir, Mike Mullane expressed fervent hopes that the explosive burn of the External Tank's propellants had been enough to completely destroy Challenger's crew cabin, barely a short distance from the astronauts, or at least breach her flight deck windows, thereby causing a rapid depressurisation and mercifully rapid death of the crew. However, tested to 140 per cent of its design strength in Lockheed's Plant 42 rig almost a decade earlier, the cabin proved extremely hardy and its wreckage showed little evidence of having experienced an explosive depressurisation. Such an eventuality would have led to an upward 'buckling' of the flight deck floor as air from the middeck rapidly expanded; no such buckling was detectable.

Additionally, wrote JSC Director of Life Sciences Joe Kerwin in a July 28th 1986 letter to NASA Associate Administrator for Spaceflight Dick Truly, the "impact damage to the windows [examined after recovery from the Atlantic] was so extreme that the presence or absence of in-flight breakage could not be determined. The estimated break-up forces would not in themselves have broken the windows. A broken window due to flying debris remains a possibility; there was a piece of debris embedded in the frame between two of the forward windows. We could not positively identify the origin of the debris or establish whether the event occurred in flight or at water impact ... Impact damage was so severe that no positive evidence for or against in-flight pressure loss could be found."

Astronauts Jim Bagian and Manley 'Sonny' Carter, both physicians, speculated that penetrations in the cabin's aft bulkhead – created by the violently severed payload

Thirty per cent of Challenger's structure was recovered from her Atlantic grave. Among the debris was the crumpled airlock hatch, originally housed in her middeck.

bay umbilical lines – could have led to a slower depressurisation and quick unconsciousness for the seven astronauts, although this proved purely conjectural. More conclusive evidence that at least some of the crew had remained alive and conscious for most of the fall to Earth came in mid-March, when four Personal Egress Air Packs (PEAPs) were recovered. These were designed to provide each astronaut with a limited amount – some six minutes' worth – of breathing air for use in emergencies. Analysis of the packs led to an announcement on May 21st that at least one had been activated in the seconds after structural break-up and, later, that this activation was not caused accidentally at water impact. Then, on June 9th, investigators revealed that one of the packs belonged to Pilot Mike Smith.

This raised an interesting scenario: Smith's PEAP was affixed to the back of his seat, placing it out of his reach, which implied that either Judy Resnik or Ellison Onizuka, seated behind him on the flight deck, had leaned forward and switched it on in a valiant effort to save his life. A second identifiable PEAP belonged to Dick Scobee and had not, apparently, been activated. The owners of the two other packs were never identified. The quantity of air remaining in Smith's PEAP, in particular, led to a suggestion that apparent 'crew inactivity' after break-up could be an indication that they had rapidly lost consciousness.

Every scrap of paper from Challenger's wreckage was analysed and it was determined that none of the astronauts had written a note; moreover, Smith's air pack was depleted by barely two and a half minutes – almost precisely the length of time it took for the cabin to fall from the fireball to the Atlantic – which suggested he had kept his helmet visor closed during the descent. If it had remained open, Mike Mullane explained, all six minutes of his PEAP air would have leaked out. Immediately after break-up, Challenger's intercom, lights, computers and electronics went dead. Bagian and Carter postulated that, in order to communicate, the crew's only option would have been to raise their visors to speak aloud. Could the fact that Smith's visor might have been closed indicate that he quickly lost consciousness?

Unfortunately, the helmets themselves were obliterated, which rendered it almost impossible to determine how, and if, the astronauts communicated during those final frantic minutes. However, Mullane believes from his own experience as a US Air Force navigator, flying in the back seat of F-4 Phantom jets in the 1960s and 1970s, that hand signals as a means of communication would have worked perfectly well. Additionally, with even the slightest possibility that the cabin's pressure integrity had been compromised, Scobee and Smith's years of experience as fighter and test pilots would have taught them to keep their visors down, rather than risk lifting them and suffocating.

One factor is almost certain: most, if not all, of the astronauts were aware of their dire predicament. Milliseconds before the External Tank disintegrated, a bright sheet of white vapour flooded across Challenger's nose – probably visible to Smith, sitting in the right-hand Pilot's seat – and may have prompted him to utter his brief "Uh, oh" comment, which turned out to be the last vocal communication from the orbiter. It is also quite possible that he saw the top of the right SRB pivot into the side of the ET. Despite hoaxed intercom 'transcripts' which later came to light, alleging that the terrified, panic-stricken crew screamed and cursed their way down to the Atlantic, Mike Mullane is confident that Scobee and Smith would have fought to the very end to regain control of their crippled ship, even though it soon became clear that they no longer had a ship to fly.

In the days after the disaster, most of the astronauts in Houston became convinced that a failure or explosion of one or more of the Shuttle's main engines was the most likely cause. Remnants of all three were dredged from the Atlantic on February 23rd, each still attached to its thrust structure, and the controllers for the Number Two (left-hand) and Three (right-hand) engines were found, disassembled, flushed with deionised water, dried, vacuum baked and their data extracted. All of the engine debris exhibited burn damage caused, according to the Rogers report, "by internal over-temperature typical of oxygen-rich shutdown". Thus, the loss of hydrogen fuel after the rupturing of the lower part of the External Tank appeared to have caused all three to begin shutting themselves down within milliseconds of each other at around T + 73.5 seconds.

Overall, the performance of the main engines was satisfactory and in line with observations from previous missions. They first exhibited "abnormal" behaviour almost exactly a second before break-up, when their fuel tank pressures dropped and the controllers responded by opening the fuel flow-rate valves. Next, turbine temperatures increased due to the leaner fuel mixture being fed into the engines' combustion chambers from the External Tank. Otherwise, the commission's report continued, "engine operation was normal" and did not contribute to the loss of STS-51L.

Nor did the gigantic ET itself, of which some 20 per cent was recovered, mostly debris from the inter-tank and the lowermost hydrogen section. Initial speculation that premature detonation of range safety explosives was quickly discounted – in part because the unexploded ordnance was among the debris – as were theories of structural imperfections in the tank's design or damage incurred at lift-off. The possibility of a liquid hydrogen leak at lift-off was also dismissed, since it would immediately have been ignited by the exhaust from the Solid Rocket Boosters or main engines and noted in the downlinked telemetry data.

In total, around 30 per cent of Challenger herself was found and subsequent inspections revealed she had disintegrated primarily as a result of massive aerodynamic overloads, with no evidence of internal burn damage or exposure to explosive forces. Chemical analyses indicated that her right side had been sprayed with hot propellant gases from the leaking SRB, but telemetry indicated that all of her systems operated normally until shortly before break-up. No problems were detected with either of her two payloads: Spartan-203 was unpowered during ascent and the deployment ordnance for the Inertial Upper Stage and TDRS-B, of which about five per cent of debris was recovered from the Atlantic, showed no indication of having prematurely activated itself. The finger of blame pointed squarely at the third component of the Shuttle system – the SRBs – and, in particular, at the leaking right-side booster.

Initial suspicion that its range safety explosive charges had been inadvertantly fired was quickly dismissed when analysis of telemetry data revealed that no such commands were sent to either booster until both were remotely destroyed by the Range Safety Officer at T + 110 seconds. For a number of engineers and managers at Morton Thiokol and within NASA, however, the cause of the disaster had been identified more than a year before Challenger's maiden voyage: the primary and secondary O-rings meant to prevent a leakage of hot gases were incapable of properly sealing the gaps between the SRB joints in extremely cold weather. Already, catastrophe had been narrowly averted on one previous cold-weather launch in January 1985 and conditions in the hours leading up to STS-51L's lift-off were colder still. Moreover, an application of zinc chromate putty, intended to act as a 'thermal barrier' and keep the combustion gas path away from the two O-rings, had been shown as early as 1984 to be susceptible to the formation of 'blow holes', which compromised its effectiveness.

"It was intended," read the Rogers report, "that the O-rings be actuated and sealed by combustion gas pressure displacing the putty in the space between the motor segments. The displacement of the putty would act like a piston and compress the air ahead of the primary O-ring and force it into the gap between the [field joint's] tang and clevis. This process is known as 'pressure actuation' of the O-ring seal. This pressure-actuated sealing is required to occur very early during the solid rocket motor ignition transient, because the gap between the tang and clevis increases as pressure

Remains of O-ring seal 'tracks' and putty in recovered debris from STS-51L's right-hand Solid Rocket Booster.

loads are applied to the joint during ignition. Should pressure actuation be delayed to the extent that the gap has opened considerably, the possibility exists that the rocket's combustion gases will blow-by the O-rings and damage or destroy the seals. The principal factor influencing the size of the gap opening is motor pressure, but gap opening is also influenced by external loads and other joint dynamics."

One of these external loads was the detrimental impact of low launch temperatures, together with the effect of water and ice, on the O-rings. In the case of STS-51L, on the night of January 27th 1986, ambient temperatures had dipped to the lowest ever recorded for a Shuttle launch: around minus 13 degrees Celsius. Indeed, at the moment of ignition the following day, the right-hand booster's aft field joint was the coldest part of the stack at minus 2.2 degrees Celsius. Ground tests had already confirmed that reduced temperatures could cause the O-rings' resiliency to degrade and, during the Rogers investigation, it was learned that a small quantity of rain water had been discovered in Columbia's SRB joints during preparations for STS-9 in November 1983. It was theorised that STS-51L, which had been sitting on Pad 39B for a total of 38 days and been exposed to significantly more rainfall than Columbia, could have suffered from the further disruption, and perhaps even 'unseating', of its O-rings by frozen water.

The observed problem with the boosters first arose in November 1981, shortly after the STS-2 mission. Routine inspections revealed significant erosion of the right-hand SRB's primary O-ring, caused by hot combustion gases, yet the secondary seal remained intact and the anomaly was not reported at the Flight Readiness Review for STS-3 in March of the following year. Morton Thiokol believed that the erosion had been caused by blow holes in the zinc chromate putty and began tests of changing the method of its application and the assembly of the booster segments. The manufacturer of the original putty, Fuller-O'Brien, discontinued its use and a new putty from the Randolph Products Company was selected in May 1982; however, after more changes, it was substituted for the original putty the following summer, shortly before the launch of STS-8.

Since December 1982, the O-rings had been designated a 'Criticality 1' item by NASA, denoting a component without a backup facility, whose failure would result in the loss of the entire Shuttle and its crew. Prior to that, during the execution of the first five missions, they had been labelled by NASA as 'Criticality 1R', meaning that, although "total element failure ... could cause loss of life or vehicle", the presence of primary and secondary O-rings lent "redundancy" to the design. The secondary seal, in effect, would expand and fill the joint if its primary counterpart failed. However, in its Critical Items List of November 1980, NASA acquiesced that "redundancy of the secondary field joint seal cannot be verified after motor case pressure reaches approximately 40 per cent of maximum expected operating pressure. It is known that joint rotation occurring at this pressure level ... causes the secondary O-ring to lose compression as a seal."

Following a series of high-pressure tests of the O-rings, conducted by Morton Thiokol in May 1982, it became clear that the secondary seal did not provide sufficient redundancy and NASA changed their criticality listing later that year. According to the agency's then-Associate Administrator for Spaceflight (Technical), Michael Weeks, who signed a waiver to accept the new criticality level in March 1983, "we felt at the time that the Solid Rocket Booster was probably one of the least worrisome things we had in the programme". It was a view shared by managers and astronauts, too. But not by Roger Boisjoly.

By the time Boisjoly, a Thiokol structural engineer, inspected severely damaged field joints from STS-51C's boosters in January 1985, a number of other missions had yielded disturbing O-ring erosion. On STS-41B almost a year earlier, the left-hand SRB's forward field joint and the nozzle joint belonging to its right-hand counterpart were found to be badly degraded, to such an extent that NASA requested Thiokol to investigate means of preventing further erosion. A week prior to the launch of STS-41C, the company concluded that blow holes in the zinc chromate putty were one "possible cause" and NASA's SRB project office at Marshall Space Flight Center (MSFC) in Huntsville, Alabama, decided that, as long as the secondary O-ring could survive gas impingement, Bob Crippen's Solar Max repair mission was safe to fly.

It was the beginning of a disturbing chain of thought within NASA and Thiokol, reflected in the Rogers report, that "there was an early acceptance of the problem" and both organisations "continued to rely on the redundancy of the secondary O-ring long after NASA had officially declared that the seal was a non-redundant, single-point [Criticality 1] failure".

One of the members of the Rogers inquiry was the celebrated physicist Richard Feynman, who judged the cavalier attitude of NASA and Thiokol as representing "a kind of Russian roulette ... [the Shuttle] flies [with O-ring erosion] and nothing happens. Then it is suggested, therefore, that the risk is no longer so high for the next flights. We can lower our standards a little bit because we got away with it last time. You got away with it, but it shouldn't be done over and over again like that." Mike Mullane later referred to it, scornfully, as the "normalisation of deviance". That deviance was still apparent, albeit in a different context, when Harold Gehman's investigative board explored the causes of the Columbia disaster 17 years later.

The STS-51C damage was among the most serious yet seen. Launched in freezing conditions of just 11 degrees Celsius on January 24th 1985, its recovered left and right SRB nozzles showed evidence of 'blow-by' between the primary and secondary O-rings and, moreover, it proved to be the first Shuttle mission in which the secondary seal displayed the effects of heat. "SRM [Solid Rocket Motor]-15," said Boisjoly of the STS-51C booster set, "actually increased concern because that was the first time we had actually penetrated a primary O-ring on a field joint with hot gas, and we had a witness to that event because the grease between the O-rings was blackened, just like coal. That was so much more significant than had ever been seen before on any blow-by on any joint."

When the blackened material was chemically analysed, Boisjoly told the Rogers hearing, "we found the products of putty in it [and] we found the products of O-ring in it". Four days after STS-51C landed, on January 31st 1985, Lawrence Mulloy, head of NASA's SRB office at MSFC, expressed concern over the impact O-ring problems may have on the next scheduled mission, STS-51E, then projected for launch in late February. One of Thiokol's conclusions before the Flight Readiness Review was that, while "low temperature enhanced probability of blow-by ... the condition is not desirable, but is acceptable". It was the first occasion on which a link between cold weather and O-ring damage had been officially acknowledged.

MISSED WARNINGS

Three months after the worrisome STS-51C boosters had drawn Boisjoly's attention, Bob Overmyer's crew lifted off on the seven-day Spacelab-3 mission. The results from their SRBs also indicated further erosion of the secondary O-ring, clearly pointing to the failure of its primary counterpart. The problem was attributed to leak check procedures. So serious was the episode, however, that "a launch constraint was placed on flight 51F and on subsequent launches," read the Rogers report. "These constraints had been imposed, and regularly waived, by the Solid Rocket Booster Project Manager at Marshall [Space Flight Center], Lawrence B. Mulloy. Neither the launch constraint, the reason for it, or the six consecutive waivers prior to 51L were known to [NASA Associate Administrator for Spaceflight Jesse] Moore or [Launch Director Gene] Thomas at the time of the Flight Readiness Review process for 51L ..."

In fact, as Overmyer would later discover, his own launch had been milliseconds from disaster. Crewmate Don Lind journeyed to Thiokol in Utah for further explanation. "The first seal on our flight had been totally destroyed," recalled Lind, "and the [other] seal had 24 per cent of its diameter burned away. Sixty-one mil[limetres] of that [last seal] had been burned away. All of that destruction happened in 600 milliseconds and what was left of that last O-ring, if it had not sealed the crack and stopped that outflow of gases, if it had not done that in the next 200 to 300 milliseconds, it would have gone [all the way]. You'd never have stopped it and we'd have exploded. That was thought provoking! We thought that was significant in our family. I painted a picture of our lift-off, then [added] two great celestial hands supporting the Shuttle and the title of that picture is 'Three-tenths of a Second'. Each of [my] children have a copy of that painting, because we wanted the grandchildren to know that we think the Lord really protected Grandpa."

Shortly after the analysis of the STS-51B boosters, on July 31st 1985, Roger Boisjoly expressed his growing concerns over the O-rings in a memorandum to Thiokol's vice-president of engineering, Bob Lund. "The mistakenly accepted position on the joint problem," he wrote, "was to fly without fear of failure and to run a series of design evaluations which would ultimately lead to a solution or at least a significant reduction of the erosion problem. This position is now changed as a result of the [STS-51B] nozzle joint erosion, which eroded a secondary O-ring with the primary O-ring never sealing. If the same scenario should occur in a field joint – and it could – then it is a jump ball as to the success or failure of the joint, because the secondary O-ring ... may not be capable of pressurisation. The result would be a catastrophe of the highest order: loss of human life."

Boisjoly recommended the establishment of a Thiokol team to investigate and resolve the problem and, on August 20th, Lund announced the formation of such a task force. However, only a day earlier, in a joint Thiokol–Marshall briefing to NASA Headquarters on the issue, programme managers concluded that the O-rings were a "critical" issue, but that, so long as all joints were leak checked with a 200 psi stabilisation pressure, were free of contamination in the seals and met O-ring 'squeeze' requirements, it was safe to continue flying. As the year wore on, Thiokol's O-ring team, which numbered only eight to ten members, found many of their efforts frustrated by senior management. "Even NASA perceives that the team is being blocked in its engineering efforts to accomplish its task," Boisjoly wrote in an October 4th memo. "NASA is sending an engineering representative to stay with us, starting October 14th. We feel that this is the direct result of their feeling that we [Thiokol] are not responding quickly enough on the seal problem."

A little over three weeks later, Challenger lifted off on her final, fully successful voyage, STS-61A, experiencing nozzle O-ring erosion and blow-by at the field joints;

neither of these problems were identified at the Flight Readiness Review for the next mission, STS-61B in November. Indeed, that flight also suffered nozzle O-ring erosion and blow-by. By December, in response to these problems, Thiokol recommended that their testing equipment needed to be redesigned. Only days later, on the 10th, the company requested closure of the O-ring critical problem issue, citing satisfactory test results, future plans and work carried out thus far by its task force. This closure request was later pounced upon by the Rogers investigators. One panel member pointed out to the Thiokol senior managers: "You close out items that you've been reviewing flight by flight – that have obviously critical implications – on the basis that, after you close it out, you're going to continue to try to fix it. What you're really saying is [that] you're closing it out because you don't want to be bothered."

Part of the problem was NASA's desire, since the mid-1970s, to create a reusable transportation system that would provide regular and routine access to low-Earth orbit. Original plans to fly the Shuttle once every fortnight, admittedly, were unrealistic because only four operational orbiters were built – rather than six or seven – but in its December 1985 launch schedule, the agency envisaged staging up to 24 missions per year from 1987 onwards. Columbia, Challenger, Discovery and Atlantis would each have to accomplish six, or perhaps more, flights apiece, shortening the turnaround time between their individual missions to as little as six or eight weeks. In correspondence with the author, one former Shuttle engineer expressed serious doubts that such flight rates could have been achieved, even with overtime and the presence of three shifts working around-the-clock in the Orbiter Processing Facility. Nine or ten missions in any 12-month period, he told me, was barely achievable and stretched resources to their limits.

Overtime and overwork presented their own problems. Numerous contract employees at KSC, the Rogers Commission heard, worked 72-hour weeks and frequently supported 12-hour shifts. "The potential implications of such overtime for safety were made apparent during the attempted launch of mission 61C on January 6th 1986," read the report, "when fatigue and shift work were cited as major contributing factors to a serious incident involving a liquid oxygen depletion that occurred less than five minutes before scheduled lift-off."

Furthermore, the commission discovered disturbing evidence that NASA's provisions to support the projected 24-flight annual rate were woefully inadequate: spares for individual orbiters were in short supply (only 65 per cent of the required parts inventory was in place by January 1986), leading to an increasingly dangerous practice of 'cannibalism' from one vehicle to equip the next, and resources were being focused primarily on 'near-term' problems, rather than longer term issues. An $83.3 million budget cut in October 1985 necessitated additional major deferrals of spare parts purchases.

The cannibalism of parts, said then-deputy chief of the astronaut office, STS-6 veteran Paul Weitz, in his Rogers testimony, "increases the exposure of both orbiters to intrusion by people. Every time you get people inside and around the orbiter, you stand a chance of inadvertent damage of whatever type, whether you leave a tool behind or, without knowing it, step on a wire bundle or a tube." Prior to the Challenger disaster, the shortage of spare parts had no serious impact on flight schedules, but, continued the Rogers report, further cannibalism was "possible only so long as orbiters from which to borrow are available. In the spring of 1986, there would have been no orbiters to use as 'spare parts bins'. Columbia was to fly in March, Discovery was to be sent to Vandenberg [Air Force Base in California] and Atlantis and Challenger were to fly in May".

Indeed, KSC Director of Shuttle Engineering Horace Lamberth predicted that, had STS-51L flown successfully, the entire schedule would have been brought to its knees that spring by the spare parts problem alone. "Compounding the problem," the report explained, "was the fact that NASA had difficulty evolving from its 'single flight' focus to a system that could efficiently support the projected flight rate. It was slow in developing a hardware maintenance plan for its reusable fleet and slow in developing the capabilities that would allow it to handle the higher volume of work and training associated with the increased flight frequency."

A FULL PLATE

Had STS-51L been completed safely, the frequency with which Challenger herself flew into space in 1986 would have greatly eclipsed her three previous years of operations. With four more missions scheduled for May, July, September and December, she would have deployed the joint US/European Ulysses probe to explore the Sun's polar regions, followed by a third Tracking and Data Relay Satellite to replace the doddery TDRS-1, retrieval of the Long Duration Exposure Facility, finally, and a secrecy-enshrouded Department of Defense assignment just before Christmas. By far the most significant of these was the launch of Ulysses, which, uniquely for the Shuttle, would have been one of two deep space missions launched within only five days. By the beginning of May, it was expected that Challenger would be installed on Pad 39B and sister ship Atlantis, carrying the Jupiter-bound Galileo explorer, on Pad 39A to support launches on the 15th and 20th of that month. Both missions were fixed within a narrow launch window which could not be slipped.

John Young referred to the missions as the 'Death Star' flights.

Behind the dark humour, however, lay real concern for the then-chief of the astronaut corps. Even with an increasingly confident outlook on the Shuttle's capabilities as it entered its sixth year of flight operations, Young instinctively knew that Challenger's STS-61F mission and Atlantis' STS-61G voyage would be two of the riskiest ever attempted. Fellow astronauts Rick Hauck and Dave Walker, who would respectively command them, echoed his concern. "As with any flight," said Hauck, who also flew aboard STS-7 and would have been Challenger's 11th Commander, "if everything goes well, it's not risky. It's when things start to go wrong that you wonder how close you are to the edge of disaster."

The loss of STS-51L and, on February 1st 2003, of Columbia upon re-entry have illustrated how fine the line is between triumph and tragedy; a line – and risk – that every astronaut knows and accepts before clambering aboard. Yet Hauck and Walker's flights, scheduled to occur just five days apart in May 1986, would have carried additional danger. This was partly due to the importance of the Ulysses and

Had STS-51L survived, these four men would have been Challenger's next crew. Scheduled for launch on May 15th 1986, STS-61F would have deployed the joint US/European Ulysses probe on a five-year mission to explore the Sun's polar regions. Seated are Pilot Roy Bridges (left) and Commander Rick Hauck and standing are Mission Specialists Mike Lounge (left) and Dave Hilmers.

Galileo payloads, both of which were equipped with controversial Radioisotope Thermoelectric Generators (RTGs). The latter were nuclear power sources, fuelled by plutonium dioxide, and the implications of a launch accident and the consequences of depositing highly radioactive material across eastern Florida did not bearing thinking about.

This risk was compounded still further by the fact that, attached to the base of each nuclear hot potato in Challenger's and Atlantis' payload bays was a thin-skinned, liquid-fed rocket that many astronauts and managers had condemned as unsafe and unacceptable for use in conjunction with a manned spacecraft. Measuring nine metres long and four metres wide, it was called the 'Centaur-G Prime' and, for Rick Hauck, it was his baby.

Just like a baby, it was both temperamental and unpredictable.

"I was assigned to be the astronaut office's project officer for Centaur," Hauck recalled two decades later. "It's pressure stabilised, which means if it's not pressurised, it's going to collapse by its own weight. If it were not pressurised, but suspended, and you pushed on it with your finger, the tank walls would 'give' and you'd see that you're flexing the metal!" Nicknamed a 'balloon' because its rigidity thus depended on full pressurisation, the Centaur had long been viewed warily by NASA's human space-flight people, whose safety rule of thumb on the Shuttle dictated that no single failure should be capable of endangering the vehicle or crew.

The Centaur-G Prime, however, did much more than that. Much of its pressure regulation hardware, disturbingly, was not redundant – it lacked a backup facility – and, worse, a failure of its internal bulkhead had the potential to rupture both its volatile liquid oxygen and hydrogen tanks. Additionally, it was recognised that the sheer mass of propellants – which totalled more than 16,500 kg – could cause 'sloshing' and a myriad of other controllability problems that could hinder Hauck or Walker if the need arose to execute an emergency landing shortly after lift-off.

In spite of the hazards, the Centaur's key advantage was that its liquid propellants provided considerably more oomph to push large payloads out of Earth orbit and onto trajectories to other planets than solid-fuelled rockets could achieve. It was also well known that liquid-fed boosters produced a much 'gentler' thrust than the notoriously harsh impulse of solids. Still, the safety concerns rightly overshadowed and ultimately overwhelmed these benefits.

"The Shuttle was obligated to launch Ulysses and Galileo," explained Hauck. "[NASA] needed the most powerful rockets they could have [and] at some point the decision was made to use Centaur, which was never meant to be involved in human spaceflight. That's important because rockets that are associated with human spaceflight have certain levels of redundancy and certain design specifications that are supposed to make them more reliable. Centaur did not come from that heritage, so, Number One, that was going to be an issue in itself, but Number Two is [that] if you've got a Return to Launch Site abort or transatlantic abort and you've got to land – and you've got a rocket filled with liquid oxygen [and] liquid hydrogen in the cargo bay – you've got to get rid of [it], so that means you've got to dump it while you're flying through this contingency abort. To make sure that it can dump safely, you need to have redundant parallel dump valves, helium systems that control the dump valves [and] software that makes sure contingencies can be taken care of. Then, when you land, you're sitting with the Centaur in the bay that you haven't been able to dump all of it, so you're venting gaseous hydrogen out this side [and] gaseous oxygen out that side. This is just not a good idea!"

To support the new rocket on STS-61F and STS-61G, both Challenger and Atlantis underwent a series of extensive modifications, costing around five million dollars apiece, which included extra plumbing to load and drain the Centaur's propellants and control panels in their aft flight decks to monitor its performance. As NASA's newest orbiter, Atlantis had been made Centaur-capable during her initial construction and was destined to spend the first part of 1986 out at Pad 39B undergoing validation tests of the new hardware. Challenger, too, had received the Centaur upgrades, which also included the addition of an S-band transmitter to handle the booster's telemetred data. During typical, pre-launch loading operations, the Centaur's liquids would have been fed through plumbing 'tapped into' the Shuttle's main propulsion system feedlines. Emergency dumping vents – capable of draining all liquid oxygen and hydrogen from the booster within 250 seconds of an abort being declared – were situated on opposite sides of the aft fuselage, just beneath the Orbital Manoeuvring System pods, none of which filled Hauck or Walker with confidence due to the risk of leakages or explosions.

As part of her validation tests, Atlantis would have been rolled to Pad 39B sometime in February 1986 – only weeks after STS-51L had vacated the same launch complex – with a 'real' Centaur-G Prime and a mock-up of Galileo in her payload bay. Whilst on the pad, the booster would have been fuelled with liquid oxygen and hydrogen and a series of tests carried out. Atlantis would then have removed from the pad, the 'real' Galileo installed and transferred to Pad 39A. By mid-April, she would have been joined on adjacent Pad 39B by Challenger, laden with Ulysses and its own Centaur.

Doubts over the reliability of the Centaur-G Prime riding the Shuttle had already, in the autumn of 1981, obliged NASA to cancel it and opt to install Ulysses and Galileo onto 'safer' – though less powerful – solid-fuelled Inertial Upper Stage boosters. For the exceptionally large Galileo, which comprised both a Jupiter orbiter and atmospheric entry probe, the swap from Centaur to IUS meant that its journey time to the giant planet would almost double to four and a half years and most likely would require the mission to be split into two 'halves'.

Predictably, its price tag soared, peaking at close to a billion dollars, until Congress pressed NASA in late 1982 to resume work on a Shuttle-borne Centaur and restore the Jupiter travel time to around two and a half years. Not only Galileo, but Challenger's Ulysses payload, required close encounters with the planet – the latter in order to alter its trajectory and rendezvous with the Sun's poles – and both missions were allocated the same, week-long launch window from May 15th to the 21st 1986. Hauck's crew would lift off from Pad 39B aboard Challenger at around 5:20 pm on the 15th, followed by Walker's team from adjacent Pad 39A aboard Atlantis five days later.

The two flights had scarcely an hour apiece available to them in which to launch and, in order to minimise weight, both would carry just four astronauts. Hauck would have been joined by Pilot Roy Bridges – a veteran of the STS-51F flight – and Mission Specialists Mike Lounge and Dave Hilmers, while Walker's crewmates were Pilot Ron Grabe and Mission Specialists Norm Thagard and James 'Ox' van Hoften. There would be no secondary experiments and their payload bays would be empty, save for the probes and their attached Centaur boosters and support structures. Even some elements of crew equipment in the middeck, including the galley, would have been eliminated to save weight. In January 1986, NASA accepted a recommendation to fly Atlantis with her main engines running at a never-before-tried 109 per cent rated thrust: launching at the standard 104 per cent, it was argued, would have meant the heavy 2,270 kg Galileo spacecraft's Centaur would have been forced to carry less propellant and limited its launch window. Ulysses, on the other hand, was considerably lighter than Galileo (at just 370 kg) and Challenger's engines for STS-61F were manifested to run at the 'standard' thrust rating.

Additionally, the two Death Star flights – scheduled to last between two and four days apiece, according to various sources – were headed for lower than normal, 168 km orbits because, said Hauck, "you need the performance to get the Centaur up because it was so heavy". Moreover, assuming an on-time lift-off of STS-61F, the astronauts would have had no more than about nine hours to get Ulysses out of Challenger's payload bay and on its way to Jupiter, since the Centaur was required to periodically dump its 'boiled-off' gaseous hydrogen to keep tank pressures within their mandated limits. After too much time, it would have 'bled' so much hydrogen that the remainder would not be sufficient to perform its trans-Jovian engine burn.

Consequently, three deployment opportunities were manifested for both missions: in the case of STS-61F, the first chance for Hauck, Bridges, Lounge and Hilmers came at 11:20 pm, some six hours after launch. Two additional options followed at 12:50 am and 2:20 am on May 16th. The Centaur-G Prime's twin Pratt and Whitney-built RL-10A-3A engines – each generating a thrust of 7,300 kg – would then have ignited about 45 minutes later and Ulysses would have been on course, first for a Jupiter rendezvous in July 1987 and ultimately for passage over the Sun's polar regions in 1989–1991.

The idea of sending a spacecraft out of the 'ecliptic' – the plane on which most of the planets circle the Sun – to investigate the mechanics of our parent star can be traced back almost half a century, although it was not until the mid-1970s that the newly-formed European Space Agency (ESA) and NASA began designing possible scenarios. One of the earliest plans was for a dual-spacecraft project, both directed initially towards Jupiter to acquire a gravity-assisted boost, after which one would head for the Sun's south pole and the other for its north pole to make simultaneous observations. This 'Out of the Ecliptic' (OOE) venture was approved in 1976, its scientific research payload defined by the following year and launch provisionally targeted aboard the Shuttle in February 1983. However, financial cutbacks obliged the cancellation of one of the two spacecraft and ESA decided to go ahead with a single probe, built in Europe, with half of its instrumentation and the RTG supplied by the United States. Ongoing development problems with the Centaur-G Prime served to push the launch back still further to the early summer of 1986.

When finally built, the boxy spacecraft was renamed 'Ulysses' and, only days before the STS-51L disaster, was shipped to Florida to commence pre-flight processing and integration with its Centaur-G Prime. Its main feature was a 1.65 m diameter high-gain antenna, through which it would communicate with ground controllers via NASA's Deep Space Network of worldwide tracking stations. Affixed to the side of Ulysses was a 5.6 m radial boom, to keep its three sets of scientific instruments well away from the main spacecraft and, in particular, from potential interference with the plutonium-fed RTG nuclear power source. These instruments included magnetometers to explore the extent of the solar field, together with plasma and ion investigations and gamma ray and X-ray detectors. Ulysses' primary scientific objective was to characterise the 'heliosphere' – a vast region of interplanetary space occupied by the Sun's atmosphere and dominated by the outflow of the solar wind – at latitudes higher than 70 degrees at both the north and south poles. Of particular interest was the behaviour of the solar wind itself, in addition to the physical properties of solar radio bursts and plasma waves, X-rays and solar and galactic cosmic rays. Near the ecliptic, the wind was known to be very turbulent, but at higher latitudes was expected to be a radial flow and to be much faster.

After deployment from Challenger, and following the nine-minute-long Centaur firing, Ulysses would also have snared another record by becoming the fastest ever man-made machine, hurtling to Jupiter at 15.9 km/sec! One can imagine that, despite their joy at getting this important international mission underway, Hauck and his crew would have been glad to see the back of both it and the Centaur. Throughout the second half of 1985 and into the spring of 1986, in addition to their rigorous training regimes, both Hauck and Walker found themselves routinely questioning their own judgement over how many potential failure modes and problems they could live with.

"In early January 1986," Hauck recalled, "we were working an issue to do with redundancy in the helium actuation system for the liquid oxygen [and] liquid hydrogen dump valves and it was clear that the [senior Shuttle management] was willing to compromise on the margins in the propulsive force being provided by the pressurised helium. We were very concerned about it. We had discussions with the technical people, but we went to a [review] board to argue why this was not a good idea to compromise on this feature. The board turned down the request. I went back to the office and said to my crew, in essence, 'NASA is doing business differently from the way it has in the past. Safety is being compromised and, if any of you want to take yourself off this flight, I will support you'. Two or three weeks later, Challenger blew up. Now, there is no direct correlation between my experience and Challenger, but it seemed to me that there was a willingness to compromise on some of the things that we shouldn't compromise on."

Years later, Hauck remained undecided as to whether he would have refused, personally, to fly STS-61F, but admitted that Shuttle programme managers were taking unacceptable risks in the months preceding Challenger's fateful launch. Only days after the tragedy, any lingering doubts were resolved for him. The Kennedy Space Center's safety office refused to approve advanced processing of the first Centaur-G Prime, citing "insufficient verification of hazard controls" from both NASA and the booster's manufacturer, General Dynamics. Additional safety concerns, and cost overruns to the tune of $100 million, ultimately led to the project's cancellation in June 1986. Fortunately, a few years later, the Galileo and Ulysses missions went ahead, reverting to the less powerful IUS to get them successfully – though not without incident and requiring longer journey times – to their celestial targets. History has shown us that both achieved considerably more than expected and truly revolutionised humanity's understanding of both our parent star and our planetary big brother.

Ulysses was finally launched by Space Shuttle Discovery's STS-41 crew in October 1990, reaching Jupiter 16 months later, thanks to the combined thrust of its IUS and a Payload Assist Module (PAM)-S booster. It reached its maximum latitude of 80.2 degrees at the Sun's south pole on September 13th 1994, then crossed the ecliptic and travelled through high northern latitudes between June and September of the following year. Both series of observations were conducted during 'quiet', or 'minimum', periods of the Sun's 11-year cycle of activity; a further set of studies at 'maximum' solar conditions were completed between November 2000 and December 2001. Additionally, Ulysses undertook serendipitous analysis of Comet Hyakutake, passing through its billion-kilometre-long tail in May 1996, and observed Comet Hale–Bopp the following year. In August 1998, employing its gamma ray experiment, it also recorded a magnetic burst from the star SGR1900 + 14 in the constellation Aquila, some 20,000 light years from Earth.

The success story which Ulysses later became – and, indeed, still is, for its operational mission has since been extended until at least March 2008 – could scarcely have been further from NASA or ESA's collective mind on January 28th 1986, as Challenger's wreckage tumbled into the Atlantic from the STS-51L fireball. All Shuttle missions, predictably, were indefinitely suspended until the Rogers Commission, whose staff panel included former astronaut Neil Armstrong and STS-7 and STS-41G veteran Sally Ride, had completed its inquiry and made recommendations.

Among its conclusions were that NASA and Thiokol's operation of the Shuttle was seriously flawed: concerns from individual engineers were not reaching appropriate managers, 'critical' items were not being given the attention they demanded and the need to stick to a 'schedule', partly in a bid to please customers, was overriding 'safety'. Not only was NASA attempting to accommodate its major customers but, evidenced in a teleconference with Marshall Space Flight Center and Kennedy Space Center managers on the evening of January 27th 1986, Thiokol showed that it was prepared to ignore the safety concerns of several of its engineers to accommodate NASA, its own major customer. Worries of potential O-ring failure under the near-freezing weather conditions predicted for the following morning, expressed by Roger Boisjoly and others, were ignored, downplayed and Thiokol collectively voted that Challenger was fit to fly, unwittingly signing the STS-51L crew's death warrants in the process.

During that fateful teleconference, Thiokol's vice-president for engineering, Bob Lund, argued that his team's 'comfort level' was not to fly SRBs at temperatures below 12 degrees Celsius – some 53 degrees Fahrenheit – for fear of catastrophic 'blow-by' of the O-rings and field joints, but he could present no evidence to Marshall that 'proved' it was unsafe to do so. In a lengthy debate, Lawrence Mulloy – based in Florida as Marshall's KSC representative manager at the time – and other NASA officials challenged Thiokol's data and questioned its logic. At one stage, the MSFC director of science and engineering, George Hardy, remarked that he was "appalled" at the company's decision. So was Mulloy, who scornfully exploded with "For God's sake, Thiokol, when do you expect me to launch? Next April?" Neither man, however, was prepared to ignore the recommendation of their major contractor. Lund stood firm and, had he continued to do so, NASA would have had little choice but to postpone the STS-51L launch. Shortly thereafter, Thiokol requested a five-minute recess from the teleconference to consider the situation. Five minutes ultimately became half an hour.

Throughout this recess, Boisjoly and fellow engineer Arnie Thompson continued to argue persuasively that it was unsafe to fly outside of their proven field joint temperature range, but the Thiokol senior executives in attendance felt the O-rings should still seat and function properly, despite the cold weather. "Arnie actually got up from his position and walked up the table, put a quarter pad down in front of the management folks and tried to sketch out once again what his concern was with the joint," Boisjoly told the Rogers Commission, "and when he realised he wasn't getting through, he stopped. I grabbed the photos and tried to make the point that it was my opinion from actual observations that temperature was indeed a discriminator and we

Members of the Rogers Commission, including chairman William Rogers (centre) arrive at the Kennedy Space Center on March 7th 1986, during the course of their inquiry into the Challenger disaster.

should not ignore the physical evidence that we had observed. I also stopped when it was apparent that I couldn't get anybody to listen."

Then, executive Jerry Mason – presumably aware of the need not to upset NASA – explicitly asked Lund to "take off your engineering hat and put on your management hat". When the teleconference resumed, Lund indeed changed his vote and Thiokol changed its position on the issue. The company's new recommendation was that, although frigid weather conditions remained a problem, their data was indeed inconclusive and the launch of STS-51L should go ahead the following morning. None of the engineers wrote out the new recommendation – "I was not even asked to participate in giving any input to the final decision charts," Boisjoly told the Rogers hearing – and none but the executive managers signed it.

However, when MSFC and KSC managers asked for any additional comments from around the Thiokol table before closing the teleconference, none of them voiced their concerns. Boisjoly, in particular, remained silent; a fact which would later lead some observers to brand him a witness who turned 'state's evidence', rather than a noble 'whistleblower'. When questioned by a Rogers panel member, he emphasised that "I never [would] take [away] any management right to take the input of an engineer and then make a decision based upon that input, and I truly believe that. There was no point in me doing anything any further than I had already attempted to do ... [but] I left the room feeling badly defeated. I personally felt that management was under a lot of pressure to launch and that they made a very tough decision, but I didn't agree with it."

Having analysed the results of the teleconference, and interviewed the participants, the Rogers report concluded that "there was a serious flaw in the decision-making process leading up to the launch ... A well-structured and managed system, emphasising safety, would have flagged the rising doubts about the Solid Rocket Booster joint seal." In fact, when brought to testify before the panel, key officials intimately involved with the decision-making process, including STS-51L Launch Director Gene Thomas, Shuttle programme manager Arnie Aldrich and NASA's Associate Administrator for Spaceflight Jesse Moore admitted that they had not been privy to the issues raised at the January 27th teleconference.

In addition to mandated changes in communication channels, such that individual engineers could express concerns more openly, the most important requirement which had to be met before the Shuttle could fly again was the redesign of the Solid Rocket Booster's field joint and O-ring seal to prevent future combustion gas leakages. In its July 1986 response to President Ronald Reagan and the Rogers Commission, NASA announced its plans: to redesign the joint's metal components, insulation and seals, thereby providing "improved structural capability, seal redundancy and thermal protection". New capture latches would reduce joint movements caused by motor pressure or structural loads and the O-rings were redesigned to not leak under structural deflection at twice the expected level. Internal insulation was modified to be sealed with a deflection relief flap, rather than putty, and new bolts, strengtheners and a third O-ring were added. External heaters with integrated weather seals would maintain future SRB joint temperatures at 24 degrees Celsius or above and prevent water from entering the seals. "The strength of the improved joint design," read NASA's reply to Reagan, "is expected to approach that of the [SRB] case walls."

Another key result of Challenger was that the Shuttle would henceforth only be employed for missions which explicitly required its unique capabilities and those of its crews. Particular focus would be granted to scientific research. More than two dozen commercial and military satellites, previously booked to fly aboard the orbiters in 1986 and beyond, were transferred to expendable rockets. Ironically, in pre-Challenger days – and in line with the 'Shuttle-only' policy inherent in the designation of the orbiter as the National Space Transportation System – these rockets were in the process of being phased out. Apart from a handful of contracts signed prior to STS-51L, including several top-secret Department of Defense payloads that had been configured to fly only aboard the Shuttle and the Italian Space Agency's second Laser Geodynamics Satellite (LAGEOS-2), no further commercial 'primary' cargoes would be trucked aloft by future crews. A deviation from that policy came in May 1992, when, on her maiden voyage, the Challenger-replacement orbiter, Endeavour, conducted a breathtaking retrieval, repair and redeployment of the stranded Intelsat-6-F3 communications satellite. Although successful and once more demonstrative of the Shuttle's unique capabilities, the STS-49 retrieval was an initial, worrying hint that the lessons from STS-51L were fading from NASA's mind.

Former astronaut Tom Henricks, who flew four missions between November 1991 and July 1996, noted that, as SRB design changes appeared to 'work' successfully, safety was once more being compromised. "The pendulum [after Challenger] had swung to as conservative as they could make it," he said, "but then that pendulum started swinging back almost immediately and it was very prevalent by the time we were going to [the Russian space station] Mir. We were still sending Americans to Mir after a fire and a collision. Near the post-Challenger timeframe, that wouldn't have happened." Henricks actually turned down the chance to command a mission to Mir in June 1998 due to these safety concerns. Ultimately, added Mike Mullane, these fears and the one-off decision to fly former astronaut and US senator John Glenn on STS-95 in October 1998 contributed to a sense of over-confidence in the Shuttle which culminated in the loss of Columbia.

Among the safety improvements made to increase the survivability of future crews in the wake of STS-51L were upgraded brakes and tyres, the development of a drag chute to support the Shuttle's high-speed touchdowns and the incorporation of an escape pole which could be used to bail out of the vehicle's middeck side hatch in the event of serious problems. It was recognised that, without a pole to provide sufficient clearance, astronauts evacuating a crippled orbiter in flight would quickly impact the left wing. Unfortunately, the pole – which was attached to the middeck ceiling during a mission – could only be used when the Shuttle was in controlled, gliding flight, and not much higher than the altitude of Challenger when she disintegrated. The seven astronauts aboard Columbia for STS-107, which broke up 61 km above Earth on February 1st 2003, stood no chance.

In the wake of Challenger, each astronaut was provided with a partial-pressure suit – later upgraded, in 1994, to a fully pressurised ensemble – which would provide hyperbaric protection during ascent and cold-water immersion protection in the event of an emergency ditching in the ocean, together with parachute and life raft. However, Mullane, who flew the second Shuttle mission after STS-51L in December 1988, commented "I was strapped into a fortress that would keep me alive long enough to watch Death's approach. If fire was to kill me, I would have time to watch the flames. If a multi-mile fall was to kill me, I would watch the Earth rushing into my face. Even a cockpit depressurisation would no longer mercifully grant us unconsciousness, as it might have spared the Challenger crew. We now wore pressure suits that would keep us alive and conscious through any cockpit rupture ..."

"THE SUN KEPT ON RISING"

Other concerns raised by the Rogers Commission were the short periods separating individual missions, which provided insufficient opportunity for flight data from one voyage to be properly analysed before the next one set off. One particular example was a potentially serious problem with Columbia's brakes during STS-61C, which was launched only 16 days before STS-51L. The Flight Readiness Review for the latter occurred on January 15th 1986, whilst Columbia was still in orbit, and the data from the brake problem – together with further, disturbing O-ring erosion – had not been analysed until the 30th. By then, of course, it was too late for Challenger and her crew. When missions resumed in September 1988, NASA mandated that a minimum of three weeks should separate every Shuttle launch, although even this was waived in November–December 1990 when two missions blasted off within 16 days of each other.

Merely casting a cursory glance over Challenger's plans for the remainder of 1986 and, especially, into 1987 indicates that individual-orbiter flight rates of six missions per annum were hopelessly optimistic and, even with crippling overtime and around-the-clock Orbiter Processing Facility operations, would probably have been unachievable. STS-61F was scheduled to land at KSC at around 3:30 pm on May 19th, completing Challenger's shortest yet flight at just under four days; interestingly, her sister ship Atlantis, which would have been sitting on Pad 39A at the time, was scheduled to lift off barely 23 hours later! Judging from Horace Lamberth's comment to the Rogers panel that a lack of spare parts would bring the Shuttle to its knees by this time, it will never be known if NASA could have succeeded in launching two missions within the same week. If either had been postponed beyond the May 1986 window, the next Jovian opportunity would not have opened until June 1987.

Most observers doubt that Ulysses and Galileo could have been launched on time, not just because of the short launch window or the lack of spare parts, but due to ongoing problems with certifying the Centaur-G Prime for its advanced processing at KSC in the spring of 1986. When the two spacecraft were finally launched several years later, *sans* Centaur, they did so in separate Jovian windows: Galileo in October 1989, Ulysses in October 1990.

After STS-61F, three further missions would have awaited Challenger in 1986. The first, a five-day flight tentatively scheduled to begin on July 22nd, was STS-61M, led by Commander Loren Shriver. He and his crew of five – Pilot Bryan O'Connor, Mission Specialists Bill Fisher, Sally Ride and Mark Lee and McDonnell Douglas Payload Specialist Bob Wood – were assigned to deploy the third Tracking and Data Relay Satellite and operate the first commercial Electrophoresis Operations in Space (EOS-1) facility. The latter was a larger, pallet-mounted payload bay variant of the continuous flow electrophoresis experiment undertaken on several previous missions. In addition to this flight, it was booked, along with TDRS-D and Wood, on Challenger's seven-day STS-71D mission in February 1987.

However, according to veteran McDonnell Douglas Payload Specialist Charlie Walker, who would have served as backup for STS-61M, Wood's training "didn't go as smoothly as hoped [and] management was considering putting me up as Number One again. I had my doubts. I was tired of the pace. Too much of a good thing too fast! Of course, it ended up being a moot issue." It is interesting to speculate that, had Walker launched a fourth time, he would have become only the second person to record so many Shuttle missions, rivalling Bob Crippen, and undoubtedly frustrating a number of 'career' astronauts, some of whom had been waiting half a decade for their first flights.

In yet another mission which Challenger herself had 'baselined' previously, an Indian National Satellite – Insat-1C – would have ridden STS-61I into orbit on September 27th 1986, together with Indian mechanical engineer Nagapathi Bhat in a middeck Payload Specialist's seat. Physically identical to the cube-shaped Insat-1B lofted during Dick Truly's STS-8 mission three years earlier, the 1,150 kg satellite would have provided telecommunications, direct-broadcast television and meteorological services to India's civilian community over a projected seven-year lifespan. Insat-1C was ultimately remanifested onto an expendable launch vehicle and successfully delivered into orbit by an Ariane-3 rocket in July 1988. Arguably the most visibly exciting aspect of the four-day STS-61I mission, however, was the retrieval of the Long Duration Exposure Facility, which Challenger herself had orbited almost two and a half years before.

This salvage operation, to which Commander Don Williams, Pilot Mike Smith and Mission Specialists Bonnie Dunbar, Sonny Carter and Jim Bagian had been assigned in the autumn of 1985, was itself overdue. Originally scheduled for February 1985, it was repeatedly postponed as commercial payloads bumped it further downstream. By the time Columbia finally approached the 12-sided satellite in January 1990, with Dunbar at the controls of the RMS mechanical arm, it had spent almost six years in space and was within weeks of re-entering Earth's atmosphere to destruction. Interestingly, Dunbar was the only member of the original STS-61I crew to actually fly the 'real' LDEF recovery mission.

In a similar vein to the Teacher in Space effort, STS-61I's second Payload Specialist seat would also have been granted to an 'ordinary' civilian – this time a journalist – and, at the time of the disaster, the applicants had been winnowed down by NASA to a list of 40 semi-finalists. These included NBC News' Theresa Anzur, Pulitzer prizewinners James Wilford and Peter Rinearson, James Asker of the 'Houston Post', freelancers Jay Barbree and Marcia Bartusiak, ABC's William Blakemore, 'Time' magazine's Roger Rosenblatt, reporter Rob Navias (later to become the Johnson Space Center spokesman for NASA) and 69-year-old CBS veteran anchorman Walter Cronkite. At the time of the disaster, the screening of those 40 semi-finalists was scheduled to take place at JSC on March 31st 1986 and the successful primary and backup candidates would have begun formal training in May. All 40 lost their chance that cold January day when not only the Shuttle was suspended, but so were its commercial ventures and any lingering illusion of it truly being the spacegoing equivalent of a passenger-carrying airliner was forever obliterated.

Challenger's final planned mission for 1986 was STS-71B on December 6th, thus designated because it was funded under NASA's budgetary allocation for the 1987 financial year. Virtually nothing is known about this flight, partly because it was a classified Department of Defense assignment and also because no crew members – apart from one Manned Spaceflight Engineer (MSE), US Air Force officer Chuck Jones, as its lone Payload Specialist – were ever named. In the years following the disaster, it has come to light that Jones and his NASA crewmates would probably have deployed an important $400 million Defense Support Program (DSP) missile early warning satellite, atop an Inertial Upper Stage. This already proven satellite system employed an array of infrared sensors to detect the heat 'signatures' of missile and booster exhausts against Earth's background.

The satellite assigned to Jones' mission came from a proud heritage: the system had undergone five upgrades since it first began operations in November 1970 and had typically exceeded its design life by around 30 per cent. Capable of being launched aboard both expendable rockets and the Shuttle, a DSP was actually inserted into orbit by Atlantis during her STS-44 mission in November 1991; the deployment scenario, leading up to a release of the payload and its IUS booster some six and a half hours after lift-off, followed an almost identical procedure to that of the Tracking and Data Relay Satellite.

The DSP itself measured almost ten metres long and seven metres wide when fully operational, weighed 2,380 kg and was dominated by its tube-shaped, wide-angle Schmidt infrared telescope. After insertion into equatorial geostationary orbit and separation from the final stage of the IUS, it would have been spin-stabilised about its Earth-facing axis, rotating at about six revolutions per minute. Detection of infrared sources was accomplished by means of the telescope and Photoelectric Cell (PEC) array portions of its subsystem. The PEC detector array, mounted in the telescope's centreline to coincide with the imaging surface of its optics, aimed off the satellite's axis, scanned Earth's surface whilst the DSP rotated. As the detector passed over an infrared source of interest, it developed an electronic signal for subsequent transmission to US Air Force-operated ground stations.

The effectiveness of these satellites, the 23rd and last of which is scheduled for launch in January 2007, was amply demonstrated during the first Gulf War, when they detected the infrared signatures of Iraqi Scud missile exhausts, allowing for timely warnings to be issued to civilian populations in Israel and Saudi Arabia. Additionally, they have seen use in the observation of volcanic eruptions and forest fires.

Military assignments would also have characterised several of Challenger's six scheduled flights during the course of 1987. Three of these – STS-71G in April, STS-81A in October and STS-81D in December – would have been mixed-cargo missions including 1,720 kg Global Positioning System (GPS) Navstar satellites. The latter project, which began in 1975, sought to supply latitude, longitude, velocity, time and altitude data to aircraft, shipping and submarines, vehicles or individuals, whether static or moving, anywhere on land, water, in the air or in space.

When fully unfurled and operational in orbit, the Navstars measured 1.2 m wide and 5.3 m long, with a pair of solar cell 'wings' providing 700 watts of electrical power. Built by Rockwell International on behalf of the Department of Defense, the 12 satellites destined to be launched by the Shuttle between 1987–1989 were part of a second-generation series boosted into 20,000 km orbits by Payload Assist Module

(PAM)-D2 upper stages. The weight of the Navstars placed them above the 1,270 kg maximum threshold of the 'standard' PAM-D boosters employed on several of Challenger's previous satellite deployment missions. For this reason, they would have been mounted atop the newer PAM-D2s, which could accommodate payloads measuring up to three metres in diameter, as opposed to just two metres for the standard variant. Indeed, thanks to its uprated Thiokol-built Star-63D motor, it could transport satellites weighing up to 1,920 kg into geosynchronous transfer orbits. In the spring of 1985, the US Air Force's Space Division awarded a $169.4 million contract to McDonnell Douglas to build a total of 28 PAM-D2s for Navstar and other purposes.

Other plans for Challenger in 1987 included STS-71J in June, which would have released the Long Duration Exposure Facility for a second period of untended free flight – lasting up to two years – with a new set of experiments, and STS-71M in August, carrying the ASTRO-3 ultraviolet observatory. The latter would also have deployed NASA's Cosmic Release and Radiation Effects Satellite (CRRES) into an initial 400 km orbit. The latter, which eventually rode an expendable rocket into space in July 1990, sought to explore fields, plasmas and energetic particles inside Earth's magnetosphere and conduct a series of chemical release experiments.

Meanwhile, ASTRO-3 was to be the last in a series of Shuttle missions to carry three sophisticated ultraviolet telescopes on a pair of Spacelab pallets in the payload bay. According to pre-Challenger plans, this third flight and its two predecessors aboard Columbia would have featured astronauts Jeff Hoffman and Bob Parker as Mission Specialists, together with a rotating system of three Payload Specialists. Sam Durrance and Ron Parise were assigned to fly ASTRO-1 – the next planned flight after STS-51L – in March 1986, after which Parise would join Ken Nordsieck for ASTRO-2 in January 1987. Durrance and Nordsieck would then be teamed for ASTRO-3. It seems quite remarkable, today, that Hoffman and Parker were scheduled to rocket into orbit on all three missions over a period of barely 17 months.

Parker, in particular, has since expressed disbelief at the sheer number of Shuttle flights planned at the time of Challenger's loss. "It's amazing, when you look back at that [schedule pressure] and the rate at which we thought we had to keep pumping this stuff out," he recalled years later. "You'd have thought the world was going to end [if we didn't meet our launch targets]. My favourite expression is: Guess what? The Sun kept on rising and setting! The Sun didn't even notice [if we missed our targets]."

The ASTRO observatory can trace its origins back to 1978, when NASA issued an Announcement of Opportunity for advanced astronomical instruments for carriage aboard future Shuttle missions. Three were ultimately chosen – the Hopkins Ultraviolet Telescope (HUT), provided by the Johns Hopkins University of Baltimore, Maryland; the Wisconsin Ultraviolet Photopolarimeter Experiment (WUPPE), built by the Space Astronomy Laboratory at the University of Wisconsin at Madison; and the Ultraviolet Imaging Telescope (UIT), sponsored by NASA's Goddard Space Flight Center. The project was to be managed by the agency's Office of Space Science.

By 1982, however, control had passed to the Marshall Space Flight Center. Two years later, the first flight of the series was tentatively scheduled for the spring of 1986 –

exactly the same time that Halley's Comet would visit the inner Solar System – and a special wide-field camera was added to permit detailed observations of the celestial wanderer. By the end of January 1986, ASTRO-1 had completed its final checkout and was ready for installation into Columbia's payload bay for STS-61E, when Challenger was lost. For the next 32 months, the Shuttle and observatory were both grounded and ASTRO-1 did not ultimately fly until December 1990, producing spectacular results and prompting calls for a second mission.

The 3.6 m long and 1.2 m wide HUT, weighing over 770 kg, was intended to explore objects such as quasars, active galactic nuclei and 'normal' galaxies at far and extreme ultraviolet wavelengths. This region of the electromagnetic spectrum was inaccessible from Earth and even to the instruments on the Hubble Space Telescope. To achieve far and extreme ultraviolet sensitivity, HUT's mirrors were coated with iridium. Meanwhile, WUPPE was designed to examine the ultraviolet polarisation of hot stars, galactic nuclei and quasars. Any star – with the obvious exception of our Sun – is so distant that it only appears as a faraway point of light in a telescope eyepiece. If its light is polarised, however, it is possible to derive more information about its geometry and physical composition. Lastly, UIT would take wide field of view images of star clusters, planetary nubulae, supernova remnants and galactic clusters. Although Hubble was expected to have higher magnification, UIT could cover larger areas of the sky at once.

Major targets for all three planned ASTRO missions included red giants, which, at the end of their lives, shrink to become dense, hot embers no bigger than Earth, known as 'white dwarfs'. Since this is believed to be the final destination for many stars, they are important areas of study; they also emit most of their radiation at ultraviolet wavelengths, thus placing them squarely within the ASTRO observatory's capabilities. Other studies would focus on 'binary systems', in which two stars reside close together and sometimes exchange material, and stellar clusters, in which anything up to a million stars reside.

In visible light, it is difficult to distinguish the light from each celestial source, but under ASTRO's ultraviolet gaze they were expected to blaze individually. More broadly, the observatory was to examine the 'interstellar medium' – the enormous expanse of gas and dust between stars – which actually provides building material for future objects. Although the interstellar medium chiefly comprises hydrogen and has a typical density no higher than one atom per thimbleful of space, it was anticipated that ASTRO would measure its physical properties more closely and explore 'pockets' of it which are much hotter than normal.

The 7,830 kg ASTRO-3 observatory would have featured a pair of Spacelab pallets and an igloo, plus the problem-prone Instrument Pointing System (IPS), marking Challenger's first 'operational' mission with this payload configuration, following the STS-51F verification flight test in July 1985. As well as the precision pointing afforded by the IPS, an additional image motion compensation device had been provided by Marshall Space Flight Center to better stabilise HUT, WUPPE and UIT during their observations. This was capable of sensing crew- or thruster-induced movements of the Shuttle and send data to the telescopes, which automatically readjusted themselves to achieve a stability finer than a single arc-second. This would have proven particularly useful for UIT, whose images were to be recorded onto sensitive astronomical film.

Eleven missions by one orbiter in a period of less than two years, even in pre-Challenger days, was overly ambitious, if not ludicrous. Moreover, all four vehicles would have been operational – Columbia and Atlantis supporting their own missions from the Kennedy Space Center and Discovery trucking exclusively polar-orbiting Department of Defense payloads aloft from her new home at Vandenberg Air Force Base in California – which meant the availability of spare parts and the option of cannibalism would have been difficult, if not impossible. By the end of 1987, had the schedule run as planned, Challenger would have roared into orbit on 20 occasions in a span of less than five years, with very little time between missions to tend to the wear and tear from those journeys. Not until after the resumption of Shuttle flights in 1988 was a co-ordinated effort set in motion to ensure that each orbiter was removed from service after every seven or eight missions for an extended period of maintenance, thorough inspection and refurbishment.

In addition to the logistical and technical nightmare of continuing to prepare and execute missions, non-stop and safely, with just six to eight weeks between each launch of individual orbiters, the production of sufficient Solid Rocket Booster hardware and enough disposable External Tanks would have driven the programme to its knees. Many observers have predicted the Shuttle flight rate to have dropped precipitously in the spring of 1986, STS-51L or no STS-51L. Ongoing problems with O-ring erosion, failures of brakes and tyres, the risk of on-the-pad main engine shutdowns, inadequate time to train crews, a multitude of niggling issues with no time to properly address them and an increasingly aggressive launch schedule would all have taken their toll. One KSC engineer from those seemingly bulletproof, strap-it-on-and-go days, when asked the question that – if Challenger had not been lost – would the reusable spacecraft have achieved even half of its projected 24-missions-per-annum flight rate today, replied with a resounding "No".

Peculiarly, as late as mid-March 1986, after the identification of the SRB problem, but still in advance of the Rogers Commission's final report, NASA was anticipating a return to flight in the spring of the following year. Furthermore, the agency envisaged no fewer than nine missions in 1987, more than a dozen in the following year and an average of between 16 and 19 per annum by the end of the decade. At around the same time, plans were still afoot to retain commercial payloads – including Westar-6S and Insat-1C – and NASA strongly wished to continue its policy of flying passengers. Its dream, even in the wake of Challenger, to continue to fly the Shuttle 'routinely' and 'regularly', was in for a rude awakening.

When Columbia disintegrated during re-entry on February 1st 2003, her mission was the first of six flights planned for that year; a pitiful figure when compared to the numbers above. The reusable fleet reached its peak of operations in 1985, when three orbiters, some of them cannibalising parts from the fourth, undertook no fewer than nine separate voyages. Although the Shuttle never again duplicated the rate at which it flew prior to STS-51L, averaging seven trips per annum throughout much of the 1990s, it scored some remarkable triumphs: longer missions of up to 18 days, four Hubble Space Telescope servicing calls, the launch of Galileo and Ulysses, the establishment of a 'full' Tracking and Data Relay Satellite network in orbit, nine visits to Mir, dozens of research flights and the ongoing construction of the International Space Station. None of these missions were carried out in the absence of risk, but thanks to the sacrifice made by Dick Scobee and his brave crew that tragic January day in 1986, coupled with the changes NASA implemented in its immediate aftermath, each Shuttle launch that followed was rendered immeasurably safer.

7

Challenger missions 1983–1986

Designation	**STS-6**
Sequence	Sixth Shuttle flight and first flight of Challenger
Milestones	First Shuttle-based spacewalk, first Shuttle mission to utilise 104 per cent-capable main engines, first use of lightweight External Tank and Solid Rocket Boosters, first Shuttle mission to employ a Heads-Up Display (HUD) in the flight deck
Launch	April 4th 1983 at 6:30:00 pm GMT from Pad 39A at the Kennedy Space Center (KSC) in Florida
Payload	Tracking and Data Relay Satellite (TDRS)-A with Inertial Upper Stage (IUS), three Getaway Special (GAS) canisters, Continuous Flow Electrophoresis System (CFES), Night/Day Optical Survey of Lightning (NOSL), Radiation Monitoring Experiment (RME) and Monodisperse Latex Reactor (MLR)
Max. altitude	285 km
Inclination	28.45 degrees
Landing	April 9th 1983 at 6:53:42 pm GMT on Runway 22 at Edwards Air Force Base in California
Orbits	80
Duration	5 days, 0 hours, 23 minutes and 42 seconds
Crew	Paul Joseph Weitz, 50, Commander Col Karol Joseph Bobko, 45, US Air Force, Pilot Donald Herod Peterson, 49, Mission Specialist 1 Dr Franklin Story Musgrave, 47, Mission Specialist 2

Designation	**STS-7**
Sequence	Seventh Shuttle flight and second flight of Challenger
Milestones	First American woman in space, first use of Ku-band communications antenna on the Shuttle, first deployment of three satellites, first photograph of the Shuttle in orbit, first time five astronauts launched in the the same spacecraft, first proximity operations and rendezvous exercises conducted with another satellite
Launch	June 18th 1983 at 11:33:00 am GMT from Pad 39A at the Kennedy Space Center (KSC) in Florida
Payload	Anik-C2 and Palapa-B1, both with Payload Assist Module (PAM)-Ds, Shuttle Pallet Satellite (SPAS)-1, Office of Space and Terrestrial Applications (OSTA)-2, seven Getaway Special (GAS) canisters, Continuous Flow Electrophoresis System (CFES), Shuttle Student Involvement Program (SSIP) and Monodisperse Latex Reactor (MLR)
Max. altitude	310 km
Inclination	28.45 degrees
Landing	June 24th 1983 at 1:56:59 pm GMT on Runway 15 at Edwards Air Force Base in California
Orbits	97
Duration	6 days, 2 hours, 23 minutes and 59 seconds
Crew	Capt Robert Laurel Crippen, 45, US Navy, Commander Capt Frederick Hamilton Hauck, 42, US Navy, Pilot Lt-Col Dr John McCreary Fabian, 44, US Air Force, Mission Specialist 1 Dr Sally Kristen Ride, 32, Mission Specialist 2 Dr Norman Earl Thagard, 39, Mission Specialist 3

Designation	**STS-8**
Sequence	Eighth Shuttle flight and third flight of Challenger
Milestones	First black American in space, oldest man in space, first nocturnal Shuttle launch and landing, first flight of improved-performance Solid Rocket Booster motors, first communications tests with Tracking and Data Relay Satellite
Launch	August 30th 1983 at 6:32:00 am GMT from Pad 39A at the Kennedy Space Center (KSC) in Florida
Payload	Indian National Satellite (Insat)-1B, Payload Flight Test Article (PFTA), Continuous Flow Electrophoresis System (CFES), Animal Enclosure Module (AEM), Incubator Cell Attachment Test (ICAT), Investigation of STS Atmospheric Luminosities (ISAL), Radiation Monitoring Experiment (RME), Shuttle Student Involvement Program (SSIP), Development Flight Instrumentation (DFI) pallet and 12 Getaway Special (GAS) canisters
Max. altitude	305 km
Inclination	28.45 degrees
Landing	September 5th 1983 at 7:40:43 am GMT on Runway 22 at Edwards Air Force Base in California
Orbits	97
Duration	6 days, 1 hour, 8 minutes and 43 seconds
Crew	Capt Richard Harrison Truly, 45, US Navy, Commander Cdr Daniel Charles Brandenstein, 40, US Navy, Pilot Lt-Cdr Dale Allan Gardner, 34, US Navy, Mission Specialist 1 Lt-Col Dr Guion Stewart Bluford Jr, 40, US Air Force, Mission Specialist 2 Dr William Edgar Thornton, 54, Mission Specialist 3

Designation	**STS-41B**
Sequence	Tenth Shuttle flight and fourth flight of Challenger
Milestones	First untethered spacewalks and first Shuttle landing back at the launch site
Launch	February 3rd 1984 at 1:00:00 pm GMT from Pad 39A at the Kennedy Space Center (KSC) in Florida
Payload	Palapa-B2 and Westar-6, both with Payload Assist Module (PAM)-Ds, Shuttle Pallet Satellite (SPAS)-1A, Inflatable Rendezvous Target (IRT), two Manned Manoeuvring Units (MMUs), five Getaway Special (GAS) canisters, Acoustic Containerless Experiment System (ACES), Isoelectric Focusing (IEF), Cinema-360 camera, Monodisperse Latex Reactor (MLR) and Radiation Monitoring Experiment (RME)
Max. altitude	320 km
Inclination	28.45 degrees
Landing	February 11th 1984 at 12:15:55 pm GMT on Runway 33 at the Kennedy Space Center (KSC) in Florida
Orbits	127
Duration	7 days, 23 hours, 15 minutes and 55 seconds
Crew	Vance DeVoe Brand, 52, Commander Lt-Cdr Robert Lee 'Hoot' Gibson, 37, US Navy, Pilot Dr Ronald Erwin McNair, 33, Mission Specialist 1 Lt-Col Robert Lee Stewart, 41, US Army, Mission Specialist 2 Capt Bruce McCandless II, 46, US Navy, Mission Specialist 3

Designation	**STS-41C**
Sequence	Eleventh Shuttle flight and fifth flight of Challenger
Milestones	First direct-insertion Shuttle launch, utilising only one Orbital Manoeuvring System burn, first satellite retrieval and repair and first operational use of the Manned Manoeuvring Unit
Launch	April 6th 1984 at 1:58:00 pm GMT from Pad 39A at the Kennedy Space Center (KSC) in Florida
Payload	Long Duration Exposure Facility (LDEF), Flight Support Structure (FSS) for Solar Max retrieval, repair and deployment, two Manned Manoeuvring Units (MMUs), Cinema-360 camera, Shuttle Student Involvement Program (SSIP), Radiation Monitoring Experiment (RME) and IMAX Camera
Max. altitude	500 km
Inclination	28.45 degrees
Landing	April 13th 1984 at 1:38:06 pm GMT on Runway 17 at Edwards Air Force Base in California
Orbits	107
Duration	6 days, 23 hours, 40 minutes and 6 seconds
Crew	Capt Robert Laurel Crippen, 46, US Navy, Commander Francis Richard Scobee, 44, Pilot Terry Jonathan Hart, 37, Mission Specialist 1 Dr James Douglas Adrianus 'Ox' van Hoften, 39, Mission Specialist 2 Dr George Driver 'Pinky' Nelson, 33, Mission Specialist 3

Designation	**STS-41G**
Sequence	Thirteenth Shuttle flight and sixth flight of Challenger
Milestones	First time seven astronauts launched in the same spacecraft, first Canadian spacefarer, first Australian-born astronaut, longest mission by Challenger, first flight of Payload Specialists aboard Challenger, first space mission to feature two women, first spacewalk by an American woman and last Kennedy Space Center landing by Challenger
Launch	October 5th 1984 at 11:03:00 am GMT from Pad 39A at the Kennedy Space Center (KSC) in Florida
Payload	Earth Radiation Budget Satellite (ERBS), Office of Space and Terrestrial Applications (OSTA)-3, Large Format Camera (LFC), Orbital Refuelling System (ORS), Auroral Photography Experiment (APE), Canadian Experiments (CANEX)-1, IMAX Camera, Radiation Monitoring Experiment (RME), Thermoluminescent Dosimeter (TLD) and eight Getaway Special (GAS) canisters
Max. altitude	350 km
Inclination	57.00 degrees
Landing	October 13th 1984 at 4:26:38 pm GMT on Runway 33 at the Kennedy Space Center (KSC) in Florida
Orbits	132
Duration	8 days, 5 hours, 23 minutes and 38 seconds
Crew	Capt Robert Laurel Crippen, 47, US Navy, Commander Cdr Jon Andrew McBride, 41, US Navy, Pilot Dr Kathryn Dwyer Sullivan, 33, Mission Specialist 1 Dr Sally Kristen Ride, 33, Mission Specialist 2 Lt-Cdr David Cornell Leestma, 35, US Navy, Mission Specialist 3 Dr Paul Desmond Scully-Power, 40, Payload Specialist 1 Cdr Dr Joseph Jean-Pierre Marc Garneau, 35, Canadian Navy, Payload Specialist 2

Designation	**STS-51B**
Sequence	Seventeenth Shuttle flight and seventh flight of Challenger
Milestones	First 'dedicated' Spacelab mission by Challenger, first mission to feature three astronauts over the age of 50, oldest man in space, first Chinese-born astronaut and first Dutch-born astronaut
Launch	April 29th 1985 at 4:02:18 pm GMT from Pad 39A at the Kennedy Space Center (KSC) in Florida
Payload	Spacelab-3 and two Getaway Special (GAS) canisters, containing North Utah Satellite (NUSAT) and Global Low-Orbiting Message Relay (GLOMR)
Max. altitude	360 km
Inclination	57.00 degrees
Landing	May 6th 1985 at 4:11:06 pm GMT on Runway 17 at Edwards Air Force Base in California
Orbits	110
Duration	7 days, 0 hours, 8 minutes and 46 seconds
Crew	Col Robert Franklyn Overmyer, 48, US Marine Corps, Commander (gold team) Col Frederick Drew Gregory, 44, US Air Force, Pilot (silver team) Dr Don Leslie Lind, 54, Mission Specialist 1 (gold team) Dr Norman Earl Thagard, 41, Mission Specialist 2 (silver team) Dr William Edgar Thornton, 56, Mission Specialist 3 (gold team) Dr Taylor Gun-Jin Wang, 44, Payload Specialist 1 (gold team) Dr Lodewijk van den Berg, 53, Payload Specialist 2 (silver team)

Designation	**STS-51F**
Sequence	Nineteenth Shuttle flight and eighth flight of Challenger
Milestones	First verification flight test of the pallet-train-and-igloo Spacelab combination, together with the Instrument Pointing System (IPS), oldest man in space, experienced Challenger's first on-the-pad main engine shutdown, followed by the programme's first (and so far only) Abort To Orbit (ATO) and first flight of carbonated beverages (Coke and Pepsi) in space
Launch	July 29th 1985 at 9:00:00 pm GMT from Pad 39A at the Kennedy Space Center (KSC) in Florida
Payload	Spacelab-2, Plasma Diagnostics Package (PDP) and Life Sciences payload in middeck
Max. altitude	230 km (390 km originally planned before ATO)
Inclination	49.50 degrees
Landing	August 6th 1985 at 7:45:26 pm GMT on Runway 23 at Edwards Air Force Base in California
Orbits	126
Duration	7 days, 22 hours, 45 minutes and 26 seconds
Crew	Col Charles Gordon Fullerton, 48, US Air Force, Commander (red/blue team) Lt-Col Roy Dunbard Bridges Jr, 42, US Air Force, Pilot (red team) Dr Karl Gordon Henize, 58, Mission Specialist 1 (red team) Dr Franklin Story Musgrave, 49, Mission Specialist 2 (blue team) Dr Anthony Wayne England, 43, Mission Specialist 3 (blue team) Dr Loren Wilber Acton, 49, Payload Specialist 1 (red team) Dr John-David Francis Bartoe, 40, Payload Specialist 2 (blue team)

Designation	**STS-61A**
Sequence	Twenty-second Shuttle flight and ninth flight of Challenger
Milestones	First Spacelab mission dedicated to West Germany, first (and so far only) time eight astronauts launched in the same spacecraft, first Shuttle flight operated from a control centre outside the United States and final fully successful mission by Challenger
Launch	October 30th 1985 at 5:00:00 pm GMT from Pad 39A at the Kennedy Space Center (KSC) in Florida
Payload	Spacelab-D1 and Global Low Orbiting Message Relay (GLOMR)
Max. altitude	320 km
Inclination	57.00 degrees
Landing	November 6th 1985 at 5:44:51 pm GMT on Runway 17 at Edwards Air Force Base in California
Orbits	111
Duration	7 days, 0 hours, 44 minutes and 51 seconds
Crew	Henry Warren Hartsfield, 51, Commander (red/blue team) Lt-Col Steven Ray Nagel, 39, US Air Force, Pilot (blue team) Dr Bonnie Jeanne Dunbar, 36, Mission Specialist 1 (blue team) Lt-Col James Frederick Buchli, 40, US Marine Corps, Mission Specialist 2 (red team) Col Dr Guion Stewart Bluford Jr, 42, US Air Force, Mission Specialist 3 (red team) Dr Reinhard Alfred Furrer, 44, Payload Specialist 1 (blue team) Dr Ernst Willi Messerschmid, 40, Payload Specialist 2 (red team) Dr Wubbo Johannes Ockels, 39, Payload Specialist 3 (red/blue team)

Designation	**STS-51L**
Sequence	Twenty-fifth Shuttle flight and tenth flight of Challenger
Milestones	First civilian passenger to fly aboard the Shuttle, first American manned mission to involve in-flight fatalities, first human space mission to launch and fail to reach its destination
Launch	January 28th 1986 at 4:38:00 pm GMT from Pad 39B at the Kennedy Space Center (KSC) in Florida
Payload	Tracking and Data Relay Satellite (TDRS)-B with Inertial Upper Stage (IUS), Shuttle Pointed Autonomous Research Tool for Astronomy (SPARTAN)-203, Fluid Dynamics Experiment (FDE), Comet Halley Active Monitoring Program (CHAMP), Phase Partitioning Experiment (PPE), Teacher In Space Project (TISP) and Shuttle Student Involvement Program (SSIP)
Max. altitude	Planned for 250 km, but vehicle destroyed at an altitude of approximately 18 km over the Atlantic Ocean
Inclination	Planned for 28.45 degrees, but failed to achieve orbit
Landing	Planned for February 3rd 1986 at 5:12:00 pm GMT on Runway 33 at the Kennedy Space Center (KSC) in Florida, but vehicle destroyed during ascent
Orbits	96 planned, but vehicle destroyed during ascent and failed to achieve orbit
Duration	Planned for 6 days, 0 hours and 34 minutes, but vehicle destroyed 1 minute and 13 seconds after lift-off
Crew	Francis Richard Scobee, 46, Commander Cdr Michael John Smith, 40, US Navy, Pilot Lt-Col Ellison Shoji Onizuka, 39, US Air Force, Mission Specialist 1 Dr Judith Arlene Resnik, 36, Mission Specialist 2 Dr Ronald Erwin McNair, 35, Mission Specialist 3 Sharon Christa McAuliffe, 37, Payload Specialist 1 Gregory Bruce Jarvis, 41, Payload Specialist 2

Bibliography

"Spacelab simulation", JSC release, January 23rd 1976.
"Shuttle Training Aircraft delivery to JSC", JSC release, June 8th 1976.
"NASA to recruit Space Shuttle astronauts", JSC release, July 8th 1976.
"Spacelab simulation crew undergoes medical tests", JSC release, December 1st 1976.
"Spacelab medical simulation starts at Johnson Space Centre", JSC release, May 13th 1977.
"NASA selects 35 astronaut candidates", JSC release, January 16th 1978.
"Shuttle orbiters named after sea vessels", JSC release, February 1st 1979.
"Space Shuttle orbiter procurement contract signed", JSC release, February 5th 1979.
"Schooling of astronauts, 35 new candidates, is varied, exciting", JSC release, April 8th 1979.
"Shuttle launch for Palapa", *Flight International*, April 28th 1979.
"RMS contract award", JSC release, May 25th 1979.
"ESA starts work on Spacelab Sled", *Flight International*, July 14th 1979.
"NASA tests new space manoeuvring backpack", JSC release, September 10th 1979.
"Astronauts may repair orbiter heat shield in flight", JSC release, September 20th 1979.
"NASA to develop Manned Manoeuvring Unit", JSC release, October 2nd 1979.
"Heads-up display for orbiter", JSC release, October 16th 1979.
"Lockheed clears Shuttle orbiter structure for flight", *Flight International*, October 27th 1979.
"Martin receives TPS repair contract", JSC release, January 22nd 1980.
"NASA signs Martin Marietta to build Manned Manoeuvring Unit", JSC release, February 29th 1980.
"NASA signs Canadians to build Shuttle robot arm", JSC release, April 14th 1980.
"Investigators file report on cause of spacesuit backpack fire", JSC release, June 10th 1980.
"Martin Marietta to build Space Shuttle orbiter tile repair kits", JSC release, July 2nd 1980.
"Two Europeans selected for Space Shuttle mission specialist training", JSC release, July 7th 1980.
"Space Shuttle to carry space toolbox", JSC release, October 14th 1980.
"Spacesuit life support system ends 14-hour test", JSC release, October 16th 1980.
"NASA names crews for three Space Shuttle missions", JSC release, March 1st 1982.
"Three Shuttle crews announced", JSC release, April 19th 1982.
"NASA names STS-10 astronaut crew", JSC release, October 20th 1982.

"Team reports on STS-5 spacesuit failures", JSC release, December 2nd 1982.
"Fifth crewmember named to STS-7 and STS-8", JSC release, December 21st 1982.
"STS-11 and STS-12 crews named", JSC release, February 4th 1983.
"Crewmembers named for STS-13, Spacelab-2 and Spacelab-3", JSC release, February 18th 1983.
"Salt spray delays Shuttle", *Flight International*, March 26th 1983.
"IUS investigation board members named", JSC release, April 7th 1983.
"Second TDRS deleted from STS-8 manifest", JSC release, May 27th 1983.
"STS flight assignments", JSC release, November 17th 1983.
"51K crew announcement", JSC release, February 14th 1984.
"The launch-and-forget satellite", *Flight International*, April 7th 1984.
"IUS on the mend", *Flight International*, April 21st 1984.
"Astronaut T.J. Hart to leave NASA", JSC release, May 10th 1984.
"NASA announces updated flight crew assignments", JSC release, August 3rd 1984.
"Making Shuttle pay its way", *Flight International*, December 15th 1984.
"NASA names crew to deploy satellites in year-end flights", JSC release, January 29th 1985.
"Boost for PAM-D2", *Flight International*, March 2nd 1985.
"NASA changes 51B landing site to Edwards Air Force Base", JSC release, April 24th 1985.
"Spacelab-3 success claimed", *Flight International*, May 18th 1985.
"NASA names astronaut crews for Ulysses, Galileo missions", JSC release, May 31st 1985.
"NASA names astronaut crew for Space Shuttle mission 61I", JSC release, June 17th 1985.
"Shuttle and Spacelab to try again", *Flight International*, July 27th 1985.
"Spacelab-2 hits trouble", *Flight International*, August 10th 1985.
"Cola duel in space", *Flight International*, August 31st 1985.
"NASA names crews for upcoming Space Shuttle flights", JSC release, September 19th 1985.
"Pointing system performance praised", *Flight International*, September 21st 1985.
"Spacelab-2 reflight studied", *Flight International*, October 26th 1985.
"Journalist to fly on Shuttle", *Flight International*, November 9th 1985.
"Spacelab-D1 controlled from Europe", *Flight International*, November 23rd 1985.
"New work for Solar Max", *Flight International*, December 7th 1985.
"Vandenberg launch schedule date slips", *Flight International*, December 21st 1985.
"NASA free-flier spies Halley", *Flight International*, January 4th 1986.
"Major changes in 1986 plan", *Flight International*, January 25th 1986.
"Teacher flies next Shuttle", *Flight International*, January 25th 1986.
"NASA assesses launcher options", *Flight International*, March 8th 1986.
"Shuttle off until spring '87", *Flight International*, March 15th 1986.
"Shuttle to lose Centaur-G Prime?", *Flight International*, June 7th 1986.
"Report of the Presidential Commission on the Space Shuttle Challenger Accident", Washington, DC, June 1986.
"Shuttle restrictions urged", *Flight International*, August 9th 1986.
"Implementation of the Recommendations of the Presidential Commission on the Space Shuttle Challenger Accident", NASA publication, June 1987.

Byars, Carlos, "Shuttle's comet tracker malfunctioning", *Houston Chronicle*, July 30th 1985.
Cooper, Henry S.F., "Before Liftoff: the making of a Space Shuttle crew". Baltimore, Maryland: The Johns Hopkins University Press, 1987.
Evans, Ben, "Space Shuttle Columbia: her missions and crews". Chichester, UK: Springer–Praxis, 2005.
Isikoff, Michael, "Remains of Shuttle crew found", *Washington Post*, March 10th 1986.

Jenkins, Dennis R., "Space Shuttle: the history of the National Space Transportation System – the first 100 missions". Hinckley: Midland Publishing, 2001.

McConnell, Malcolm, "Challenger: a major malfunction". New York: Doubleday, 1987.

Mullane, Mike, "Riding Rockets: the outrageous tales of a Space Shuttle astronaut". New York: Scribner, 2006.

Reichardt, Tony, (ed.) "Space Shuttle: the first 20 years". Washington, DC: Smithsonian Institution, 2001.

Savage, P.D., G.C. Jahns, B.P. Dalton, R.P. Hogan and A.E. Wray, "The Rodent Research Animal Holding Facility as a barrier to environmental contamination", NASA Technical Memorandum 102237, September 1989.

Shapland, David and Michael Rycroft, "Spacelab: research in Earth orbit". Cambridge: Cambridge University Press, 1984.

Trento, Joseph J., "Prescription for disaster". New York: Crown, 1987.

Vaughan, Diane, "The Challenger launch decision: risky technology, culture and deviance at NASA". Chicago: University of Chicago Press, 1996.

Index

Note: page numbers in **bold** refer to illustrations.

Acton, Loren 198–200
Airborne Support Equipment 32, **33**
Aldrich, Arnie 22
Allen, Joe 2, 5

Blaha, John 41
Bluford, Guy 73, 80–82, 85, 93
  as first black American spacefarer 73, 82–83
Bobko, Karol 'Bo' 1–2, 10, 12
  and training for STS-6 28
  role in STS-6 spacewalks 10, 14
Brand, Vance 103–104, 113
Brandenstein, Dan 48, 80, 82, 85, 91, 93

Challenger
  as lighter orbiter than Columbia 22
  flight deck of **21**
  Flight Readiness Firing of 23–25, **24**
  Heads-Up Display of 41
  hydrogen leaks prior to STS-6 25–26
  main engines of 23–26
  middeck of 27
  modification from Structural Test Article to orbiter 8–10, **9**
  origin of name of 8
  planned missions after STS-51L 256–262, 266–270
  records set by 227
  rollover to Vehicle Assembly Building of **6**
  test equipment aboard 21–22
Colonna, Richard 5

Electrophoresis Operations in Space 38
Enterprise
  approach and landing runs of 7
  as testbed for Columbia and Challenger 7
  ejection seats of 7
  structure of 7

Greene, Jay 26, 97
Gregory, Fred 175–176, 189–190

Haise, Fred 7
Hart, Terry 97, 114, 126–129
Hauck, Rick 45–46, 48–49, 54, 60, 63

Inertial Upper Stage 31–36, **33**, 73–74, 95–96, 114

Lenoir, Bill 2–3
Lind, Don 39, 188
Lunney, Glynn 35

McBarron, Jim 5
McCandless, Bruce 3, 29, 98–104, 134
McCann, Joe 13–14

Mission Specialists 22
Musgrave, Story 1–2, 3, **4**
 and preparations for STS-6 spacewalk 10–13
 first experience of spaceflight of 29–30, 36–37
 role in Tracking and Data Relay Satellite deployment 31–36
 standing during STS-6 re-entry 41

Nelson, George 'Pinky' 4, 12, 14, 16, 114, 121, 126–129, 134
North, Warren 22

Payload Assist Module 62
Peterson, Don 1–2, 3, 4
 and preparations for STS-6 spacewalk 10–13
 and training for STS-6 26–27
 role in Tracking and Data Relay Satellite deployment 31–36
Pohl, Henry 19

Ride, Sally 46–47, **47**, 59–60, 63, 137–139
 as first American woman in space 52–53
Ross, Jerry 28–29

Scott, Dave 4
Shriver, Loren 91–92
Shuttle Landing Facility 69–70
Shuttle Training Aircraft 40–41
Solar Max 15, 63, 104–106, **105**, 114–115, 123–129, **125**, 131–133
Spacelab 161–230
 development and history of 161–165, **164**
 igloo 201
 Instrument Pointing System 166, 200, 202–209, **205**
 pallets 200–202, **201**
 pressurised module 165–166, 167
Space Shuttle
 ability to land at night of 91–93
 airlock of 11
 and satellite repair missions 115–116, 133–134
 approach and landing runs of 7
 as operational 'space liner' 5
 assembly of 20
 astronauts of 45–46
 ejection seats of 20–21, 22–23
 flight deck of **21**
 flight deck windows of 68–69
 External Tank of 18–19, 23
 juggling of orbiters and missions 139–140, 167–170
 landing of 41–43
 launch of 27–29, **28**
 main engines of 19
 manoeuvrability of 107
 middeck of 27
 mission training for 140
 Payload Specialists on 137
 physical appearance of 10
 Remote Manipulator System of 50–52, **52**, 86–88, 104–106
 sleeping aboard **90**
 Solid Rocket Boosters of 18, 19–20, 23, 76
 spacesuit of 10–17, **4**, **15**
 STS-41D main engine shutdown and 135–137
 testing of 7
 thermal protection system of 42–43
space sickness 53–56, 76-78, 83, 121–123
STS-5
 spacesuit failures on 2–3, 5
STS-6 1–44
 as oldest astronaut crew to date 1–2
 Continuous Flow Electrophoresis System on 37–39, **37**
 EVA on 13–17, **15**
 EVA training for 3–5
 F-Troop nickname of crew 1
 Getaway Special experiments on 40
 landing of 41–43
 launch of 27–29, **28**
 long wait of crew to fly 18
 Monodisperse Latex Reactor on 39–40
 Night/Day Optical Survey of Lightning on 39
 thermal protection system damage suffered by 42–43
 Tracking and Data Relay Satellite and 26, 30–36, **33**, 43–44
 wheel damage suffered by 42
STS-7 45–71
 Anik-C2 and 57–59, **58**, 60–62, **61**
 as first five-person Shuttle crew 49–50, **50**, 53–56

brake damage on 71
Continuous Flow Electrophoresis System on 68
Getaway Special experiments on 67
landing of 69–71, **70**
launch of 59–60
micrometeroid impact on 68–69
Monodisperse Latex Reactor on 68
Office of Space and Terrestrial Applications payload 65, 66–67
Palapa-B1 and 57–59, **58**, 62–63
preparations for 56–58, **58**
Remote Manipulator System on 50–52, **52**
role of Ku-band antenna on 65–66
Shuttle Pallet Satellite aboard 52, **58**, 63–65, **64**
space 'Olympics' on 70
STS-8 73–93
and first nocturnal Shuttle landing 91–93, **92**
and first nocturnal Shuttle launch 78, **79**, 80–82, **81**
Animal Enclosure Module on 86
Continuous Flow Electrophoresis System on 85–86
Evaluation of Oxygen Interaction with Materials on 88–89
Getaway Special experiments on 89–91
heat pipe investigation on 89
high-performance Solid Rocket Boosters on 76
Insat-1B and 73, **74**, 75, 83–85, **84**
Payload Flight Test Article and 73, 75, 86–88, **87**
removal of Tracking and Data Relay Satellite from 36, 73–74
STS-41B 95–114
EVAs on 102–107
Inflatable Rendezvous Target and 110–112, **111**
landing of 112–114, **113**
Manned Manoeuvring Unit and 98–104, **99**, **101**, **106**
new numbering system and 95–97
Palapa-B2 and 109–110
Shuttle Pallet Satellite aboard 104–106, **105**
Westar-6 and 107–109, 110
STS-41C 114–134
'direct insertion' ascent of 121
EVAs on 123–129
landing of 129–131
launch of 120–121
Long Duration Exposure Facility and 117–119, **118**, 122-123
Manned Manoeuvring Unit and 123–129, **125**
Solar Max repair 123–129, 125
'triskaedekaphobia' and 97, 114
STS-41G 134–159
crew of 137–139
discovery of Ubar 149–150
Earth Radiation Budget Satellite and 141–145, **144**
EVA on 151–155, **154**
Ku-band antenna problems on 146–147, 155
Office of Space and Terrestrial Applications payload 150–151
Payload Specialists on 155–158, **156**
Shuttle Imaging Radar and 146–151, **148**, 155
STS-51B 166, 170–193
as 'dual-shift' mission 181
ATMOS observations 179–181
'gravity gradient' attitude adopted by **180**, 182
landing site changed to Edwards Air Force Base 176
landing of 192–193
launch of 178–179
microgravity research aboard 175, 182–188
NUSAT deployment and 190–191
squirrel monkeys aboard 170–175
STS-51F 193–217
Abort To Orbit on 193–197, **195**
as first pallet-only Spacelab mission 209–210, **210**
Coke and Pepsi aboard 198–200, **199**
life sciences experiments on 216–217
main engine shutdown and 197–198, **197**
Plasma Diagnostics Package on 211–214, **212**
STS-61A 218–230
as first eight-person Shuttle crew 218
life and microgravity sciences research on 219–222

STS-61A (*cont.*)
operated from outside the United States 222
training for 218–219
STS-51L 231–272
cause of the disaster 249–256
cold weather and 232, **233**
destruction of 241–247, **243**, **245**, **246**
fate of the crew on 247–249
first 'private' citizen in space on 231–232, 238–239
Greg Jarvis and 239–241
planned Halley's Comet observations by 232–236
Tracking and Data Relay Satellite and 236–238

T-38 jet trainers 78–80
Thagard, Norm 53–56, **55**
Thirty Five New Guys astronaut class 45–49, 75–76
Thornton, Bill 76–78, **77**
Tracking and Data Relay Satellite 17–18, 30–36, **33**, 43–44, 73–74, 88, 114, 146–147, 167–168
Truly, Dick 93

Vaughan, Chester 14
van Hoften, James 'Ox' 126–129
Vidrine, Dave 115

Walker, Charlie 21, 38, 54, 68, 120
Weightless Environment Training Facility 3, **4**
Weitz, Paul 1–2, 41

Printing: Mercedes-Druck, Berlin
Binding: Stein+Lehmann, Berlin